METHODS IN MOLECULAR BIOLOGY™

For other titles published in this series, go to
www.springer.com/series/7651

Serum/Plasma Proteomics

Methods and Protocols

Edited by

Richard J. Simpson and David W. Greening

Ludwig Institute for Cancer Research Ltd,
Royal Melbourne Hospital, Parkville, Victoria, Australia

Humana Press

Editors
Richard J. Simpson, Ph.D.
Ludwig Institute for Cancer Research Ltd
Royal Melbourne Hospital
Parkville, Victoria
Australia
richard.simpson@ludwig.edu.au

David W. Greening, Ph.D.
Ludwig Institute for Cancer Research Ltd
Royal Melbourne Hospital
Parkville, Victoria
Australia
david.greening@ludwig.edu.au

ISSN 1064-3745 e-ISSN 1940-6029
ISBN 978-1-61779-067-6 e-ISBN 978-1-61779-068-3
DOI 10.1007/978-1-61779-068-3
Springer New York Dordrecht Heidelberg London

Library of Congress Control Number: 2011923966

Printed on acid-free paper

Humana Press is part of Springer Science+Business Media (www.springer.com)

Preface

Since its early beginnings of religious and "medicinal" purposes, blood science has become a cornerstone of medicine – especially, in the areas of clinical chemistry, disease diagnosis, and therapeutic monitoring. Because blood constituents, primarily proteins, reflect diverse physiological, pathological, and pharmacological states they are of great clinical significance – i.e., blood can be considered the "window of disease." The ease with which blood (especially, its plasma/serum components) can be sampled in a noninvasive manner makes it a logical choice for diagnostic screening applications.

Over the past decade, we have witnessed the advent of more powerful proteomics technologies that allow deeper drilling of the blood proteome. These technological improvements have, in part, fuelled the quest for the discovery of novel blood-based biomarkers of disease.

This volume describes recent developments in the relatively new area of blood proteomics. One significant technical challenge of any blood proteome analysis is the issue of sample complexity – for example, the dynamic range of protein concentrations in blood is in the order of 10^9–10^{10}, which is beyond any single protein fractionation method. Thus, in order to examine low-abundance proteins in blood, it is necessary to perform serial fractionation. Part I of this volume comprises seven chapters devoted to fractionation strategies for in-depth blood proteome analysis. This section includes chapters devoted to enriching glycoproteins and low-molecular weight classes of proteins, many of which are of very low abundance not typically seen in whole blood proteome analyses.

Another major challenge facing serum-based biomarker discovery, and blood proteomics in general, relates to blood collection and storage. Not to put too fine point on it, if screening blood samples for the purpose of discovering novel biomarkers, one needs to be sure that "normal" and "disease" blood samples are in fact being compared, and not variances in blood collection/handling/storage protocols. These vexed issues are covered in three chapters in Part II. To aid blood proteome researchers, we also include current standard operating procedures (SOPs) for plasma collection for the purpose of clinical research, the measured concentrations of many plasma proteins from quantitative assays, and reference ranges for blood tests. We also include a reference summary of the international collaborative effort, involving 38 laboratories, conducted by the Human Proteome Organization (HUPO) in 2005 to address the effects of preanalytical variance and different proteomic protocols on acquiring a comprehensive blood proteome. These aids are appended in detail at the end of the volume.

Needless to say, quantitative assaying of blood-based biomarkers is an important aspect of clinical chemistry. Part III of this volume includes four chapters with detailed protocols for performing both antibody-based (e.g., multiplex fluorescent microsphere-based assay using the BioPlex system) and nonantibody based (e.g., mass spectrometry-based multiple reaction monitoring, MRM) quantitative assays. Additionally, an updated protocol for analyzing glycated proteins in human plasma of patients presenting with glucotoxicity is presented.

Part IV of this volume includes five chapters focusing on proteome analysis of blood cell compartments (e.g., platelet concentrates, platelet membranes, cyroprecipitate, and cyrodepleted plasma), circulating nanomembranous vesicles (exosomes) and blood-related fluids, including tissue interstitial fluid. Any text covering blood proteome analysis would be found wanting if data management, statistical design, and bioinformatic challenges were not covered. This topic is detailed through four chapters in Part V, along with a featured protocol for using PeptideAtlas, an essential adjunct to any MRM assay design.

Serum/Plasma Proteomics is a comprehensive resource of protocols for areas, preanalytical through to analytical, of plasma and serum proteomics. This book, contributed by leading experts in the field, provides a valuable foundation for the development and application of blood-based proteomics.

The editing of this book was supported by National Health and Medical Research Council (Australia) Program Grant 487922.

Parkville, Australia

Richard J. Simpson
David W. Greening

Recommended Reading

Starr, D., (2002). *Blood: An Epic History of Medicine and Commerce.* 2nd ed. New York: HarperCollins Publishers. p464. ISBN 0-688-17649-6.

Contents

Part III Methods for Quantitative Assaying of Blood-Based Biomarkers

Part IV Proteome Analysis of Blood Cell Components, Vesicles, and Blood-Related Fluids

Part V Bioinformatic Analysis of Blood Proteins and Peptides

Contributors

RUEDI AEBERSOLD • *Institute of Molecular Systems Biology, ETH Zurich, Zurich, Switzerland; Institute for Systems Biology, Seattle, WA, USA*

THOMAS W. BARNES • *Ludwig Institute for Cancer Research, Royal Melbourne Hospital, Parkville, Australia*

KURT T. BARNHART • *University of Pennsylvania, Philadelphia, PA, USA*

IHOR BATRUCH • *Department of Pathology and Laboratory Medicine, Mount Sinai Hospital, Toronto, ON, Canada*

LYNN A. BEER • *Molecular and Cellular Oncogenesis Program, The Wistar Institute, Spruce Street, Philadelphia, PA, USA*

OLIVER K. BERNHARD • *Ludwig Institute for Cancer Research, Royal Melbourne Hospital, Parkville, Australia*

SIEGMUND BRAUN • *Deutsches Herzzentrum, Munich, Germany; First Medizinische Klinik der Technischen Universitt Munchen, Munich, Germany*

TIMOTHY CLOUGH • *Department of Statistics, Purdue University, West Lafayette, IN, USA*

DAVID CRAFT • *BD Diagnostics, Franklin Lakes, NJ, USA*

ERIC W. DEUTSCH • *Institute for Systems Biology, Seattle, WA, USA*

ELEFTHERIOS P. DIAMANDIS • *Department of Laboratory Medicine and Pathobiology, University of Toronto, Toronto, ON, Canada; Department of Pathology and Laboratory Medicine, Mount Sinai Hospital, Toronto, ON, Canada; Department of Clinical Biochemistry, University Health Network, Toronto, ON, Canada*

ALLEN EVERETT • *Department of Pediatrics, Johns Hopkins University, Baltimore, MD, USA*

TERRY FARRAH • *Institute for Systems Biology, Seattle, WA, USA*

VLADIMIR FOKIN • *Department of Mathematics, Indiana University, Bloomington, IN, USA*

JOSEPH M. FOSTER • *EMBL Outstation, European Bioinformatics Institute, Wellcome Trust Genome Campus, Hinxton, Cambridge, UK*

QIN FU • *Department of Medicine, Johns Hopkins University, Baltimore, MD, USA*

CRAIG A. GELFAND • *BD Diagnostics, Franklin Lakes, NJ, USA*

SCOTT GERBER • *Norris Cotton Cancer Center, Dartmouth-Hitchcock Medical Center, Rubin BuildingLebanon, NH, USA*

CICEK GERCEL-TAYLOR • *James Graham Brown Cancer Center, University of Louisville School of Medicine, Louisville, KY, USA*

DAVID W. GREENING • *Ludwig Institute for Cancer Research, Royal Melbourne Hospital, Parkville, Australia*

SAMIR HANASH • *Fred Hutchinson Cancer Research Center, North Seattle, WA, USA*

FUCHU HE • *Beijing Proteome Research Center, Beijing, China; China National Center of Biomedical Analysis, Beijing, China*
RÜDIGER HESS • *PXBioVisioN GmbH, Hannover, Germany*
DENIS HOCHSTRASSER • *Department of Genetics and Laboratory Medicine, Clinical Proteomics Laboratory, Geneva University Hospitals, Geneva, Switzerland*
YING JIANG • *Beijing Proteome Research Center, Beijing, China; China National Center of Biomedical Analysis, Beijing, China*
KIAN KANI • *University of Southern California, 90033, Los Angeles, CA, USA*
JONATHAN E. KATZ • *University of Southern California, 90033, Los Angeles, CA, USA*
VATHANY KULASINGAM • *Department of Laboratory Medicine and Pathobiology, University of Toronto, Toronto, ON, Canada; Department of Clinical Biochemistry, University Health Network, Toronto, ON, Canada*
URS LEWANDROWSKI • *Leibniz – Institute for Analytical Sciences – ISAS – e.V., Dortmund, Germany*
YAN LI • *Department of Pathology, Division of Clinical Chemistry, John Hopkins University, Baltimore, MD, USA*
ZHAOXIA LIU • *BD Diagnostics, Franklin Lakes, NJ, USA*
PARAG MALLICK • *University of Southern California, University of California, Los Angeles, CA, USA*
LENNART MARTENS • *Department of Medical Protein Science, Universiteit Gent – VIB, B-9000, Gent, Belgium; Department of Biochemistry, Universiteit Gent – VIB, B-9000, Gent, Belgium*
ROMMEL A. MATHIAS • *Ludwig Institute for Cancer Research, Royal Melbourne Hospital, Parkville, Australia*
ANDREW MENG • *Department of Medicine, Johns Hopkins University, Baltimore, MD, USA*
RETO OSSOLA • *Institute for Molecular Systems Biology and Biognosys AG, ETH Zurich, Zurich, Switzerland; Biognosys AG Zurich, Switzerland*
ILKA OTT • *Deutsches Herzzentrum, Munich, Germany; First Medizinische Klinik der Technischen Universitt Mnchen, Munich, Germany*
PAOLA PICOTTI • *Institute for Molecular Systems Biology, ETH Zurich, Zurich, Switzerland*
THOMAS PREMSLER • *Leibniz – Institute for Analytical Sciences – ISAS – e.V., Dortmund, Germany*
FELICIANO PRIEGO-CAPOTE • *Department of Structural Biology and Bioinformatics, Biomedical Proteomics Research Group, University Medical Centre, University of Geneva, Geneva, Switzerland*
SUSANNE RAGG • *Indiana University School of Medicine, Indianapolis, IN, USA*
MARIA RAMÍREZ-BOO • *Department of Structural Biology and Bioinformatics, Biomedical Proteomics Research Group, University Medical Centre, University of Geneva, Geneva, Switzerland*
LUKAS REITER • *Institute of Molecular Biology, University of Zurich, 8057, Zurich, Switzerland*

Oliver Rinner • *Institute for Molecular Systems Biology and Biognosys AG, ETH Zurich, Zurich, Switzerland*
Jean-Charles Sanchez • *Department of Structural Biology and Bioinformatics, Biomedical Proteomics Research Group, University Medical Centre, University of Geneva, Geneva, Switzerland*
Gunther Schadow • *Indiana University School of Informatics, Indianapolis, IN, USA*
Ralph Schiess • *Institute for Molecular Systems Biology and ProteoMediX AG, ETH Zurich, Zurich, Switzerland*
Simon Sheng • *Department of Medicine, Johns Hopkins University, Baltimore, MD, USA*
Albert Sickmann • *Leibniz – Institute for Analytical Sciences – ISAS – e.V., Dortmund, Germany; Medizinisches Proteom-Center (MPC), Ruhr-Universität, Bochum, Germany*
Richard J. Simpson • *Ludwig Institute for Cancer Research, Royal Melbourne Hospital, Parkville, Australia*
Helena Skalnikova • *Department of Medicine, Johns Hopkins University, Baltimore, MD, USA; Institute of Animal Physiology and Genetics, Academy of Sciences of the Czech Republic, Libechov, Czech Republic*
Christopher R. Smith • *Department of Clinical Biochemistry, University Health Network, Toronto, ON, Canada*
Rosemary L. Sparrow • *Research Unit, Australian Red Cross Blood Service, South Melbourne, VIC, Australia*
David W. Speicher • *Molecular and Cellular Oncogenesis Program, The Wistar Institute, Philadelphia, PA, USA*
Wei Sun • *Beijing Proteome Research Center, Beijing, China; China National Center of Biomedical Analysis, Beijing, China*
Harald Tammen • *PXBioVisioN GmbH, Hannover, Germany*
Hsin-Yao Tang • *Molecular and Cellular Oncogenesis Program, The Wistar Institute, 3601 Spruce Street, Philadelphia, PA, USA*
Douglas D. Taylor • *James Graham Brown Cancer Center, University of Louisville School of Medicine, Louisville, KY, USA*
John Tra • *Department of Medicine, Johns Hopkins University, 21224, Baltimore, MD, USA*
Jennifer Van Eyk • *Department of Medicine, Johns Hopkins University, Baltimore, MD, USA*
Olga Vitek • *Department of Statistics, Purdue University, West Lafayette, IN, USA; Department of Computer Science, Purdue University, West Lafayette, IN, USA*
Maryann S. Vogelsang • *University of Southern California, Los Angeles, CA, USA*
Hong Wang • *Fred Hutchinson Cancer Research Center, North Seattle, WA, USA*
Jizu Yi • *BD Diagnostics, Franklin Lakes, NJ, USA*

WOLFGANG ZACHARIAS • *James Graham Brown Cancer Center, University of Louisville School of Medicine, Louisville, KY, USA*
RENÉ PEIMAN ZAHEDI • *Leibniz – Institute for Analytical Sciences – ISAS – e.V., Dortmund, Germany*
HUI ZHANG • *Department of Pathology, Division of Clinical Chemistry, John Hopkins University, Baltimore, MD, USA*

Part I

Fractionation Strategies for In-depth Blood Proteome Analysis

Chapter 1

Plasma Biomarker Discovery Using 3D Protein Profiling Coupled with Label-Free Quantitation

Lynn A. Beer, Hsin-Yao Tang, Kurt T. Barnhart, and David W. Speicher

Abstract

In-depth quantitative profiling of human plasma samples for biomarker discovery remains quite challenging. One promising alternative to chemical derivatization with stable isotope labels for quantitative comparisons is direct, label-free, quantitative comparison of raw LC–MS data. But, in order to achieve high-sensitivity detection of low-abundance proteins, plasma proteins must be extensively pre-fractionated, and results from LC–MS runs of all fractions must be integrated efficiently in order to avoid misidentification of variations in fractionation from sample to sample as "apparent" biomarkers. This protocol describes a powerful 3D protein profiling method for comprehensive analysis of human serum or plasma proteomes, which combines abundant protein depletion and high-sensitivity GeLC–MS/MS with label-free quantitation of candidate biomarkers.

Key words: Proteomics, Major protein depletion, Plasma biomarkers, Label-free quantitation

1. Introduction

The human plasma proteome is a major focus of proteomic studies directed toward disease biomarker discovery, because blood is a routinely collected biological fluid that is likely to contain multiple novel biomarkers for many physiological conditions and diseases (1). It is believed that most cells in the body secrete proteins into the blood and, therefore, plasma is very likely to contain information concerning the physiological status of all tissues and organs in the body – and is potentially the most informative sample to describe an individual's current health state (1–4). Although plasma is a potential source of disease biomarkers, systematic discovery of potential clinical biomarkers is greatly complicated by the wide dynamic range of known blood proteins, which spans 9–12 orders of magnitude (3, 5–8). The huge complexity and dynamic nature

Richard J. Simpson and David W. Greening (eds.), *Serum/Plasma Proteomics: Methods and Protocols*, Methods in Molecular Biology, vol. 728, DOI 10.1007/978-1-61779-068-3_1, © Springer Science+Business Media, LLC 2011

of the plasma proteome present great technological challenges for proteomic analyses, specifically when trying to identify and quantitate low-abundant (<100 ng/mL) proteins (9, 10).

Recent advances in proteomic technologies, specifically high-resolution, high-mass-accuracy, mass spectrometry (MS), have made quantitative comparisons of proteins in biospecimens more robust and reproducible. However, the wide dynamic concentration range of plasma proteins greatly exceeds the analytical capabilities of single LC–MS/MS runs and other 1D and 2D proteomic methods. One strategy to increase dynamic range of the plasma proteome is to use multi-dimensional separation strategies that involve orthogonal protein/peptide fractionation steps (5, 11–13).

The first step in our multi-dimensional plasma profiling scheme, outlined in Fig. 1, is major protein depletion. Until approximately 2003, the most effective technology available for depletion of abundant blood proteins was represented by kits that combined Cibacron® blue dye for removal of albumin, which

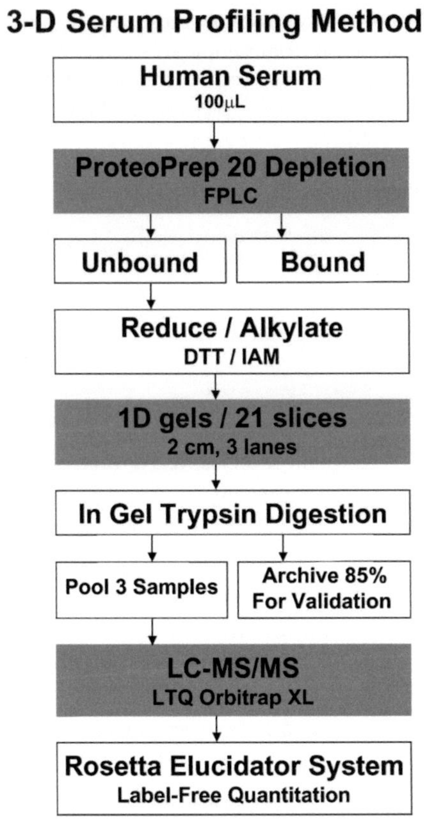

Fig. 1. Scheme for comprehensive protein profiling using major protein depletion and GeLC–MS/MS prior to label-free quantitation. To identify a broad range of proteins in plasma in a clinically relevant setting (ectopic pregnancy compared with intrauterine pregnancy), we utilize "Top 20" depletion and fractionation on short SDS-PAGE gels to improve detection of low-abundance proteins.

constitutes ~65% of total plasma protein, with Protein A or G for removal of immunoglobulins, which constitute an additional 15% of plasma protein. Unfortunately, these columns incompletely removed albumin, which remained as one of the most abundant proteins. Furthermore, many lower abundance proteins were also removed by the dye columns due to nonspecific binding as visualized by 2D gels (14). The Multiple Affinity Removal System (Agilent Technologies, Santa Clara, CA) developed in 2003 was the first commercial product that simultaneously depleted several abundant proteins (15). Specifically, this technique effectively depleted six abundant proteins and reduced the protein content of human plasma or serum by ~85%. This substantially improved protein profiling capacities, but the next tier of abundant proteins continues to limit volumes of plasma that can be analyzed. Ideally, additional medium-abundant proteins should be depleted so that at least 98–99% of total protein content is initially removed (5, 14). Other manufacturers have recently produced antibody columns to target 7, 12, 14, or 20 abundant plasma proteins in the next phase of immunodepletion. In addition, the next generation of immunodepletion methods will likely target more medium-abundance proteins. For instance, the recently introduced Seppro® Supermix column (Sigma–Aldrich, St. Louis, MO) is an immunoaffinity column designed to be used in tandem with IgY columns and targets medium-abundance proteins. A recent report showed that the tandem IgY columns were reproducible and greatly increased detection of lower-abundance proteins by removing ~60 medium- to high-abundance proteins (16).

Following major protein depletion, there are multiple alternative protein and peptide separation techniques for further fractionation of the proteome. Our strategy uses GeLC–MS/MS, an efficient strategy to identify a broad dynamic range of protein abundances in complex proteomes. Samples are separated using 1D SDS-PAGE, and gel slices are excised and typically digested followed by nano-LC–MS/MS. The advantages of GeLC–MS/MS approaches are that the molecular weight (MW) of the proteins of interest are known, providing information on possible protein processing or posttranslational modifications with relative changes in MW distribution (17). In addition, the routine method combines reproducible protein separation methods with well-established sensitivity of in-gel digestions, followed by LC–MS/MS for protein identifications (18).

Finally, we have chosen to use comparative LC–MS pattern analysis for label-free quantification as an alternative approach to chemical derivatization with stable isotope labels for several reasons. First, most methods used to compare proteins quantitatively select either portions of the total proteome, e.g., cysteine-containing peptides or glycosylated peptides, or label proteins after trypsin digestion. Such strategies make separations at the protein level prior to labeling impractical due to potential misinterpretation of technical variations as biological differences. Second, the cost of isotopic labeling reagents is a disadvantage for large-scale experiments with high numbers of samples. Third, direct, label-free, quantitative comparison of raw LC–MS data is a promising alternative approach based on the observation that ion peak intensity in ESI-MS is usually proportional to the concentration of a peptide in a sample (9, 19). By comparing ion intensities between LC–MS runs of multiple samples, comprehensive quantitation of each peptide between samples is possible (9, 10). However, one limitation of the label-free method is that in order to achieve accurate, relative quantitation of peak intensities in multiple LC–MS datasets, it is necessary to have high-resolution and high-mass-accuracy mass spectrometers (LTQ-FT or LTQ-Orbitrap™ instruments). Furthermore, dedicated software for identification, matching, and alignment of peaks, as well as for integration of data across fractions is essential. High-performance MS has become more routinely available in the past few years, and it has been shown that precise and comprehensive analysis of complex proteomes is best achieved using high-resolution proteomics technologies (20).

2. Materials

2.1. FPLC Affinity Depletion of Plasma or Serum

1. Human plasma or serum.
2. HPLC or FPLC system capable of operating at low pressure (<30 psi).
3. ProteoPrep® 20 LC Plasma Immunodepletion Column kit (#PROT20LC-1EA, Sigma–Aldrich). Individual components are as follows:
   (a) ProteoPrep® 20 LC Plasma Immunodepletion Column, 9.4 mL resin, 100 μL plasma (50–70 mg) capacity, storage: 2–8°C.
   (b) ProteoPrep® 20 equilibration buffer, 10× concentrate, storage: 2–8°C.
   (c) ProteoPrep® 20 elution solution, 10× concentrate, storage: 2–8°C.
   (d) ProteoPrep® preservative concentrate, storage: 2–8°C.
4. Stericup® GP Express Plus membrane (Millipore, Billerica, MA), 0.22-μm filter units.
5. Amicon Ultrafree®-MC (Millipore), 0.22-μm microcentrifuge filters.
6. Milli-Q® (Millipore) water or equivalent.
7. 20% (v/v) Ethanol in water.

2.2. Ethanol Precipitation

1. Unbound fraction from ProteoPrep® 20 LC depletion.
2. Ethanol (e.g., molecular biology grade, 200 proof, Sigma–Aldrich), chilled to –20°C.
3. SpeedVac® (Thermo Scientific) centrifuge.

2.3. Reduction and Alkylation of Samples Prior to 1D SDS-PAGE

1. Depleted and ethanol-precipitated pellet of human plasma or serum.
2. SDS-resuspension buffer: 1.0–1.2% (w/v) SDS in 100 mM aqueous Tris–HCl, pH 8.6.
3. 1 M Aqueous dithiothretiol.
4. 0.5 M Iodoacetamide in 100 mM aqueous Tris–HCl, pH 8.6.
5. 37°C Thermostatically controlled incubator/shaker.

2.4. 1D SDS-PAGE

1. Reduced and alkylated ProteoPrep® 20 depleted sample.
2. 2× Protein solubilizing buffer: 0.4 M sucrose, 6% (w/v) SDS, 125 mM Tris–HCl, 4 mM Na_2EDTA, 2% (v/v) 2-mercaptoethanol, and 2% (v/v) saturated bromophenol blue solution.
3. NuPAGE® MES running buffer 20× (Invitrogen, Carlsbad, CA), storage: 2–8°C.
4. NuPAGE® 12% Bis–Tris gel, 1 mm, 10-well (Invitrogen), storage: 2–8°C.
5. XCell *SureLock*™ Mini-Cell (Invitrogen).
6. Heat block set to 90°C.
7. BenchMark™ (Invitrogen) molecular weight marker, storage: –20°C.
8. 1 cc Insulin syringe.
9. India ink, several drops.
10. Novex® Colloidal blue staining kit (#LC6025, Invitrogen):
    (a) Stainer A.
    (b) Stainer B.
11. Fixing solution: 50% (v/v) methanol/10% (v/v) acetic acid in water.
12. Methanol (Optima® 0.2-μm filtered, Fisher Scientific, Pittsburgh, PA).
13. Glacial acetic acid.
14. Milli-Q water or equivalent.
15. Staining trays.

2.5. In-Gel Trypsin Digestion

1. PCR hood with built-in laminar flow and equipped with a HEPA filter and a lightbox.
2. 0.1% (v/v) trifluoroacetic acid/50% (v/v) methanol in water.

3. Methanol (Optima® 0.2-µm filtered, Fisher Scientific).
4. 96-well pierced V-bottomed plate (e.g., Thermo Fisher Scientific, Waltham, MA, or BioMachines, Research Triangle Park, NC).
5. V-bottomed, 96-well storage plates (e.g., Thermo Fisher Scientific).
6. Polystyrene plate covers (e.g., Thermo Fisher Scientific).
7. Glass plate (cut to the same size as a 96-well plate, 8 × 12 cm).
8. Gel-cutting device: e.g., MEG-1.5 Gel Cutter (The Gel Company, San Francisco, CA) or custom razor-blade array.
9. Stainless steel razor blades.
10. Ziploc bags.
11. Jewelers microforceps (e.g., Fisher Scientific).
12. SpeedVac® centrifuge equipped with 96-well plate centrifuge rotor.
13. Eight-channel pipette.
14. 37°C thermostatically controlled incubator/shaker.
15. Destain solution: 50% (v/v) acetonitrile in 0.2 M aqueous ammonium bicarbonate.
16. Sequencing grade-modified trypsin (Promega, Madison, WI).
17. Trypsin working solution: 0.02 µg/µL trypsin in 40 mM aqueous ammonium bicarbonate.
18. Trypsin wash buffer: 0.3% (v/v) formic acid in 40 mM aqueous ammonium bicarbonate.
19. Autosampler sample tubes.

2.6. LC–MS/MS

1. Solvent A: Milli-Q water containing 0.1% (v/v) formic acid (LC–MS grade, Sigma–Aldrich).
2. Solvent B: Acetonitrile (HPLC grade, Thomas Scientific, Swedesboro, NJ) containing 0.1% (v/v) formic acid (LC–MS grade, Sigma–Aldrich).
3. 180 µm i.d. ×2 cm UPLC symmetry trap column packed with 5 µm C18 resin (Waters, Milford, MA).
4. Reversed-phase 75 µm i.d. ×25 cm length analytical column packed with 1.7-µm C18 resin [e.g., nanoACQUITY UPLC BEH C18 Column (Waters, Milford, MA)].
5. UPLC equipped with a chilled microvolume autosampler, a 10-µL injection loop, and a column heater maintained at 40°C.
6. Hybrid linear ion trap-Fourier transform (LTQ FT ICR Ultra) or LTQ Orbitrap™ XL mass spectrometer.

2.7. Label-Free Quantitation

1. Label-free analysis software (e.g., Rosetta Elucidator® System, Rosetta Biosoftware, Cambridge, MA).
2. Database search engine (e.g., BioWorks v. 3.3.1, Thermo Scientific).

3. Methods

3.1. FPLC Affinity Depletion of Plasma or Serum

The following protocol describes HPLC/FPLC immunoaffinity depletion using the ProteoPrep® 20 column (21), which currently depletes the largest number of proteins and removes approximately 95% of total plasma protein (see Note 1). The ProteoPrep® 20 LC column packing has a rated pressure maximum of 30 psi, which is much lower than the pressure system requirements of most HPLCs (typically, at pressures of 100 psi or greater). Hence, an FPLC-type system, capable of operating at low pressures, is strongly preferred. If the available HPLC requires higher operating pressures, the necessary backpressure can be generated in front of the column with a flow restrictor, such as an appropriate length of small internal diameter PEEK tubing (e.g., 1 m of 0.005 or 0.007 in. i.d.), and after the column with a larger ID PEEK tubing (0.030 or 0.040 in. i.d.) to avoid exposing the column to high pressures. The maximum pressure limit must be set only slightly above the operating pressure to reduce the risk of compressing the column bed:

1. Prepare Buffer A (1× equilibration buffer): Dilute one part of ProteoPrep® 20 equilibration buffer, 10× concentrate with nine parts of room temperature ultrapure water. Filter the diluted buffer through a 0.22-μm filter unit and degas for 5 min at room temperature using vacuum filtration with stirring. The final volume needed for depletion is ~55 mL; buffers should be prepared fresh each day.
2. Prepare Buffer B (1× elution solution): Dilute one part of ProteoPrep® 20 elution solution, 10× concentrate with nine parts of room temperature ultrapure water. Filter the diluted buffer through a 0.22-μm filter unit and degas for 5 min at room temperature using vacuum filtration with stirring. The final volume needed for depletion is ~30 mL; buffers should be prepared fresh each day.
3. Before connecting the column, flush Buffer A through the system to the column inlet line for 15 min at 1–3 mL/min. Transfer the ProteoPrep® 20 LC plasma immunodepletion column from 2–8°C to room temperature at this time.
4. Connecting the column: To prevent air from being introduced into the column, after flushing the system as described in

step 3, stop flow, remove the plug from the top of the column only, start the pump at 0.5 mL/min, and while holding the column inlet tubing over the top end of the column, fill the column end fitting with Buffer A. Stop flow and connect the tubing to the column inlet.

5. Remove the screw plug from the bottom of the column, connect the detector inlet solvent line to the column bottom, and equilibrate with Buffer A for 5 min at 1 mL/min (see Note 2).
6. Filter undiluted plasma (generally 100 μL) with a pre-rinsed 0.22-μm microcentrifuge filter (see Note 3). Keep the filtered sample on ice until ready for use (see Note 4).
7. Prior to injection of the first sample of the day, run a blank run (no sample injection) and re-equilibrate the column using Buffer A (see Note 5).
8. Inject 100 μL of filtered plasma at 0.3 mL/min (see Note 6).
9. Elute the depleted plasma (unbound fraction) from the column with Buffer A at 0.3 mL/min. Collect the flow-through fractions and pool the unbound fractions containing protein as it comes off the column (see Note 7). Keep the pooled sample on ice.
10. Elute bound proteins with Buffer B at 3 mL/min. Collect and pool bound proteins and store on ice (see Note 8).
11. Re-equilibrate column with Buffer A for 10 min at 3 mL/min.
12. Inject the next sample and repeat steps 8–11 until all samples are separated or a sufficient volume of a single sample is depleted.
13. Disconnect and store the column at 4°C in Buffer A (see Note 9).
14. Replace Buffers A and B with ultrapure water and flush the system, including detector, for 30 min at 1 mL/min followed by 20% (v/v) ethanol in water for 30 min at 1 mL/min (see Note 10).

3.2. Ethanol Precipitation

Precipitation of proteins with ethanol is an advantageous alternative to ultra-filtration for concentrating unbound and bound fractions. Ethanol precipitation requires less operator time and more effectively removes salts and detergent. The ethanol should be of high quality, chilled to –20°C, and added quickly for efficient protein precipitation. If precipitating the bound fraction, the pooled sample should be neutralized by adding 1 M NaOH to bring the pH to ~8.0 prior to precipitation:

1. Quickly add nine volumes of 200-proof ethanol (–20°C) to the pooled unbound fraction.
2. Vortex the sample thoroughly.

3. Incubate overnight at 0°C.
4. Centrifuge the sample for 30 min at 1,500 × *g*, 4°C.
5. Carefully remove the supernatant and carefully air-dry pellet under a gentle stream of argon (see Note 11).
6. Freeze pellets and store the samples at −20°C for short-term (less than 1 week) or −80°C for long-term storage.

3.3. Reduction and Alkylation of Samples Prior to 1D SDS-PAGE

Reducing and alkylating the unbound sample prior to running gels save time in the downstream in-gel digestion protocol, especially for large-scale experiments with many fractions and/or samples. This protocol assumes that 100 μL of human plasma has been depleted, precipitated with ethanol, and will be re-suspended in 100 μL (original plasma volume) of SDS re-suspension buffer. Reduction, alkylation, and quenching buffer volumes should be calculated and subtracted from the final sample volume (in this case, 100 μL) to determine the volume of re-suspension buffer to be used:

1. Thaw ethanol-precipitated plasma pellet from −20°C.
2. Re-suspend pellet in 81 μL of 1.2% SDS and 100 mM Tris–HCl, pH 8.0 (final SDS concentration: ~1%).
3. Reduce samples by adding 2 μL of 1 M DTT (final concentration: 20 mM). Incubate for 1 h at 37°C with shaking.
4. Alkylate proteins by adding 12 μL of 0.5 M iodoacetamide/100 mM Tris–HCl, pH 8.6 (final concentration: 60 mM). Incubate for 1 h at 37°C in the dark, with shaking.
5. Quench the reaction by adding an additional 5 μL of 1 M DTT (final concentration: 50 mM) to the sample and incubate for 15 min at 37°C with shaking.

3.4. 1D SDS-PAGE

The following instructions are for SDS-PAGE with precast NuPAGE® gels, buffers, and electrophoresis equipment, followed by gel staining with colloidal Coomassie® blue. To prevent keratin contamination, prior to electrophoresis all gel units and staining containers should be thoroughly washed with a 1% solution of mild laboratory detergent, followed by rinses with neat methanol and ultrapure water. It is also important to wear gloves that have been rinsed with ultrapure water, and staining solutions should be removed with vacuum aspiration to avoid any contact with the gel:

1. Prepare reduced and alkylated plasma samples in protein solubilizing buffer (mix 1:1 v/v with 2× solubilizing buffer).
2. Mark a precast 1-mm, 12% Bis–Tris, 10-well NuPAGE® gel cassette at a point 2–4 cm from the bottom of the sample wells (see Note 12).

3. Prepare 1× MES running buffer: Add 50 mL of 20× NuPAGE® MES buffer to 950 mL of Milli-Q water.
4. Heat the samples at 90°C for 2 min.
5. Assemble the gel in the XCell *SureLock*™ Mini-Cell unit and add the running buffer to the upper and lower chambers. Load samples into each lane, with a MW standard in the first lane (see Note 13).
6. Run gels at 200 V using constant voltage.
7. Stop the gels when the dye front has reached the desired separation distance.
8. Disassemble the gel unit and remove the gel from plastic plates. Cut the left, bottom corner of the gel for proper orientation and, using a syringe, carefully mark the outer edges of the dye front with India ink (see Note 14).
9. Prepare fixing solution (100 mL per gel): 50% (v/v) methanol/10% (v/v) acetic acid in water.
10. Place the gel in fixing solution for 10 min, with gentle agitation on a platform shaker.
11. Prepare staining solution (95 mL per gel): 20% (v/v) methanol/20% (v/v) Stainer A in water.
12. Remove the fixing solution with vacuum aspiration and add the staining solution to the gel container. Shake the gel in the staining solution (without Stainer B) for 10 min.
13. Add the colloidal Coomassie® Stainer B (5 mL per gel container) to the existing staining solution. Shake the gel in the staining solution between 3 and 12 h.
14. Remove the staining solution with vacuum aspiration and replace with 200 mL of ultrapure water to destain. The gel will have a clear background after approximately 7 h, but the water can be changed several times to expedite the destaining process.

3.5. In-Gel Trypsin Digestion

To minimize airborne keratin contamination, we generally perform the in-gel digestion in a PCR hood equipped with laminar flow and a HEPA filter. For higher throughput, large-scale digestions are done in 96-well plates with laser-cut holes allowing for centrifuge-aided removal of liquids. Prior to cutting gels, all 96-well plates and autosampler tubes and caps should be washed twice with 0.1% (v/v) trifluoroacetic acid/50% (v/v) methanol in water, followed by two rinses with 100% methanol, and allowed to air-dry in the PCR hood. All reagents are added with an eight-channel pipette and removed with centrifugation for 1 min in a bench-top centrifuge equipped with a 96-well plate rotor. A clean polystyrene plate cover should be used to cover the 96-well plate any time it is removed from the PCR hood for centrifugation or incubations:

1. Turn on the fan in the PCR hood approximately 15 min prior to cutting gels to achieve optimal air flow and to remove any dust particles that may be inside the hood.
2. Slice lanes of each gel into uniform 1 × 4 mm slices using a gel-cutting device (see Note 15). Several adjacent lanes (2–3) of the same sample may be combined to maximize protein load (see Note 16).
3. Using microforceps, transfer the excised gel slices into separate wells of a pre-cleaned, pierced, 96-well plate that has been stacked on top of a clean, V-bottomed, 96-well collection plate.
4. Destain the gel slices for 30 min with 100 μL of 200 mM ammonium bicarbonate/50% acetonitrile at 37°C, with gentle shaking. Centrifuge the plates for 1 min to remove the buffer into the collecting plates. Discard the destaining extract after centrifugation.
5. Dry gel slices for approximately 20–30 min using a SpeedVac® evaporator.
6. Re-hydrate gel bands by adding either 20 μL (one to two 1 × 1 × 4 mm bands per well) or 40 μL (three bands per well) of trypsin working solution (see Note 17). Incubate for 16–18 h at 37°C in a thermostatically controlled incubator (see Note 18).
7. Collect the digested protein extract into a clean, 96-well collecting plate by centrifuging for 1 min.
8. Add 15 μL (for one to two bands per well) or 30 μL (for three bands per well) of wash buffer and incubate for 30 min at 37°C.
9. Collect the second extract into the 96-well collecting plate containing extract 1 by centrifuging for 1 min.
10. Transfer the samples into pre-cleaned autosampler sample vials. If desired, the samples may be pooled at this point (see Note 19).

3.6. LC–MS/MS

Label-free quantitation through comparative LC–MS requires data acquisition with high-resolution and high-mass-accuracy mass spectrometers. Analysis of complex peptide mixtures on lower-resolution instruments is not sufficient to resolve co-eluting peptides with similar m/z ratios (20). MS-based proteomics has seen great improvements in the past 5–6 years, and extremely high-performance MS is now routinely available. The evolution of hybrid linear ion trap-FT and Orbitrap™ instruments has improved not only resolution, but also ion capacity, scan speed, and mass resolving power to reduce false-positive peptide identifications dramatically (20). Our analysis was performed using an LTQ-Orbitrap™ XL instrument:

1. Connect a 75-μm i.d. × 25 cm column packed with BEH C18 resin, 1.7 μm particle size, to the mass spectrometer. To avoid

temperature fluctuations, the column should be contained in a thermostat-controlled compartment maintained at 40°C.

2. Set the flow rate at 200 nL/min and equilibrate the column with 3% Solvent B.
3. Inject 4–8 μL (see Note 20) of each trypsin digest at 5 μL/min over a 4-min period with 100% Solvent A. The HPLC gradient is as follows (see Note 21):

   3–28% Solvent B over 42 min.

   28–50% Solvent B over 25.5 min.

   50–80% Solvent B over 5 min.

   Hold at 80% Solvent B for 5 min.

   Return to 3% Solvent B over 1 min.
4. The following parameters are used for MS/MS data acquisition and have been optimized for downstream analysis with the Elucidator® system (see Note 22):
   (a) Scan ranges: 400–2,000 *m/z*.
   (b) Full MS scan: FT at 60,000 resolution in profile mode.
   (c) Number of MS/MS: three most abundant with data-dependant scans.
   (d) Charge state screening enabled.
   (e) Monoisotopic precursor selection enabled.
   (f) Charge state rejection enabled to reject $z=1$.

3.7. Label-Free Quantitation in the Elucidator® System

There are a number of label-free quantitation software programs available as open-source or commercial packages. While the freely available packages are in various stages of development, the commercial packages seem to have more advanced data-processing algorithms and user-friendly interfaces (9). A critical distinguishing feature is whether the software can reliably combine quantities of a given peptide if it is distributed across fractions. This protocol describes label-free quantitation with the Rosetta Elucidator® System. The basic principles and typical workflow of the Elucidator® System are common to most comparative LC–MS packages, and this protocol can be used as a guideline for most available label-free quantitation programs:

1. Import RAW MS spectra into the Elucidator® System.
2. Define each experiment's treatment groups and corresponding technical and/or biological replicates. Multi-fraction experiments should be analyzed in a multidimensional workflow (see Note 23). If the dataset is too large for the software program to handle in a single analysis, the proteome may be

divided into subsets based on gel migration for quantitative comparisons (see Note 24).

3. Perform retention time (RT) alignment and extract features, i.e., unique time and m/z boundaries for peaks in individual runs, over the entire chromatographic time window. In the Rosetta Elucidator® System, this step is executed using the Peak Teller algorithm (see Note 25).
4. Acquire high-quality features: filter out any features that have $z \leq 1$ or $z \geq 5$, or peak time score or peak m/z score less than 0.7 and 0.8, respectively (see Note 26).
5. Create DTA files for all features that pass the above quality filters. DTAs should be created for the monoisotopic peak using the scan closest to the center of the chromatographic peak only.
6. Search MS/MS data from features of interest using SEQUEST or other search engines to determine peptide sequences and associated proteins.
7. Assign peptide and protein information to features using SEQUEST Annotation. For high-quality peptide and protein identifications, search results should be filtered at mass error ≤10 ppm and dCN ≥0.05 prior to annotation.
8. Find peptides or proteins of interest whose expression levels change significantly across treatment groups (see Note 27).

4. Notes

1. The current commercially available antibody columns deplete 6, 7, 12, 14, or 20 abundant plasma proteins in both HPLC/FPLC and spin column formats. In our experience, most commercial multi-protein immunodepletion columns perform similarly, in that they efficiently remove >98–99.5% of most targeted abundant proteins, while not removing significant amounts of untargeted minor proteins (14). However, if serum albumin is present at 40–50 mg/mL and 99% is removed, the residual 0.4–0.5 mg/mL of albumin is still a predominantly abundant remaining major component. Therefore, for studies directed toward low-abundant protein analyses, it is often necessary to remove greater than 99%, at least for the most abundant proteins.

   There are several factors to consider when choosing a depletion column. First, whether to use an FPLC or spin column format. FPLC columns require relatively expensive instrumentation not available in every laboratory, and antibody LC columns are usually larger and more expensive than spin columns.

However, the FPLC columns are highly robust, more completely remove the targeted proteins, and require less of the operator's time because they usually remove larger volumes of plasma or serum in a single pass.

Other factors that may influence decisions concerning the type of column to use include (1) column size and capacity, i.e., spin column capacities typically range from 8 to 20 μL of serum or plasma, while LC columns can be purchased with standard capacities ranging from 20–80 μL or 100–300 μL for high capacity columns, (2) price-to-capacity differences, (3) compositions of buffers and whether they are defined or proprietary, and, most importantly, (4) the amount of the total plasma protein removed by the column. In general, it is most beneficial to deplete as many high- to medium-abundant proteins as possible with the aim of improving the identification of proteins of lower abundance.

2. The ProteoPrep® LC column comes packed in a storage solution containing glycerol. To remove the glycerol solution prior to the first use of the column, attach the top of the column to the tubing, as described in Subheading 3.1, step 4. Do not connect the column outlet line to the bottom of the column. Instead, place the column over a beaker to collect the glycerol solution. Allow the column to equilibrate with 20–25 mL of 1× equilibration buffer at no higher than 0.5 mL/min. It is important to monitor pressure and decrease the flow rate carefully as to not exceed 30 psi. Increased pressures at this step may cause bed compression and a void at the top of the resin bed, which can be readily observed because the column is made of glass. When the resin has changed from opaque to white and the buffer runs clear, the column outlet line may be connected. The column should be conditioned with two complete blank cycles prior to injecting a sample. It is also preferable to run several noncritical control plasma samples before depleting plasma samples that will be used in comparative studies in case there is minor irreversible absorption of low-abundance proteins or antibody bleed during the first several runs.
3. The manufacturer originally recommended diluting the plasma sample 5× with Buffer A prior to filtering. This dilution step has subsequently been omitted to enhance depletion of apolipoproteins and recovery of unbound proteins. We have not observed any major changes in apolipoprotein removal, and a fivefold dilution of the plasma did not significantly increase the final volume of the unbound pool. Also, diluting the sample makes it more convenient to add protease inhibitors to the sample before injection, if desired.
4. Several considerations to note during sample preparation are (1) the number of depletions and/or (2) different plasma

samples that will be run in a single day. If multiple runs of the same sample are being depleted, it is best to thaw and filter the entire sample at once and store it on ice until all depletions are complete for optimal reproducibility. If multiple samples from different donors are being depleted, it is recommended that a single sample is thawed and filtered shortly before its injection. To minimize proteolysis, thaw and prepare the next sample while the previous sample is running. Alternatively, another option is to add protease inhibitors to the plasma sample prior to injection; however, the value of adding protease inhibitors is unclear. Instead of adding protease inhibitors to plasma or serum, we prefer to minimize exposure of samples to conditions that favor proteolysis. However, the one step where it is not practical to avoid protease-favorable conditions is during binding to the antibody resin. This is because in order to optimize binding affinity, conditions close to physiological pH without denaturants must be used, and in order to achieve reasonable binding kinetics, low temperatures should be avoided. However, it should be noted that plasma and serum are far more stable to proteolysis than cell lysates and intrinsically contain many proteases and protease inhibitors. Moderate-length exposure to physiological conditions at room temperature appears to have only very minor effects on plasma proteomes (22). Regardless of the exact sample preparation and handling methods to be used, it is critical to ensure that all plasma samples to be compared are processed in a consistent manner.

5. Running an initial "quick blank gradient" at 3 mL/min for the entire gradient with no sample injection is useful for several reasons. First, it will ensure that the column is equilibrated to room temperature prior to the plasma sample being injected. Also, as the column warms up, solvent out-gassing may occur; but the increased flow rate forces any small air bubbles out of the column before large pockets can form and disrupt the column media packing. Finally, a blank cycle run as the first run of the day removes any residual protein that may be bound to the column from previous depletions. Any protein peaks observed during the blank run should be run on 1D SDS-PAGE and analyzed with silver stain.

6. The manufacturer describes this column's capacity as 100 μL of plasma with a concentration of 40–50 mg/mL. However, this is a very low estimate for protein concentration. For example, the range for total protein in normal human serum is quite broad, ~60–85 mg/mL, and plasma samples will have similar or slightly higher values. Furthermore, in our experience, plasma concentration ranges in patients with diseases are often even more variable. Total protein concentration may

also be affected by the use of different anticoagulants and collection tubes. Therefore, it is recommended to determine the protein concentration of each sample prior to depletion and base injection loads upon total protein, not volume.

7. To monitor depletions and analyze individual fractions, it is useful to reserve a small aliquot of each fraction before pooling the unbound and bound samples for analysis on 1D SDS-PAGE.

8. Bound proteins should ideally be collected and stored at –20°C for potential future analysis. Prior to freezing, the bound pool should be neutralized by adding 1 M NaOH to bring the pH to approximately 8.0.

9. It is essential to store and monitor antibody LC columns properly to maintain optimal column performance. Storage in equilibration buffer at 2–8°C is suitable for short-term storage (less than 1 week). Long-term column storage (1 week or more) requires the addition of a preservative solution, included with the ProteoPrep® 20 LC columns. Otherwise, flushing the column with equilibration buffer containing 0.01% sodium azide should prevent any microbial growth in the column. It is important to keep track of the total number of blank and depletion cycles run on each column. Periodically, and especially when the number of recommended uses has been reached, column performance should be critically evaluated. If column capacity has inevitably declined, it is usually adequate to reduce protein loads slightly to achieve a degree of depletion similar to that observed with a new column. To track antibody bleed and/or proteolysis of the coupled antibodies from the column, bound fractions from plasma depletions and bound fractions from blank cycles after extended column storage can be monitored by digesting an aliquot of the bound pool with trypsin, followed by LC–MS/MS. The resulting data ideally should be searched against an all-species database. Alternatively, a human sequence database supplemented with all immunoglobulin sequences from the species of origin for the antibodies on the column should be used.

10. After the final run of each day, the FPLC or HPLC system should be thoroughly flushed with water, followed by 20% ethanol. This will prevent corrosion due to halogen salt-containing buffers or clogging of lines due to bacterial growth.

11. It is preferable to dry the pellet completely under a gentle stream of argon (this may be done at room temperature) or, alternatively, in a SpeedVac® for a few minutes to prevent airborne contamination or sample oxidation. The pellet is then resuspended in the desired resolubilization buffer, or the dried pellet may be stored at –20 to –80°C until it is needed.

To check for potential sample loss during the ethanol precipitation, dry 10% of the ethanol supernatant in a SpeedVac®, and run this sample in parallel with proportional volumes of sample before and after precipitation on 1D SDS-PAGE.

12. 12% Bis–Tris gels combined with NuPAGE® MES buffer effectively separate complex serum proteins and provide high resolution of small- and medium-sized proteins. Ten-well gels allow larger volumes to be loaded, and the wider gel lane is advantageous for in-gel digestions, as any protein smearing at the lane edges can be removed with a razor blade while keeping an optimal lane width of ~4 mm. Gel length can differ per experiment based on sample complexity, but in general, running plasma samples for a short distance (2 or 3 cm) will effectively separate the proteins while reducing the total number of fractions to be analyzed by LC–MS/MS.

13. Protein loads on gels should be maximized to increase the total sample digested and analyzed by LC–MS/MS. We typically load the equivalent of 10 μL of original serum per lane and run triplicate lanes per sample, for a total of 30 μL of original serum analyzed. When running more than one gel for quantitation, it is critical to measure the separation distance precisely to eliminate gel-to-gel migration differences. Therefore, we usually only run one gel per chamber because two gels in the same electrophoresis unit can migrate slightly differently. While some gel-to-gel variability is inevitable, tracking protein migration with MW standards will help to point out any major gel separation discrepancies. Figure 2

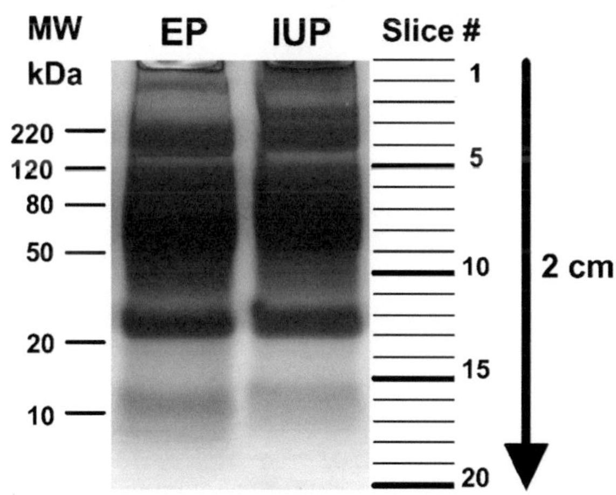

Fig. 2. Representative 1D SDS-PAGE gel used for GeLC–MS/MS. Depleted and concentrated unbound serum from patients with ectopic pregnancy (EP) and normal intrauterine pregnancy (IUP) was run for a short distance (2 cm) on SDS-PAGE. Each lane was sliced into 20 fractions and each fraction was digested with trypsin for further analysis on LC–MS/MS.

shows a typical 2-cm gel used for a serum biomarker discovery study; specifically, detection of protein changes between ectopic (EP) and normal intrauterine (IUP) pregnancy.

14. Marking the tracking dye with India ink is helpful for measuring the exact gel running distance. This is important because the low MW region of plasma may be sparse with few stained proteins, and it can be difficult to determine where to cut the bottom of the gel to ensure that these low-abundance proteins are included in the subsequent in-gel digestion. Also, some gel swelling may occur after staining and can increase gel volume slightly, thereby artifactually shifting gel migration. MW standards should be carefully examined to correct for any migration differences.
15. If using the gel cutter tool, which is currently the only commercial gel cutting device for 1D SDS-PAGE, gel lanes will have to be sliced individually. This tool is convenient, but some mashing of the gel occurs. Hence, a custom razor blade array constructed of stainless steel razor blades separated by 1-mm spacers is preferred. The custom array is beneficial because the width of the blades (4 cm) allows up to three lanes to be sliced simultaneously, preventing sample mix-up that can occur when lanes are sliced separately. Also, the sharpness of the razor blades, compared to the thin grid-like design of the gel cutter, prevents mashing of the gel, which can cause small particles of gel to contaminate other protein fractions.
16. To combine three lanes of a sample, the entire lanes of the gel can be cut with a stainless steel razor, removing any portion with no protein, i.e., loading wells, outer edges, gel region between lanes, or the bottom of the gel. We typically cut along the inner edges of the gel lane to eliminate any edge effects on the sides of the gel lane. Adjacent lanes can be aligned by matching up major proteins horizontally prior to cutting the gel.
17. A 96-well pierced plate will easily hold up to three $1 \times 1 \times 4$ mm gel slices. However, increasing the number of gel slices, and consequently gel volume, will absorb more trypsin solution as the gels re-hydrate during the overnight incubation. To prevent significant buffer losses, we increase the trypsin extract volumes in Subheading 3.5, steps 5 and 7 to account for the increased gel volume if three slices are combined. Using more than three $1 \times 1 \times 4$ mm gel slices is not recommended due to the need for excessive reagent volumes and increased trapping of peptides in the large gel volume.
18. To prevent sample evaporation during overnight incubation, place the pierced plate (top) and collection plate (middle) on

top of a third 96-well plate with ~80 μL of Milli-Q® water in each well. Place all stacked plates into a Ziploc® bag. The humidity created in the bag will prevent sample evaporation. To prevent sample leakage from the pierced plate, cover the top (pierced) plate with a glass plate placed under the polystyrene plate cover. The glass plate acts as a seal to prevent leakage of trypsin solution into the collecting plate. Shaking the samples during overnight incubation should be avoided, as the agitation may promote sample leakage.

19. The effect of pooling samples in biomarker discovery remains unclear. Pooling samples reduces cost and analysis time associated with multi-dimensional experiments, which may include tens to hundreds of patient samples. However, pooling averages the concentration of individual proteins across all pooled samples and, in some cases, outliers skew the results and overall intensities of some low-abundance proteins may fall below the detection threshold in pooled samples (23). Alternatively, pooling may help to reduce sample variability by minimizing individual variation. This is suitable for analyses where the focus is not specific to the individual, but rather on global changes in expression patterns (24). Pooling can be done in two ways: complete pools or sub-pools. Complete pooling of all samples in a treatment group reduces biological variation, and any replicates analyzed are technical rather than biological. Sub-pooling randomly combines several individual samples into multiple pools per experimental group, thus providing biological replicates and still allowing identification and removal of anomalous samples, if necessary (24, 25).

    We typically create sub-pools of samples after proteins are digested with trypsin. Tryptic digests of individual samples are then archived at –20°C for future use. Specifically, from nine serum samples from patients with ectopic pregnancy and nine gestational-stage matched controls, we pooled corresponding fractions from three patients each, for a total of six sub-pools. The sub-pools provide biological replicates for our analyses, yet reduce the total number of fractions to be analyzed by LC–MS/MS. In future validation stages of biomarker discovery, the individual samples can be easily thawed and analyzed in a highly reproducible manner, because the same samples used for discovery are being used for initial validation.

20. The injection volume can be adjusted based on the intensity of protein staining in individual digests to prevent overloading of the column.

21. The sample run time is 78 min. To help minimize carryover between samples, a blank should be run between samples.

A reasonable blank cycle time is 20 min, giving a total analysis time per sample of 98 min.

22. To facilitate the label-free pattern analysis in the Elucidator® system, it is essential that rigorous controls be maintained to ensure that HPLC column performance (resolution and peak widths), peak RTs, and mass spectrometer performance are maintained within restricted tolerances. Specifically, the same column (ideally temperature controlled) and gradients should be used, RTs of trypsin autolytic peptides should be monitored via extracted ion chromatograms and remain within a ±1.0-min window, and injection volumes – measured by weighing the sample vials before and after injection – should remain within ±10%.

23. One of the benefits of using the Elucidator® system is the ability to utilize a multi-dimensional workflow. While there are a number of LC–MS and LC–MS/MS pattern analysis software programs available, most approaches and current commercial software can only handle data where an entire proteome is represented by a single LC–MS/MS run. Unfortunately, without some fractionation of the proteome, only a modest number of plasma proteins can be identified, which will include few, if any, low-abundance proteins. Alternatively, the Elucidator® system effectively integrates multiple LC–MS/MS datasets such as those resulting from separating a single proteome into 20 fractions (gel slices). The system initially aligns RT and *m/z* patterns from corresponding slices, e.g., all slice 1 samples or all slice 2 samples. Therefore, the initial comparison is within corresponding gel slices; but importantly, the software adjusts for variable distributions of some proteins between adjacent slices. This is critical because such complex samples represent a continuum of proteins on a 1D gel (or any other protein or peptide fractionation method), and there will always be some proteins that are variably distributed across adjacent slice boundaries in different gel lanes. After initial comparisons within gel slices are completed, the entire dataset is integrated by matching corresponding *m/z* values with the same RT between slices. Intensities for matching *m/z* values in adjacent slices are summed if they are identified as matching the same peptide, or if the *m/z* in one slice was identified and the *m/z* with matching mass and RT did not trigger an MS/MS. Because matching peptide intensities are summed even if they do not result in a high-quality MS/MS spectra, the major source of variation often encountered due to undersampling in MS/MS analyses of complex mixtures is avoided.

Summed signal intensity over the entire gel lane is obtained for every identified protein in the proteome by combining the intensities for all peptides contained within that protein. Average intensities and standard deviations for each protein are calculated for the control and experimental samples. The variance in these data points reflects a combination of experimental and biological variation. An advantage of this analysis is that signals arising from very low-level peptides that produce a detectable *m/z* peak, but are too low to trigger MS/MS analyses consistently in all samples, can still be quantitatively compared across all samples.

24. It was our experience with an earlier version of the Elucidator® system that software limitations prevented the analysis of entire proteomes from six pools with a total of 252 RAW files in a single quantitative comparison. Hence, we divided the data into four subsets based on gel migration (slices 1–4, 5–10, 11–15, and 16–21).

25. The Peak Teller algorithm is a complex processing step that corrects for local shifts in RT across all runs in the experiment and removes noise and background. In addition, the algorithm creates peak regions, or features, to match peaks across runs and treatment groups correctly. The algorithm summarizes features into isotope groups, i.e., groups of features originating from the same peptide and an ion of one charge state, and isotope groups are associated into charge groups representing all detected charge states from a given peptide. Finally, the algorithm generates expression profiles, which include peak intensities and associated error model statistics. Figure 3 shows peak intensities at the isotope group level for three peptides of a candidate biomarker.

26. During RT alignment and feature (RT × *m/z* surface) extraction, the Peak Teller algorithm performs background subtraction and smoothing across the alignment window (both in RT and *m/z* dimensions) and uses a maximum alignment time correction of ±4 min. Each feature is then scored for quality, where a score of 0 indicates uncorrelated intensities and a score of 1 represents ideal peak shape (26). In these experiments, only features with scores greater that 0.7 and 0.8 were included.

27. The Elucidator® system has a number of system tools to filter full sets of peaks and return a subset relevant to the specific experiment. Differentially expressed peaks across treatment groups can be detected using significance testing methods, such as ANOVA, as well as by applying fold-change criteria. We find that there are some advantages in using the trend analysis tool to detect increases or decreases in intensities at

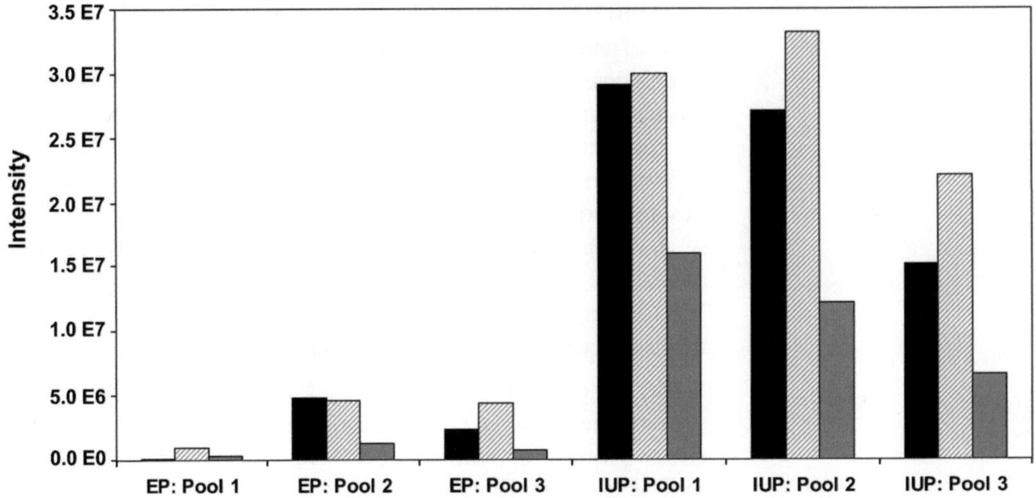

Fig. 3. Differential peak intensity. Differential peak intensities quantitated using the Rosetta Elucidator® system. Peak intensities shown are summarized as isotope groups, i.e., groups of isotopic mass peaks determined to originate from the same charge state of one peptide. The isotope groups represent three different peptides of a candidate biomarker, which is decreased in ectopic pregnancy across all pooled samples ($n = 3$ in each pooled group).

the peptide level between experimental groups. Trends were created for each segment of the gel proteome (see Note 24) and intensities for each protein were then manually combined across the four subsets. As expected, individual peptides were spread across several gel slices, and minor variations occurred from sample to sample. These gel migration differences caused medium- to high-abundant proteins within one region of the gel to look artifactually interesting in a flanking segment of the gel. When peptide intensities for these multi-fraction proteins were combined, the net fold change was often not significant. Variable intensity distributions across fractions may be due to a number of factors, as illustrated in Fig. 4. These include possible extensive aggregation/glycosylation (*Panel A*), and proteolysis (*Panel B*). These MW differences may also reflect an isoform shift or a posttranslational modification that could distinguish between normal and ectopic pregnancy. GeLC–MS/MS provides the advantage of identifying these potential MW shifts, which would be missed in most other fraction schemes.

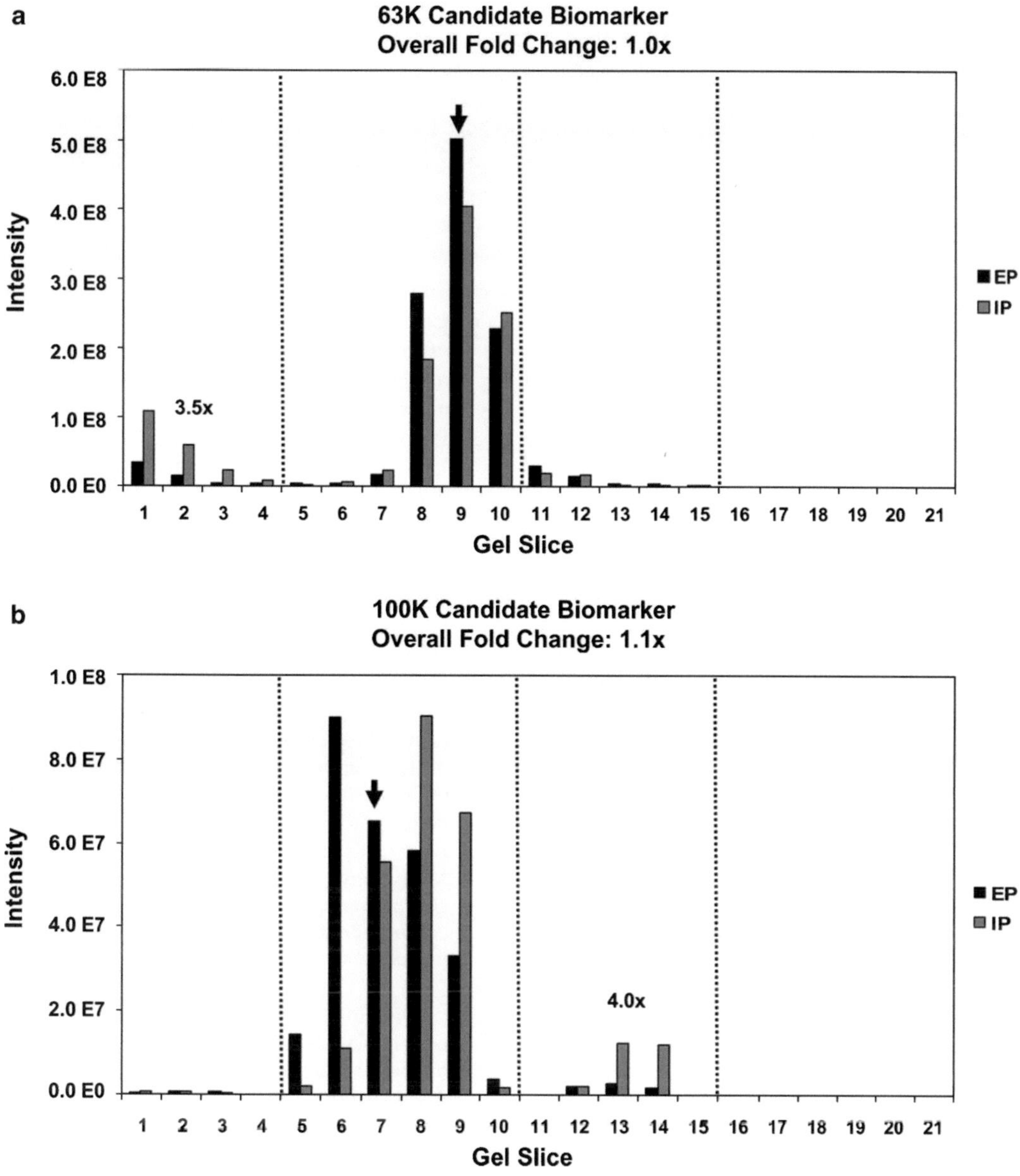

Fig. 4. Edge effects caused by dividing the proteome into multiple groups due to *software* capacity limitations. Individual peptides are spread across multiple gel slices as expected, and minor variations occur from sample to sample. These gel migration differences cause proteins within one region of the gel (*vertical dashed lines*) to exhibit an apparently significant fold change (values indicated above intensity *bars*). When peptide intensities for these multi-fraction proteins were combined across all gel fractions, the net fold change was not significant (values indicated below *panel title*). Variable intensity distributions across fractions are due to a number of factors, including extensive aggregation or glycosylation (**a**) and proteolysis (**b**). *Arrows* indicate expected MW (*gel slice*) of protein based on amino acid sequence.

Acknowledgments

This work was supported by NIH grants HD036455 (K.T.B.), CA120393 and CA131582 (D.W.S.), and by an institutional grant to the Wistar Institute (NCI Cancer Core Grant CA10815).

References

1. Jacobs, J. M., Adkins, J. N., Qian, W. J., et al. (2005) Utilizing human blood plasma for proteomic biomarker discovery, *J Proteome Res* **4**, 1073–1085.
2. Anderson, N. L., and Anderson, N. G. (2002) The human plasma proteome: history, character, and diagnostic prospects, *Mol Cell Proteomics* **1**, 845–867.
3. Hanash, S. M., Pitteri, S. J., and Faca, V. M. (2008) Mining the plasma proteome for cancer biomarkers, *Nature* **452**, 571–579.
4. Hu, S., Loo, J. A., and Wong, D. T. (2006) Human body fluid proteome analysis, *Proteomics* **6**, 6326–6353.
5. Hoffman, S. A., Joo, W. A., Echan, L. A., and Speicher, D. W. (2007) Higher dimensional (Hi-D) separation strategies dramatically improve the potential for cancer biomarker detection in serum and plasma, *J Chromatogr B Analyt Technol Biomed Life Sci* **849**, 43–52.
6. Omenn, G. S., States, D. J., Adamski, M., et al. (2005) Overview of the HUPO Plasma Proteome Project: results from the pilot phase with 35 collaborating laboratories and multiple analytical groups, generating a core dataset of 3020 proteins and a publicly-available database, *Proteomics* **5**, 3226–3245.
7. States, D. J., Omenn, G. S., Blackwell, T. W., et al. (2006) Challenges in deriving high-confidence protein identifications from data gathered by a HUPO plasma proteome collaborative study, *Nat Biotechnol* **24**, 333–338.
8. Tirumalai, R. S., Chan, K. C., Prieto, D., et al. (2003) Characterization of the low molecular weight human serum proteome, *Mol Cell Proteomics* **2**, 1096–1103.
9. America, A. H., and Cordewener, J. H. (2008) Comparative LC-MS: a landscape of peaks and valleys, *Proteomics* **8**, 731–749.
10. Bantscheff, M., Schirle, M., Sweetman, G., Rick, J., and Kuster, B. (2007) Quantitative mass spectrometry in proteomics: a critical review, *Anal Bioanal Chem* **389**, 1017–1031.
11. Faca, V., Pitteri, S. J., Newcomb, L., et al. (2007) Contribution of protein fractionation to depth of analysis of the serum and plasma proteomes, *J Proteome Res* **6**, 3558–3565.
12. Lee, H. J., Lee, E. Y., Kwon, M. S., and Paik, Y. K. (2006) Biomarker discovery from the plasma proteome using multidimensional fractionation proteomics, *Curr Opin Chem Biol* **10**, 42–49.
13. Pernemalm, M., Orre, L. M., Lengqvist, J., et al. (2008) Evaluation of three principally different intact protein prefractionation methods for plasma biomarker discovery, *J Proteome Res* **7**, 2712–2722.
14. Echan, L. A., Tang, H. Y., Ali-Khan, N., et al. (2005) Depletion of multiple high-abundance proteins improves protein profiling capacities of human serum and plasma, *Proteomics* **5**, 3292–3303.
15. Pieper, R., Su, Q., Gatlin, C. L., et al. (2003) Multi-component immunoaffinity subtraction chromatography: an innovative step towards a comprehensive survey of the human plasma proteome, *Proteomics* **3**, 422–432.
16. Qian, W. J., Kaleta, D. T., Petritis, B. O., et al. (2008) Enhanced detection of low abundance human plasma proteins using a tandem IgY12-SuperMix immunoaffinity separation strategy, *Mol Cell Proteomics* **7**, 1963–1973.
17. Steen, H., and Mann, M. (2004) The ABC's (and XYZ's) of peptide sequencing, *Nat Rev Mol Cell Biol* *5*, 699–711.
18. Schirle, M., Heurtier, M. A., and Kuster, B. (2003) Profiling core proteomes of human cell lines by one-dimensional PAGE and liquid chromatography-tandem mass spectrometry, *Mol Cell Proteomics* **2**, 1297–1305.
19. Old, W. M., Meyer-Arendt, K., Aveline-Wolf., et al. (2005) Comparison of label-free methods for quantifying human proteins by shotgun proteomics, *Mol Cell Proteomics* **4**, 1487–1502.
20. Mann, M., and Kelleher, N. L. (2008) Precision proteomics: the case for high resolution

and high mass accuracy, *Proc Natl Acad Sci U S A* **105**, 18132–18138.

21. Echan, L. A., and Speicher, D. W. (2009) Immunoaffinity depletion of high abundance plasma and serum proteins, In *The Protein Protocols Handbook* (Walker, J. M., Ed.) 2nd ed., pp 139-153, Humana Press, New York.
22. Rai, A. J., Gelfand, C. A., Haywood, B. C., et al. (2005) HUPO Plasma Proteome Project specimen collection and handling: towards the standardization of parameters for plasma proteome samples, *Proteomics* **5**, 3262–3277.
23. Sadiq, S. T., and Agranoff, D. (2008) Pooling serum samples may lead to loss of potential biomarkers in SELDI-ToF MS proteomic profiling, *Proteome Sci* **6**, 16.
24. Karp, N. A., and Lilley, K. S. (2009) Investigating sample pooling strategies for DIGE experiments to address biological variability, *Proteomics* **9**, 388–397.
25. Peng, X., Wood, C. L., Blalock, E. M., Chen, K. C., Landfield, P. W., and Stromberg, A. J. (2003) Statistical implications of pooling RNA samples for microarray experiments, *BMC Bioinformatics* **4**, 26.
26. Neubert, H., Bonnert, T. P., Rumpel, K., et al. (2008) Label-free detection of differential protein expression by LC/MALDI mass spectrometry, *J Proteome Res* **7**, 2270–2279.

Chapter 2

Intact Protein Separation by One- and Two-Dimensional Liquid Chromatography for the Comparative Proteomic Separation of Partitioned Serum or Plasma

Simon Sheng, Helena Skalnikova, Andrew Meng, John Tra, Qin Fu, Allen Everett, and Jennifer Van Eyk

Abstract

Serum- and plasma-based biomarker discovery requires technologies with specific capabilities: sufficient proteome coverage and depth, technical reproducibly, and the scalability to enable analysis on a large number of samples at reasonable cost. We have shown that plasma samples processed using IgY LC10 Proteome Partitioning kits to remove the most highly abundant proteins selectively, followed by intact protein separation by two-dimensional liquid chromatography (2DLC, chromatofocusing, and reversed phase) can uniquely enrich for middle to lower-abundant proteins. Equally, 1DLC (reversed phase) separation of intact proteins is complementary to 2DLC. The serial use of a single piece of equipment can be prohibitively time consuming and thus, this chapter also describes the harmonization of multiple LC instruments in order to minimize technical variation and ensure reproducibility. These technical improvements allow large numbers of individual clinical samples to be analyzed with multiple instruments in a timely manner, while retaining optimal reproducibility and allowing precise differential analysis at the proteome scale.

Key words: Liquid chromatography, Chromatofocusing, Reversed phase, Reproducibility, Protein load, Plasma, Serum, Proteomics, Biomarkers

1. Introduction

Serum/plasma biomarker discovery requires technology platforms that are capable of sufficient proteome coverage and depth, along with technical reproducibly and the scalability to analyze significant volumes of a large number of samples at a reasonable cost and within an acceptable time frame. Liquid chromatography (LC) of intact proteins has developed into a sophisticated separation

Richard J. Simpson and David W. Greening (eds.), *Serum/Plasma Proteomics: Methods and Protocols*,
Methods in Molecular Biology, vol. 728, DOI 10.1007/978-1-61779-068-3_2, © Springer Science+Business Media, LLC 2011

technique in the fast-growing field of proteomics (1, 2). Protein pre-fractionation of a complex mixture is able to reduce complexity prior to mass spectrometry (MS) analysis, leading to increased proteomic coverage and deeper identification of post-translational modifications (PTMs), isoforms, and potentially of single nucleotide polymorphisms (3, 4). Our laboratory has optimized the 2DLC system that combines chromatofocusing (CF) and reversed-phase (RP) HPLC for serum and plasma analysis, using the ProteomeLab™ PF 2D Protein Fractionation system, a commercially available Beckman Coulter instrument (4) (Fig. 1). Moreover, we have found that intact protein separation by RP alone (1DLC) is complementary to 2DLC.

Timeliness, however, requires simultaneous use of multiple, harmonized systems. It is especially crucial to ensure reproducibility in analyzing samples between multiple PF 2D instruments when a large number of individual clinical samples are analyzed for biomarker discovery. However, harmonization of multiple PF 2D systems is particularly difficult because it involves two sequential separation steps: high performance CF and RP chromatography. Any variation in first-dimension fraction collection (by pH range) will alter the subsequent overall profile of second dimension hydrophobicity chromatograms, which complicates

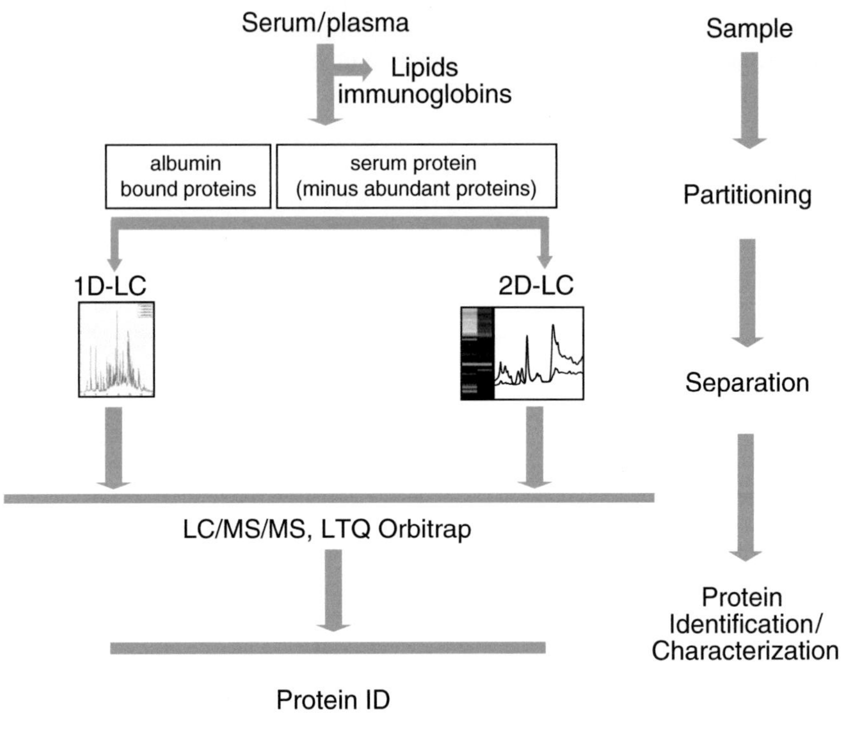

Fig. 1. Schema of 1DLC and 2DLC serum/plasma intact protein separation pipeline.

both chromatogram alignment and differential comparison of the multiple samples in a clinical sample set. In a 2008 publication, Suberbielle et al. (5) used hardware and software improvements to provide a comprehensive assessment of PF 2D performance. Their results demonstrated a high level of reproducibility of a single PF 2D system. In this report, we outline the parameters needed to harmonize multiple 2DLC systems for the separation of intact proteins – in particular, for plasma partitioning of albumin, IgG, or the 12 most highly abundant proteins. This required the precise alignment of four PF 2D systems, with improvements in sample preparation, HPLC tubing and columns, running conditions, sample load, and instrument hardware and software.

2. Materials

2.1. Equipment

1. For affinity chromatography use to remove the high(est) abundant protein, use ProteomeLab™ PPS Proteome Partitioning System (Beckman Coulter Inc., CA) or any 1D HPLC (e.g., System Gold system with an auto-injector, Beckman Coulter) system consisting of an isocratic pump, a UV detector, and a fraction collector.
2. For 1DLC RP intact protein separations, use any 1D HPLC (e.g., System Gold system with an auto-injector, Beckman Coulter) or second dimension of the 2D LC systems described below.
3. For 2DLC intact protein separation, use any 2DLC system (e.g., ProteomeLab PF 2D, Beckman Coulter). The first dimension of PF 2D consists of a single piston pump (HPCF Module), a manual injector for sample introduction, a pH monitor, and a UV detector. The first-dimension separation consists of CF based on charge. Fractions from the first dimension are collected into 96-deepwell plates (Nunc, plates; #278752, well cap; #276002) in the fraction collector/injector (FC/I Module), which is the interface between the first and second dimensions. The fractions are automatically introduced into the second-dimension RP chromatography, which separates on the basis of hydrophobicity. The second dimension consists of a binary pump system (HPRP Module), a column heater, a UV detector, and a fraction collector. Fractions are collected into 96-well plates for mass spectrometry analysis. The hardware is controlled by a 32-Karat software (Beckman Coulter). With this system, the first and second dimensions occur sequentially in an automatic manner.

2.2. Plasma/Serum Partitioning

2.2.1. Immunoglobulin (IgG) Depletion

1. Reagent: HiTrap protein G HP column (GE Healthcare, #17-0404-01).
2. 10 mM HEPES is made from 1 M HEPES, pH 7.4 stock: To 119.15 g of HEPES, add dH_2O and adjust the volume to 300 mL, then add concentrated NaOH dropwise to adjust the pH to 7.4, bring up the volume to 500 mL with dH_2O, filter with a 0.45-μM filter, and store at room temperature (RT; 25°C).
3. 0.10 M NaCl is made by diluting a 5 M NaCl stock: To 292.2 g of NaCl, add dH_2O and make the volume to 1 L, filter with a 0.45-μM filter, and store at RT.
4. Mini-spin column (Pierce, #69705).

2.2.2. Chemical Removal of Albumin

1. Ethanol is refrigerated in a flame-retardant container in an explosion-proof refrigerator or cold room (4°C).
2. 0.15 M NaCl is made by diluting a 5 M NaCl stock: To 292.2 g of NaCl, add dH_2O and make the volume to 1 L, filter with a 0.45-μM filter, and store at room temperature.

2.2.3. IgY Immunodepletion of Top Abundant Plasma Proteins

1. Dilution buffer is proprietary but includes 150 mM or 1.5 M NaCl and 10 or 100 mM Tris–HCl, pH 7.4, and is provided as 10× concentrates with the column (Beckman Coulter, #A24355; Sigma–Aldrich, #S4199) (see Note 1).
2. Striping buffer is proprietary but includes 100 mM-1 M glycine-HCl, pH 2.5, and is provided as 10× concentrates with the column (Beckman Coulter, #A24355; Sigma–Aldrich, S4324) (see Note 1).
3. Neutralization buffer is proprietary but can includes 100 mM-1 M Tris–HCl, pH 8.0, and is provided as 10× concentrates with the column (Beckman Coulter, #A24355; Sigma–Aldrich, S4449) (see Note 1).
4. 0.45-μm centrifugation filters (Costar® Spin-X® cellulose acetate filters, CTA, Corning, #8163).
5. IgY_{12}-LC10 column (Beckman Coulter, #A24355), or IgY_{14}-LC10 (Seppro®, #S5074-1EA, Sigma–Aldrich, St. Louis) bovine serum albumin (BSA) Sigma–Aldrich, St. Louis.

2.3. Concentration of Depleted Sample Using Ultra Centrifugal Filtration

1. PF 2D Start buffer, pH 8.5, is proprietary and is provided in the PF 2D buffer kit (Beckman Coulter, #A51678).
2. Plasma/serum denaturing buffer consists of 7.5 M urea, 2.5 M thiourea, 12.5% glycerol, 2.5% *n*-octylglucoside, 62.5 mM Tris–HCl, pH 8.0, and 1.25 mM EDTA.
   To make 12.5% glycerol, add 12.5 mL glycerol to a 100-mL volumetric flask and fill up to 100 mL by adding dH_2O.
   To a 100-mL beaker, add 20 mL of 12.5% glycerol and under stirring continously add 0.123 g Tris base, 0.333 g Tris–HCl,

9.52 g thiourea, 22.52 g urea, 1.25 g *n*-octylglucoside, and 0.023 g EDTA. Stir until the components are dissolved completely, then transfer the buffer to a 50-mL volumetric flask, and make up the volume to 50 mL by adding 12.5% glycerol. Aliquot into 2-mL Eppendorf tubes and store up to 6 months at –20°C.

3. Human serum albumin (Sigma–Aldrich, #A9511 St. Louis).
4. Amicon Ultra-15 centrifugal filtration units with 5 kDa molecular weight cutoff (MWCO) (Millipore, #UFC900524) (see Note 2). Another option is to concentrate the flow through directly on the 1DLC reversed phase column.

2.4. Reversed-Phase (1DLC) Separation of Intact Protein-Depleted Sample

1. Solvent A consists of 0.1% HPLC-grade trifluoroacetic acid (TFA) in HPLC-grade water (1 mL of TFA in 1,000 mL dH_2O). Solvents are stable at RT for 2 weeks.
2. Solvent B consists 0.08% TFA in HPLC-grade acetonitrile (800 μL of TFA in 1,000 mL acetonitrile). Solvents are stable at RT for 2 weeks.
3. Test kit solvent 50% methanol in HPLC-grade water (v/v).
4. Isocratic test mix (PN 238447 Beckman Coulter). It is a three-component QC test mix and contains methyl *P*-hydroxybenzoate, propyl *P*-hydroxybenzoate, and ethyl paraben.
5. Nonporous C18 RP column (Beckman Coulter) or a Jupiter-5 micron C18 300 Å 250 × 4.6 mm RP column (Phenomenex, Torrance, CA, PN 006-4053-E0).
6. 96-Deepwell plates (Nunc deepwell, U96 PP2).

2.5. 2DLC Separation of Intact Protein-Depleted Samples

1. PF 2D Start buffer, pH 8.5, is a proprietary buffer provided in the PF 2D buffer kit (Beckman Coulter, #A51678). The buffer includes urea and a reducing agent to maintain protein solubility and other proprietary reagents.
2. Elution buffer, pH 4.0, is a proprietary buffer provided in the PF 2D buffer kit (Beckman Coulter, #A51678). The buffer includes urea and a reducing agent to maintain protein solubility and other proprietary reagents.
3. Saturated solution (50 mg/mL) of iminodiacetic acid and 1 M NH_4OH solution is necessary for precise pH adjustment.
4. PF 2D high-salt wash (HISS) buffer is a proprietary buffer provided in the PF 2D buffer kit (Beckman Coulter, #A51678) that contains 1 M NaCl in 30% isopropanol (prepared from 58.44 g NaCl dissolved in 30% HPLC-grade isopropanol in HPLC-grade water and filtered through 0.2-μm filter), stored at +4°C for 1 week.
5. Solvent A comprises 0.1% TFA in HPLC-grade water (1 mL of TFA in 1,000 mL of water). Solvents are stable at RT for 2 weeks.

6. Solvent B consists of 0.08% TFA in HPLC-grade acetonitrile (800 μL of TFA in 1,000 mL of acetonitrile). Solvents are stable at RT for 2 weeks.
7. PD-10 column (GE Healthcare, #17-0851-01) packed with Sephadex™ G-25 medium for sample desalting and buffer exchange prior to CF.
8. High-performance chromatofocusing column (CF column, 250×2.1 mm) included in the PF 2D column kit (Beckman Coulter, #A51680).
9. High-resolution nonporous C18 reverse-phase column (RP column, 33×4.6 mm) included in the PF 2D column kit (Beckman Coulter, #A51680).

2.6. Preparation for Reversed-Phase Fractions for Subsequent MS Analysis

1. 20 mM NH_4HCO_3 is made from a 1 M NH_4HCO_3 stock. 1 M stock is made fresh every 2 weeks: add 3.95 g of NH_4HCO_3 to 40 mL of H_2O and then make up the volume to 50 mL. The 20 mM solution is made fresh prior to tryptic digestion.
2. HPLC-grade TFA (Pierce).
3. Sequencing-grade modified trypsin (Promega, #V511A).
4. Speed vacuum system.

2.7. Analysis by LC–MS/MS

1. Solvent A, which is comprised of 4% (v/v) aqueous acetonitrile with 0.1% formic acid.
2. Solvent B, which is comprised of 90% (v/v) aqueous acetonitrile in 0.1% formic acid.
3. BioBasic C18 RP nanocolumn for LTQ-Orbitrap. PicoFrit™ nanocolumn packed with 5 μm BioBasic™ C18, 300 Å pore size column (New objective, #PF7515-100H052-3P).
4. LTQ-Orbitrap (ThermoFisher, CA) with an Agilent 1,200 nano-LC system, although other high-end LC–MS/MS instrumentation are acceptable for label-free quantification.

3. Methods

3.1. Plasma/Serum Preparation

The first step in a biomarker development pipeline is often the partitioning of the high(est) abundant proteins (6). In our laboratory, we have found that delipidation is required prior to partitioning of plasma/serum samples. We routinely deplete IgG using Protein G affinity chromatography, albumin using a chemical-based method (albumin remains soluble and majority of the other proteins are precipitated), or the top 12 or 14 abundant plasma proteins using IgY affinity chromatography.

3.1.1. Delipidation of Plasma/Serum

1. Plasma/serum is centrifuged at 15,000 × *g* at 4°C for 15 min (see Notes 3 and 4).
2. The samples are kept on ice, and the insoluble matter and the top lipid layer removed and discarded (see Note 5).

3.1.2. Immunoglobulin (IgG) Depletion

1. Protein G beads are equilibrated with 100 mM NaCl and 10 mM HEPES, pH 7.4, and washed three times with this buffer.
2. The protein G beads are transferred to a Mini-spin column.
3. Spin at 3,000 × *g* for 3 min at 4°C.
4. 100 μL of (delipidated) sample is diluted with 200 μL of 0.15 M NaCl and added to a Mini-spin column containing 200 μL of protein G beads and incubated with gentle rocking for 1 h at RT.
5. IgG-depleted plasma/serum is separated from protein G beads by centrifugation (1,500 × *g* for 3 min).
6. The supernatant (~300 μL, IgG-depleted serum) is collected.

3.1.3. Chemical Removal of Albumin

1. Plasma, serum, or the IgG-depleted serum/plasma is incubated at 4°C with gentle stirring for 1 h (see Note 6).
2. 95% (v/v) cold aqueous ethanol is added until a final concentration of 42% (e.g., add 252 μL of 95% ethanol to 300 μL of IgG-depleted serum/plasma, making the final concentration of 42% ethanol) (see Note 7).
3. The mixture is incubated for one additional hour at 4°C with gentle stirring.
   Albumin remains in the supernatant and the albumin-depleted low-abundant proteins precipitate. Low-abundant proteins are pelleted by centrifugation at 16,000 × *g* for 45 min at 4°C (see Note 8).

3.1.4. IgY Immunodepletion of Top Abundant Plasma Proteins

High-capacity LC10 affinity IgY_{12} (Beckman) or IgY_{14} LC10 affinity column (Sigma) utilizes IgY-conjugated microbeads packed into a 10-mL column bed. A total of 14 high-abundant plasma proteins (including albumin, total IgG, α1-antitrypsin, IgA, IgM, transferrin, haptoglobin, α1-acid glycoprotein (orosomucoid), α2-macroglobulin, fibrinogen, complement C3, HDL (apolipoproteins A-I and A-II), and LDL (mainly apolipoprotein B)) are depleted from human serum/plasma in a single step (7) (see Note 9). In all, 20 mg of plasma/serum proteins loaded on the IgY-LC10 column can result in about 750 μg of depleted plasma proteins. Thus, four IgY-LC10 runs may be necessary to ensure 2.5 mg of low-abundant proteins for subsequent 1DLC or 2DLC (PF2D) analysis. The whole immunodepletion is performed at RT (25°C) and the dilution, stripping, and neutralization buffers as well as column should be equilibrated at RT before use (see Note 10):

1. Equilibrate the IgY column on a ProteomeLab™ PPS Proteome Partitioning System or any 1D HPLC with the dilution buffer for 30 min at a flow rate of 2 mL/min and 5 min at a flow rate of 0.5 mL/min (see Notes 11 and 12).
2. Dilute 250–300 μL of (delipidated) serum/plasma with 450–750 μL of dilution buffer, filter through 0.45-μm Spin-X centrifugation filter, and load the diluted sample onto the IgY column.
3. Collect the flow-through fraction in a 50 mL tube (depleted sample, monitored by UV detector) at a flow rate 0.5 mL/min of dilution buffer. Load capacity is 1.25 mL of serum or plasma.
4. Elute the bound fraction (the 12 or 14 most abundant serum proteins) by stripping buffer at a flow rate of 2 mL/min for 15 min. Collect the bound fraction into a new 50-mL tube. Neutralize the bound fraction with 0.1× fraction volume of 10× neutralizing buffer.
5. Neutralize the column by adding neutralization buffer for 6 min at a flow rate 2 mL/min.
6. Re-equilibrate the column for 5 min at 2 mL/min and 2 min by addition of dilution buffer at 0.5 mL/min. Inject the next sample. Under these conditions, up to 100 depletion runs can be performed on a single LC10 affinity IgY column.
7. The flow-through (~8–12 mL per run, depleted sample) is eluted and the sample is immediately frozen at −80°C. The flow-through is analyzed in the subsequent steps, however. The bound fractions also have the potential to deliver biomarkers (8). The collected samples are concentrated by ultrafiltration before use for 1D or 2DLC (see Note 13).

3.1.5. Concentration of Depleted Sample Using Ultracentrifugation Filtration

1. To minimize nonspecific protein absorption to the ultrafiltration membrane, Amicon Ultra 15 filter devices with 5 K MWCO are seasoned by passing 3 mL of the PF 2D start buffer and centrifuged using swinging bucket rotor at 3,900×*g* at 4°C for 10 min (see Note 14).
2. Flow-through fractions (up to 15 mL) are placed in pre-wet Ultra-15 filter devices according to the manufacturer's instructions and centrifuged at 3,900×*g* at 4°C for 30 min. Usually three to four centrifugation steps are required to concentrate all of the flow-through fractions from four IgY-LC10 runs. The entire flow-through volume is concentrated to 0.5 mL.
3. To the concentrated solution, add 2 mL of plasma/serum denaturing buffer and mix for 30 min at RT. The mixed solution is added onto an equilibrated PD-10 gel filtration column for buffer exchange. Proteins are eluted with 3.5 mL of PF 2D start buffer. The sample is transferred into two centrifugation microtubes and insoluble material pelleted by centrifugation at 20,000×*g*, 18°C for 60 min.

4. Protein concentration of the supernatant after ultrafiltration is determined by Micro BCA™ Protein Assay kit (Pierce).
5. The samples are frozen at −80°C or analyzed directly. It is strongly recommended not to re-freeze protein solutions containing start buffer.

3.2. Reversed-Phase (1DLC) Separation of Intact Protein-Depleted Sample

3.2.1. 1DLC Fractionation

1. A total of 100–500 μg of partitioned plasma samples can be separated on a RP column provided with the PF 2D chemistry kit or other RP column of interest.
2. A linear 3.33% B/min gradient is used in which solvent A is 0.1% aqueous TFA and solvent B is 0.08% TFA in acetonitrile. Flow rate is 0.75 mL/min at RT.
3. Between 16 and 32 fractions are collected into 96-deepwell plates and stored at −80°C or analyzed directly.

3.2.2. Testing Reproducibility and Column Performance of Reversed-Phase 1DLC

1. Isocratic Test Mix containing methyl *P*-hydroxybenzoate, ethyl paraben, and propyl *P*-hydroxybenzoate is injected and run using an isocratic gradient comprised of Solvent Test. Flow rate is 0.75 mL/min and monitored at 210–214 nm (see Note 15).

3.3. 2DLC Separation of Intact Protein-Depleted Sample

3.3.1. 2DLC Fractionation

1. For buffer exchange, 2.5 mL of the prepared depleted samples dissolved in plasma/serum denaturing buffer is loaded onto a PD-10 gel filtration column equilibrated with start buffer. The PD-10 flow-through is discarded (see Note 16).
2. The sample is eluted with 3.5 mL of PF 2D Start buffer into a clean tube (see Note 17).
3. Protein concentration is determined (Subheading 3.1.5), verifying the amount of sample to be loaded onto the PF 2D.
4. CF is carried out at RT at the flow rate of 0.2 mL/min. The CF column is equilibrated for 130 min with the proprietary PF 2D Start buffer at pH 8.50 (see Note 18).
5. 2.5–10 mg of total protein is injected on the equilibrated CF column in start buffer alone or with 20% isopropanol. A linear pH gradient from pH 8.5 to 4.0 is achieved by mixing two buffers with different pH (PF 2D Start buffer and PF 2D Eluent buffer). The pH gradient is achieved by introducing increasing amounts of the PF 2D Eluent buffer (pH 4.0) alone or 20% isopropanol (see ref. 4) for over 95 min. The separation is monitored by an inbuilt pH probe and UV detector at 280 nm.
6. The CF column is washed with high salt wash (HISS) for 45 min. Furthermore, 30% isopropanol is added to increase protein elution and column cleanup.
7. The CF column can be re-equilibrated or stored in water (see Note 19).

8. Fractions from the first dimension are collected in the fraction collector/injector (FC/I Module) cooled to 10°C, which is the interface between the first and second dimensions. The fractions are initially collected every 8.5 min until the pH gradient portion of the run, when fractions were collected at 0.3-pH intervals between pH 8.3 and 4.0, followed by a HISS wash (1 M NaCl in 30% isopropanol) to elute acidic and hydrophobic proteins.
9. 250 μL of each first-dimension fraction is automatically introduced into the second-dimension RP chromatography column. A linear 0–100% B gradient is used, in which solvent A is 0.1% aqueous TFA and solvent B is acetonitrile with 0.08% TFA. The duration of the gradient is 35 min, with the flow rate of 0.75 mL/min. The column is run at 55°C with UV detection at 214 nm. The RP column is re-equilibrated for 10 min by 100% solvent A, with the next 1D fraction automatically injected.
10. Fractions from the second dimension are automatically collected into 96-well plates by a fraction collector.
11. The fractions are stored at −80°C prior to carrying out the steps required for mass spectrometry analysis.

3.3.2. Testing Reproducibility and Column Performance of 2DLC on Same Instrument

To test reproducibility of the first dimension (CF step), the peak retention time and the elution pH of two selected and well-resolved peaks (Fig. 2.) are measured across multiple runs as outlined below. RP column is analyzed using the method described above:

1. Pooled human depleted sample was run according to the method described (Subheading 3.4) (see Note 20).
2. Determine relative standard deviation (RSD%) for retention time and elution pH (see Note 21).

3.4. Preparation of Reversed-Phase Fractions for Subsequent MS Analysis

1. The fractions are concentrated to dryness using a SpeedVac concentrator (see Note 22).
2. 20 mM NH_4HCO_3 is added to resuspend proteins from each of the RP fractions obtained from the 1DLC or 2DLC.
3. Sequencing-grade modified trypsin at an enzyme-to-substrate ratio of 1:50 is added and incubated overnight (>12 h) at 37°C.
4. 10% (v/v) aqueous TFA is added to stop the digestion.
5. The digested peptide solution is concentrated to dryness and stored at −80°C until mass spectrometry analysis.

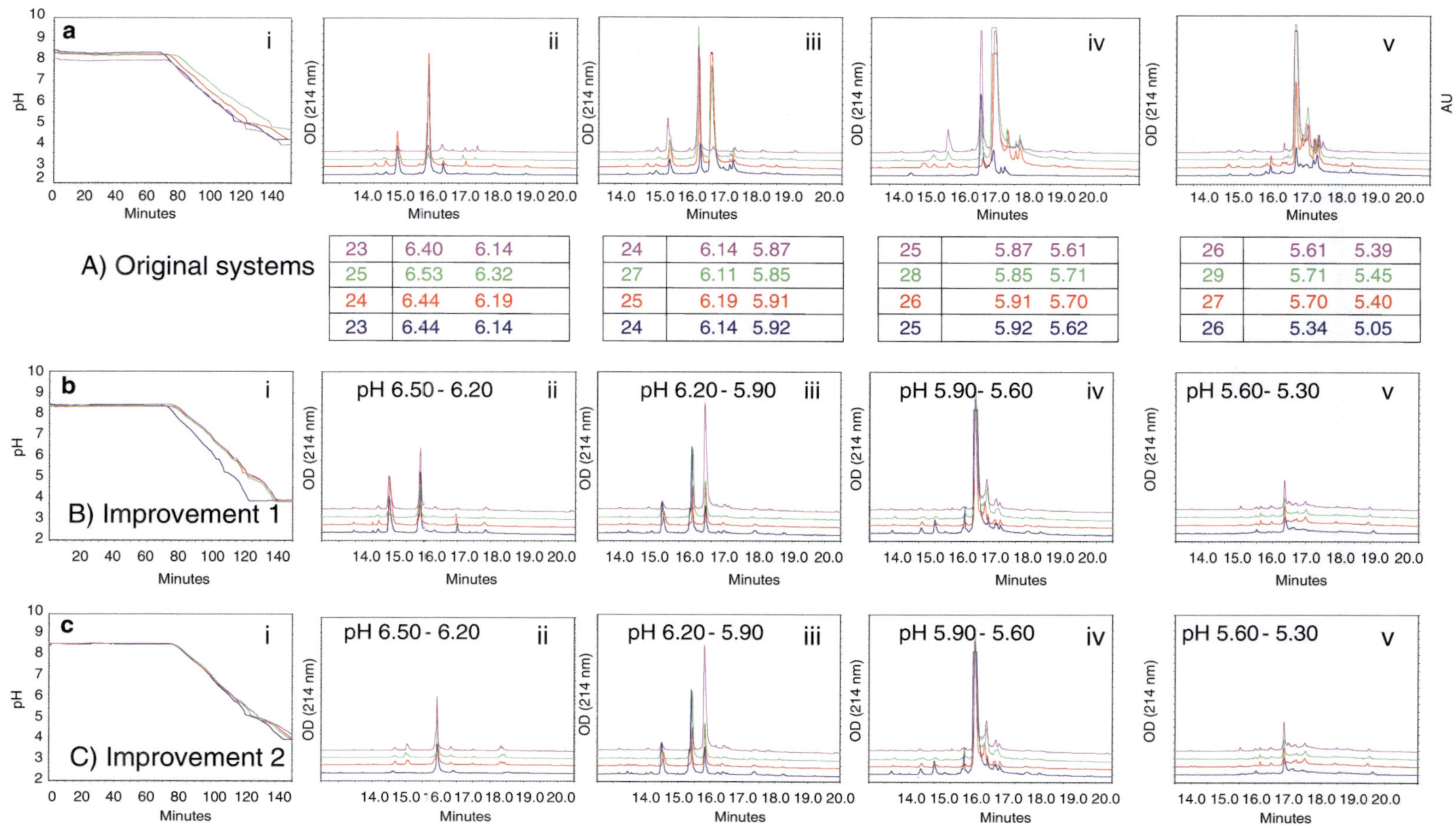

Fig. 2. Harmonization of four PF 2D systems: (**a**) Original commercial systems; (**b**) Improvement 1; (**c**) Improvement 2. In each series, (i) shows first-dimension pH profiles (pH 8.3-4), while (ii)–(v) show selected second-dimension chromatograms corresponding to four pH ranges: 6.50–6.20, 6.20–5.90, 5.90–5.60, and 5.60–5.30. All traces from the four systems were overlaid. Note that the harmonization in (**b**) and (**c**) allowed reproducible comparisons of second-dimension chromatograms across four different instruments. However, in order to overlay four chromatograms in (**a**), different fractions (Tables 1 and 2 under chromatograms of Aii–Av) were selected based on chromatogram similarities rather than fraction number.

3.5. Analysis by LC–MS/MS

For more details on LC–MS/MS experiments on the LTQ-Orbitrap MS instrument, see ref. 9:

1. Peptides from the trypsin-digested LC fractions are resuspended in 6 μL of resuspension buffer consisting of 4% (v/v) aqueous acetonitrile, with 0.1% formic acid.
2. Samples (3 μL) are loaded onto a 75 μm × 10 cm BioBasic C18 reversed-phase column (see Note 23).
3. Peptides are eluted into an LTQ-Orbitrap using an nano-LC system. The linear HPLC gradient of 5–60% B comprises 90% acetonitrile/water in 0.1% formic acid over 30–60 min (see Note 24).
4. Data from the LTQ-Orbitrap were searched against the IPI or Uniprot human databases using Sorcerer Sequest (Sagen, CA). The search results were validated and displayed using Sorcerer-integrated Trans-Proteomic Pipeline (TPP). BLAST (http://www.ebi.ac.uk/Tools/blast2/index.html) searching of the amino acid sequence for each protein was conducted to remove protein name redundancy. The cut-off criteria were minimum probabilities of 0.5 for both protein and peptide (0.93 sensitivity and 0.06 error).
5. Each protein isoform or processed protein was confirmed by matching a tryptic peptide fragment to a unique amino acid sequence on the isoform or intact protein, and the peptides were carefully handled through manual inspection of the tandem MS (MS/MS) spectra.
6. To create a nonredundant database, the protein identifications were examined manually in the database for possible redundancies, including multiple names and homologies, because numerous instances exist where the same protein contains multiple database protein identifications.

3.6. Harmonization of Multiple PF 2D (2DLC) Systems

In our laboratory, we have aligned four PF 2D systems to allow for harmonization among the systems. This was required due to the large differences in their gradient and fraction collection. Figure 2 shows these four systems, comparing the pH curves of the first dimension (CF) and selected second-dimension (RP) chromatograms the pH curves of the first dimension (CF) for analysis. Panel A shows the analysis on four unaltered commercial systems (Panel Ai–Av), while Panels B and C show equivalent samples separated on the same systems, but after system modification to improve performance and harmonization (Panels Bi–Bv and Ci–Cv). In Panel A (original commercial systems), the variance in fractions collected during the pH gradient illustrates a lack of first-dimension alignment across the four instruments (online Table 1). As a second-dimension example, fraction 21 was collected at four different pH ranges (such as 7.01–6.71, 7.29–6.99, 7.60–7.35, and 6.99–6.70) on the four different systems,

Table 1
High reproducibility of reverse-phase technique, using a three-compound standard mix

Reversed-phase test mix	Retention time (RSD%)	Area (RSD%)	Height (RSD%)
Methyl *P*-hydroxybenzoate	0.6	7.5	6.9
Ethyl paraben	0.8	8.3	7.3
Propyl *P*-hydroxybenzoate	1.1	7.6	7.6

The standard mix was separated on six different, aligned PF 2D systems over several days. The standard was analyzed two to six times per instrument. Retention time, area under the peak, and peak height were determined for each analysis and used to calculate percent relative standard deviation (RSD%)

resulting in significant variation in the chromatograms produced by the various systems (Fig. 2a (ii–v)), and making comparisons of the resulting second-dimension chromatograms difficult.

3.7. Improvement to the PF 2D Systems

1. Refit each instrument with identical tubing diameter and length for each component of the various systems.
2. A software package (SP1, Beckman Coulter) was installed that enforces initiation of the fraction collection protocol (used during the pH gradient, and collecting every 0.3 pH units) immediately when the pH drops below a fixed pH of 8.3, rather than waiting to finish the preset 8.5-min fraction collection protocol used during the void volume (>pH 8.3). This ensures that fractions collected during the pH gradient were aligned at the same starting pH, independent of the slight variations in elution time.
3. The pH electrodes used for each of the four 2DLC systems must be selected based on their similar pH responsiveness during initial pH testing (all electrodes measured within pH ±0.1) (see Note 25).
4. Flat-bottom electrodes and a corresponding flow cell (#PN A48657 Beckman Coulter) were installed. Flat-bottom electrodes were specifically designed for increased accuracy by better withstanding the high back pressures that can occur in the LC system (Fig. 3, Table 2) (see Note 26).

4. Notes

1. 10× Buffer concentrates should be stored at 4°C. Once diluted, the buffers should be stored at 4°C for maximum 1 week.

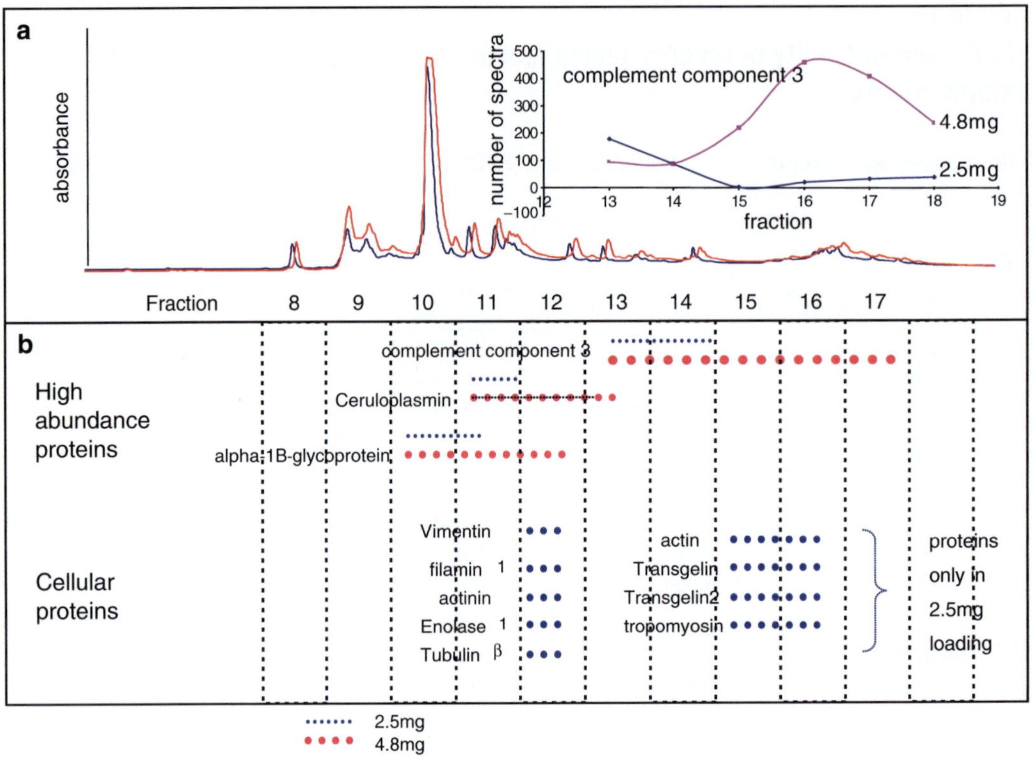

Fig. 3. Differential loading of IgY-12 partitioned plasma in the PF 2D system. Displayed by varying protein load (2.5, 4.8, or 10 mg); the optimal protein load was 2.5 mg based on the total number of proteins identified by LC–MS/MS. (**a**) Overlaid second-dimension chromatograms of fraction 21 (pH 4.69–4.99) from two different protein load: 2.5 and 4.8 mg. The insert is the plot of number of spectra of complement component 3 vs. fraction number. Note that CC3 spreads over a large number of multiple fractions in the 4.8-mg load. (**b**) Selected proteins identified by MS in corresponding fractions. High-abundance plasma proteins such as CC3, ceruloplasmin, and alpha-1β-glycoprotein spread over more fractions in the 4.8-mg load (*large dotted line*) than in the 2.5-mg load (*small dotted line*). Cellular proteins (defined by subcellular location) were only identified in the 2.5-mg load.

2. The 5 K MWCO membrane minimizes the losses of serum/plasma proteins and also prevents membrane clogging.
3. Carrying out an initial delipidation will often reduce clogging of the affinity columns and increase both capacity and longevity of the columns.
4. Centrifugation mostly removes triglyceride-rich chylomicrons and low-density lipoproteins.
5. This method has been previously reported (6). The standard serum/plasma delipidation method using butanol/di-isopropyl ether extraction, methanol/chloroform extraction, and a serum/plasma lipid and lipoprotein absorbent, PHM-L LOPOSORB absorbent (EMB Biosciences).
6. If IgG depletion step is not performed, the addition of 0.15 M NaCl to a final concentration of 100 mM in the serum or plasma is required.

Table 2
First-dimension fraction collection tables for four PF 2D systems (S1–S4)

S-1		S-2		S-3		S-4 (8)	
F#	pH range	F#	pH range	F#	pH range	F#	pH range
Panel A							
21	7.01–6.71	21	7.29–6.99	21	7.60–7.35	21	6.99–6.70
22	6.71–6.44	22	6.99–6.69	22	7.35–7.10	22	6.70–6.40
23	6.44–6.14	23	6.69–6.44	23	7.10–6.82	23	6.40–6.14
24	6.14–5.92	24	6.44–6.19	24	6.82–6.53	24	6.14–5.87
25	5.92–5.62	25	6.19–5.91	25	6.53–6.32	25	5.87–5.61
26	5.62–5.34	26	5.91–5.70	26	6.32–6.11	26	5.61–5.39
27	5.34–5.05	27	5.70–5.40	27	6.11–5.85	27	5.39–5.12
28	5.05–5.00	28	5.40–5.10	28	5.85–5.71	28	5.12–4.81
29	5.00–4.91	29	5.10–4.94	29	5.71–5.45	29	4.81–4.63
30	4.91–4.79	30	4.94–4.87	30	5.45–5.25	30	4.63–4.57
31	4.79–4.53	31	4.87–4.78	31	5.25–5.04	31	4.57–4.46
32	4.53–4.29	32	4.78–4.60	32	5.04–4.74	32	4.46–4.25
33	4.29–4.16	33	4.60–4.36	33	4.74–4.69	33	4.25–3.96
Panel B							
vv	8.38–8.39	vv	8.32–8.32	vv	8.35–8.34	vv	8.38–8.37
16	6.49–6.19	16	6.49–6.19	16	6.49–6.19	16	6.50–6.20
17	6.19–5.89	17	6.19–5.89	17	6.19–5.89	17	6.20–5.90
18	5.89–5.59	18	5.89–5.59	18	5.89–5.59	18	5.90–5.59
19	5.59–5.29	19	5.59–5.28	19	5.59–5.29	19	5.59–5.29
20	5.29–4.99	20	5.28–4.99	20	5.29–4.99	20	5.29–4.99
21	4.99–4.69	21	4.99–4.69	21	4.99–4.69	21	4.99–4.69
22	4.69–4.39	22	4.69–4.39	22	4.69–4.39	22	4.69–4.39
23	4.39–4.09	23	4.39–4.09	23	4.39–4.09	23	4.39–4.09
Panel C							
vv	8.53–8.52	vv	8.55–8.55	vv	8.57–8.56	vv	8.56–8.56
15	6.80–6.49	16	6.80–6.50	16	6.80–6.49	16	6.80–6.50
16	6.49–6.19	16	6.50–6.20	17	6.49–6.19	17	6.50–6.20
17	6.19–5.89	17	6.20–5.90	18	6.19–5.89	18	6.20–5.90
18	5.89–5.59	18	5.90–5.60	19	5.89–5.59	19	5.90–5.60
19	5.59–5.29	19	5.60–5.29	20	5.59–5.29	20	5.60–5.30
20	5.29–4.99	20	5.29–4.99	21	5.29–4.99	21	5.30–4.99
21	4.99–4.69	21	4.99–4.69	22	4.99–4.69	22	4.99–4.77
22	4.69–4.39	22	4.69–4.39	23	4.69–4.39	23	4.77–4.47
23	4.39–4.09	23	4.39–4.09	24	4.39–4.09	24	4.47–4.17

Panel A (original commercial systems) shows mismatched fractions in pH range; fractions in Panel B (Subheading 3.7, step 1) and Panel C (Subheading 3.7, step 2) are perfectly aligned for each PF2D system (S1 to S4)

7. It is very important to maintain NaCl concentration at 100 mM and EtOH concentration at 42% to enhance the efficiency of precipitation.
8. Over 95% of the albumin and IgG are removed, as shown in a previously reported publication (6).

9. IgY antibodies are produced by injecting hens with highly purified mammalian protein antigens, and then antigen affinity purifying the polyclonal antibodies recovered from egg yolks (10).
10. For regeneration of the column, one can carry out wash steps; however, some bound proteins will remain. Therefore, randomization of the patient samples being depleted is needed.
11. Please see ref. 11 for methods that can be used to clean up the column between subsequent patient samples. However, these methods most likely will reduce longevity of the affinity column.
12. For large-scale studies and for maximum efficiency, two LC systems can be used by a single investigator. In a normal day, four runs per instrument can be carried out, allowing collectively eight samples to be depleted per day.
13. In some cases, it was necessary to combine the flow-through fractions from four IgY affinity chromatography runs to obtain the 2.5 mg of partitioned sample for downstream analysis on the PF 2D system. When more than one run is required for a sample, then the flow-through is collected and pooled prior to the next steps. This results in ~40–50 mL of flow-through prior to concentration.
14. If the only downstream fractionation method is RP, then this pre-concentrating step is not required. In this case, one can load the flow-through directly onto the 1DLC column. We use a 5- or 10-mL injection loop and load using isocratic start conditions. As needed, sequential loading of the RP column can be done until all of the flow-through has been loaded. At this point, the gradient can be started.
15. See Table 1 for expected performance (relative standard deviation (RSD%) for retention time, peak height, and peak area) of each component of the test mix on six different aligned PF 2D systems using only the second-dimension RP columns. The standards were analyzed two to six times per instrument.
16. We have determined the capacity of the PF 2D system using 2.5, 4.8, and 10 mg of IgY-12 partitioned plasma (all originating from the same pooled sample). All fractions from the first dimension were analyzed by second-dimension fractionation, with fraction 21 (pH 4.69–4.99) selected for tryptic digestion and MS analysis with the LTQ-Orbitrap. The optimal protein load was found to be 2.5 mg, based on the total number of proteins identified compared to the 4.8- or 10-mg loads; this is most likely the result of less peak overlap among adjacent fractions (Fig. 2c). Note that with increased protein loads at 4.8 and 10 mg, higher-abundance proteins remaining after IgY-12 partitioning (such as complement C3, ceruloplasmin, and alpha-1B-glycoprotein) spread over a large

number of multiple fractions. At a 2.5-mg load, these proteins were observed by MS in only one or two adjacent pH fractions. Similarly, proteins such as actin, transgelin, alpha tropomyosin, filamin 1, enolase 1, and tubulin were found in one or two fractions (Fig. 3), with less interference from higher-abundance proteins.

17. It is strongly recommended not to re-freeze protein solutions containing Start buffer as considerable protein loss and precipitation appear upon thawing.
18. We have found that adding 20% isopropanol reduces nonspecific binding to the column (4).
19. The first dimension (CF) of PF 2D consists of a single piston pump (HPCF Module), a manual injector for sample introduction, a pH monitor, and a UV detector at 280 nm.
20. We have found that with a single system and using the same column on three different days, the RSD% was 0.67% for retention time and 1.02% for elution pH.
21. Aliquots of pooled IgY-12 partitioned human plasma were kept as standards for testing the CF column.
22. Storing the RP fractions as intact proteins rather than as digested peptides seems to reduce loss. Comparison by MS/MS analysis of the same freshly digested sample to those digested and frozen shows a loss of ~50% in the latter case.
23. Normally, we run each peptide sample in duplicate sequentially. We have found that freezing the digested peptides results in loss of peptide upon thawing and, hence, reduction in protein coverage.
24. The mass spectrometer was operated in the data-dependent mode, in which every FT–MS scan (survey 350–2,000 Da) was followed by MS/MS scans of the five most abundant ions.
25. With the first three adjustments, reproducibility of the first- and second-dimension chromatograms was dramatically improved (Fig. 2b, Table 1). Table 1 shows that all fractions from the pH gradient were collected at virtually identical pH intervals (0.3 pH units), with the same starting point of pH 6.8. The second-dimension chromatograms in Panel B show identical overlap between the same pH fractions among all four systems. This harmonization allowed reproducible comparisons of many samples across different instruments.
26. The flat-bottom electrodes provide even closer matching of pH profiles among the different PF 2D systems, compared to original electrodes (Fig. 2b, c, Panel b and c); this is very beneficial for comparing fractionation profiles of complex proteomes. However, these replacement kits are not necessarily readily available and are not necessarily needed in order to achieve harmonization of multiple 2DLC instruments.

Acknowledgments

The authors would like to acknowledge Beckman Coulter for their support. H.S. fellowship in the J.E. Van Eyk's laboratory was partially supported by Czech Ministry of Education project No. LC07017. The study was funded by the NHLBI Proteomics Innovation Contract N01-HV-28180 (JEV) and the Institute for Clinical and Translational Science Award (Grant NO 1U54RR023561-01A1).

References

1. Fu Q, Sheng S, Van Eyk JE (2007) Development of biomarker development pipeline: search for myocardial ischemia biomarkers. Book Chapter, *Clinical Proteomics: From Diagnosis to Therapy*, J. Van Eyk and M.J. Dunn (Eds), Wiley-VCH.
2. Schoenhoff FS, Fu Q, Van Eyk JE (2009) Cardiovascular proteomics: implications for clinical applications. *Clin. Lab. Med.* 29:87–99.
3. McDonald T, Sheng S, Stanley B. et al. (2006) Expanding the subproteome of the inner mitochondria using protein separation technologies: one and two-dimensional liquid chromatography and two-dimensional gel electrophoresis. *Mol. Cell. Proteomics*, 5:2392–2398.
4. Sheng S, Chen, D, Van Eyk JE (2006) Multidimensional liquid chromatography separation of intact protein by chromatographic focusing and reversed phase of the human serum proteome: optimization and protein database. *Mol. Cell. Proteomics* 5:26–34.
5. Suberbielle E, Gonzalez-Dunia D, Pont F (2008) High reproducibility of two-dimensional liquid chromatography using pH-driven fractionation with a pressure-resistant electrode. *J. Chromatography B* 871(1):125–9.
6. Fu Q, Garnham CP, Elliott ST, Bovenkamp DE, Van Eyk JE (2005) A robust, streamlined, and reproducible method for proteomic analysis of serum by delipidation, albumin and IgG depletion, and two-dimensional gel electrophoresis. *Proteomics* 5:2656–64.
7. Liu T, Qian WJ, Mottaz HM. et al. (2006) Evaluation of Multiprotein Immunoaffinity Subtraction for Plasma Proteomics and Candidate Biomarker Discovery Using Mass Spectrometry. *Mol. Cell. Proteomics* 5(11): 2167–74.
8. Gundry RL, Fu Q, Jelinek CA, Van Eyk JE, Cotter RJ (2007) Investigation of an albumin-enriched fraction of human serum and its albuminome. *Proteomics Clin. Appl.* 1:73–88.
9. Gundry RL, White MY, Murray CI, Kane LA, Fu Q, Stanley BA, Van Eyk JE (2009) Preparation of proteins and peptides for mass spectrometry analysis in a bottom-up proteomics workflow. *Curr. Protoc. Mol. Biol.* Chapter 10:Unit10.25
10. Huang L, Harvie G, Feitelson J. et al. (2005) Immunoaffinity separation of plasma proteins by IgY microbeads: Meeting the needs of proteomic sample preparation and analysis. *Proteomics* 5(13), 3314–28.
11. Gundry RL, White MY, Nogee J, Tchernyshyov I, Van Eyk JE (2009) Assessment of albumin removal from an immunoaffinity spin column: critical implications for proteomic examination of the albuminome and albumin-depleted samples. *Proteomics* 9:2021–8.

Chapter 3

In-Depth Analysis of a Plasma or Serum Proteome Using a 4D Protein Profiling Method

Hsin-Yao Tang, Lynn A. Beer, and David W. Speicher

Abstract

Comprehensive proteomic analysis of human plasma or serum has been a major strategy used to identify biomarkers that serve as indicators of disease. However, such in-depth proteomic analyses are challenging due to the complexity and extremely large dynamic range of protein concentrations in plasma. Therefore, reduction in sample complexity through multidimensional pre-fractionation strategies is critical, particularly for the detection of low-abundance proteins that have the potential to be the most specific disease biomarkers. We describe here a 4D protein profiling method that we developed for comprehensive proteomic analyses of both plasma and serum. Our method consists of abundant protein depletion coupled with separation strategies – microscale solution isoelectrofocusing and 1D SDS-PAGE – followed by reversed-phase separation of tryptic peptides prior to LC–MS/MS. Using this profiling strategy, we routinely identify a large number of proteins over nine orders of magnitude, including a substantial number of proteins at the low ng/mL or lower levels from approximately 300 μL of plasma sample.

Key words: Plasma proteome, Low-abundance protein, Immunoaffinity depletion, Biomarker, MicroSol IEF, SDS-PAGE, LC–MS/MS

1. Introduction

Human plasma (or serum) is one of the most valuable specimens for protein biomarker discovery because it is easily collected and contains thousands of proteins, including those secreted or leaked into the blood by normal cell and tissue processes or cell damage (1). Thus, the presence or change in expression of proteins might potentially indicate the onset and progression of most disease states. However, systematic analysis of the plasma proteome is extremely challenging due to its complexity, the extremely wide dynamic range of protein concentrations that span more than ten

Richard J. Simpson and David W. Greening (eds.), *Serum/Plasma Proteomics: Methods and Protocols*, Methods in Molecular Biology, vol. 728, DOI 10.1007/978-1-61779-068-3_3,

orders of magnitude, and the extensive physiological variation among patient samples. The plasma proteome is also dominated by a small number of high-abundance proteins, such as albumin and immunoglobulins, which together constitute more than 80% of the total plasma proteins. In contrast, known disease biomarkers, such as prostate-specific antigen (PSA) and carcinoembryonic antigen (CEA), are low-abundance proteins that are typically found in the low ng/mL to pg/mL range, and most other specific disease biomarkers are likely to be present at similar concentration levels (1–3).

The strategies that have been used most frequently to overcome the dynamic range and sample complexity problems of the plasma proteome are to deplete the major plasma proteins and subject the remaining proteome to multiple fractionation steps for further reduction of sample complexity (4–6). The fractionation steps employed should exploit the orthogonal physicochemical properties of the molecules as a basis for separation. One of the most common multidimensional approaches involves separation of tryptic peptides by strong cation exchange (SCX) followed by reversed-phase (RP) liquid chromatography (7, 8). However, tryptic peptides have limited orthogonal physicochemical properties; therefore, separation methods that involve both proteins and peptides can maximize the effectiveness of fractionation strategies. We have developed several multidimensional protein profiling strategies for a more efficient and comprehensive analysis of the human plasma and serum proteomes. One strategy uses three dimensions of separation coupled with a label-free quantitative comparison of proteomes (9). For an even greater depth of analysis, a 4D method has been developed (5). This strategy utilizes three sequential, and substantially orthogonal, protein pre-fractionation steps followed by online, reversed-phase, tryptic peptide separation prior to mass spectrometry analysis (LC–MS/MS). It is particularly well suited for in-depth cataloging of as many proteins as possible in very complex proteomes, such as human or mouse plasma, rather than for quantitative comparisons of multiple samples. We previously applied this protein profiling method for comprehensive analysis of serum and plasma proteomes and showed that it could detect a large number of proteins from a serum sample in the pilot phase of the Human Proteome Organization Plasma Proteome Project, including 14 of the 20 known proteins in the 1–100 ng/mL range and several proteins in the pg/mL range (5, 10).

2. Materials

2.1. Immunoaffinity Depletion of Major Plasma Proteins from Mouse Plasma or Serum

1. Mouse plasma or serum (see Note 1).
2. Milli-Q® water (Millipore, Bedford, MA) or equivalent. "Water" in this text refers to Milli-Q® water unless otherwise specified.
3. ÄKTA™ Purifier fast protein liquid chromatography (FPLC) system (GE Healthcare, Piscataway, NJ) with 1-mL sample loop.
4. Multiple Affinity Removal System® (MARS) for mouse serum proteins, 4.6 × 100-mm column (Agilent Technologies, Santa Clara, CA). Store at 2–8°C when not in use.
5. MARS buffer A. Before use, add the following final concentration of protease inhibitors: 5 mM ethylenediaminetetraacetic acid (EDTA), 0.15 mM phenylmethylsulfonyl fluoride (PMSF), 1 μg/mL leupeptin, and 1 μg/mL pepstatin A.
6. MARS buffer B. The solution is stable for at least 1 year at 25°C.
7. 50% (v/v) 2-propanol (Optima grade, Fisher Scientific, Pittsburgh, PA).
8. Stericup® GP Express Plus membrane (Millipore), 0.22-μm filter units, 500 mL capacity.
9. Amicon® Ultrafree®-MC 0.22-μm microcentrifuge filters (Millipore).
10. 5-kDa MWCO spin separators (Amicon® Ultra 4, Millipore).
11. 10 mM Sodium phosphate buffer, pH 7.4: Prepare 0.1 M stock buffer by dissolving 3.1 g of NaH_2PO_4 H_2O and 10.9 g of anhydrous Na_2HPO_4 in water to a final volume of 1 L. The stock buffer will have a pH of 7.4 and can be stored for up to 1 month at 4°C. Prepare the working solution by diluting one part with nine parts water.
12. Beckman J-6B centrifuge with JS-4.2 swinging bucket rotor (Beckman Coulter, Fullerton, CA).

2.2. Microscale Solution Isoelectrofocusing Fractionation

1. Urea (PlusOne™; GE Healthcare).
2. 3 M Tris–HCl, pH 8.5: Dissolve 36.3 g of Tris in 90 mL water, adjust pH to 8.5 with 1 N HCl, and add water to a final volume of 100 mL.
3. 3 M dithiothreitol (DTT) stock solution: Prepare 10 mL by dissolving 4.63 g in water. Stored in single-use aliquots at −20°C.
4. *N*,*N*-Dimethylacrylamide (DMA) solution (Sigma–Aldrich, St. Louis, MO). The concentration of this reagent is approximately 9.7 M.

5. Thiourea (Sigma–Aldrich).
6. 3-[(3-Cholamidopropyl)dimethylammonio]-1-propanesulfonic acid (CHAPS) (GE Healthcare).
7. MicroSol IEF apparatus (ZOOM® IEF Fractionator, Invitrogen, Carlsbad, CA). Reagents for operating the ZOOM® IEF Fractionator were obtained from Invitrogen unless otherwise specified.
8. ZOOM® focusing buffers, pH 3–7 and pH 7–12.
9. Electrophoresis power supply (EPS 3501 XL, GE Healthcare) with a capacity of at least 1,000 V that can operate at currents below 1 mA.
10. pH membranes (ZOOM® disks): 3.0, 4.6, 5.4, 6.2, and 12.0. ZOOM® disks are stable for 3 months when stored at 4°C.
11. Anode buffer: Prepare 20 mL by mixing 3.5 mL Novex® IEF anode buffer (50×), 9.6 g urea, 3.0 g thiourea, and 7 mL water. Adjust to pH 3.0 with Novex® IEF buffer (50×) and add water to a final volume of 20 mL. Novex® IEF anode buffer (50×) is stored at 25°C and is stable for 1 year.
12. Cathode buffer: Prepare 20 mL by mixing 2 mL Novex® IEF cathode buffer, pH 9–12 (10×), 9.6 g urea, 3.0 g thiourea, and 7 mL water. Adjust to pH 12.0 with 1 N NaOH, and add water to a final volume of 20 mL. Store Novex® IEF cathode buffer (10×) at 4°C.
13. MicroSol sample buffer: 8 M urea, 2 M thiourea, 4% (w/v) CHAPS, 1% (w/v) DTT, and 1% (v/v) pH 3–7 and 1% (v/v) pH 7–12 of ZOOM® focusing buffers. Prepare by mixing 4.8 g urea, 1.5 g thiourea, 0.4 g CHAPS, 0.1 g DTT, and 100 μL of each ZOOM® focusing buffers in water to a final volume of 10 mL.
14. Falcon 3047 24-well, flat-bottom plate (Becton, Dickinson and Company, Franklin Lakes, NJ).
15. 10% NuPAGE® Bis–Tris precast gels (Invitrogen).
16. Centrifugal filter devices (0.5 mL, 0.22 μm, Millipore).

2.3. 1-D SDS-Polyacrylamide Gel Electrophoresis

1. 1 N NaOH, to neutralize acidic samples.
2. 200-Proof ethanol (Sigma–Aldrich), chilled to −20°C.
3. SDS-solubilizing buffer: 63 mM Tris–HCl, pH 6.8, 0.2 M sucrose, 3% (w/v) SDS, 2 mM Na_2EDTA, 1% (v/v) 2-mercaptoethanol, and 1% (v/v) saturated bromophenol blue solution.
4. XCell SureLock™ Mini-Cell (Invitrogen).
5. 1× NuPAGE® MES running buffer (Invitrogen): Prepare fresh by mixing 50 mL of 20× buffer with 950 mL of Milli-Q® water.

6. NuPAGE® 12% Bis–Tris, 1-mm thick, ten-well precast gel (Invitrogen). Store at 4°C.
7. Benchmark protein ladder (Invitrogen).
8. 1-cc Insulin syringe (Becton, Dickinson and Company).
9. Higgins waterproof India ink (Sanford Corporation, Oak Brook, IL).
10. Novex® Colloidal Blue Staining kit (Invitrogen).

2.4. High-Throughput In-Gel Trypsin Digestion

1. Methanol (Optima grade, Fisher Scientific).
2. Wash solution: 0.1% (v/v) aqueous trifluoroacetic acid, 50% (v/v) aqueous methanol.
3. PCR cabinet with built-in laminar flow (e.g., Captair®Bio, Erlab) equipped with a light box.
4. 96-Well pierced plates (BioMachines Inc., Carrboro, NC).
5. V-bottom, 96-well collecting plates (AB-1058, Thermo Fisher Scientific, Waltham, MA). Also used as a trypsin collecting plate and a humidifier plate.
6. Plate covers for 96-well plates (AB-0751, Thermo Fisher Scientific).
7. Clean glass plates: 20 × 20 cm and 8 × 12 cm.
8. Custom-made gel-cutting device consisting of multiple stainless steel razor blades separated by 1-mm Teflon spacer. Alternatively, the MEG-1.5 Gel Cutter (The Gel Company, San Francisco, CA).
9. Microforcep (Thermo Fisher Scientific).
10. SpeedVac® Concentrator with 96-well plate centrifuge rotor (Thermo Fisher Scientific).
11. Eight-channel motorized pipette (Rainin Instrument, Bristol, PA).
12. 37°C Micro-incubator shaker (M-36, Taitec Corporation, Cupertino, CA).
13. 25-mL Reagent reservoirs (Diversified Biotech, Boston, MA) for pipetting with an eight-channel pipette.
14. Destain solution: 0.2 M ammonium bicarbonate and 50% acetonitrile.
15. Trypsin solution: 0.02 μg/μL trypsin (Sequencing-grade Modified Trypsin, Promega Corporation, Madison, WI) in 40 mM ammonium bicarbonate, pH 8.0.
16. Trypsin wash buffer: 40 mM ammonium bicarbonate, 3% formic acid.
17. Autosampler sample tubes (12 × 32 mm polypropylene vials, Thermo Fisher Scientific).

2.5. Liquid Chromatography–Tandem MS Analysis

1. Formic acid (LC/MS grade), 10 × 1 mL sealed ampoules (Fluka Analytical, Seelze, Germany).
2. Hamilton 500-μL glass syringe (GASTIGHT® #1750, Hamilton Company, Reno, NV).
3. Solvent A: Milli-Q® water with 0.1% formic acid. Add acid using a glass syringe.
4. Solvent B: Acetonitrile with 0.1% formic acid. Add acid using a glass syringe.
5. Trap column: NanoACQUITY™ UPLC Symmetry 5-μm C18 trap, 180 μm i.d. × 2 cm long (Waters Corporation, Milford, MA).
6. Analytical column: NanoACQUITY™ UPLC BEH 1.7-μm C18 column, 75 μm i.d. × 25 cm long (Waters); or self-packed, 25-cm long PicoFrit® column (75 μm i.d./15 μm tip i.d., New Objective, Woburn, MA) with 3-μm 100 Å Magic C18AQ™ resin (Michrom Bioresources, Auburn, CA).
7. High-performance or ultra-performance LC (UPLC) system, capable of nanoliter flow rates, with a chilled autosampler (e.g., NanoACQUITY™ UPLC, Waters).
8. Mass spectrometer with fast scan rate (e.g., LTQ Orbitrap XL™, Thermo Fisher Scientific).
9. MS/MS data analysis software (e.g., Sequest or Mascot).

3. Methods

The protein profiling strategy described here consists of three sequential protein separation steps, i.e., major protein depletion, microscale solution isoelectrofocusing (MicroSol IEF), and 1D SDS-polyacrylamide gel electrophoresis (SDS-PAGE), followed by online, reversed-phase, LC peptide separation prior to mass spectrometry (MS) analysis. Each of these separation steps is designed to reduce the complexity of the plasma proteome incrementally and thus allow a higher amount of sample to be processed or analyzed in subsequent steps. Maximizing the protein load at each separation step is critical to obtain comprehensive, in-depth proteome coverage.

While this multidimensional protein profiling method enables in-depth proteome analysis, it also will increase the total number of samples (and analysis time) for downstream analysis. Therefore, a major consideration before beginning the protocol is to determine the total number of fractions per sample for LC–MS/MS analysis that is desired. In this protocol, the sample number is affected mainly by the number of MicroSol IEF fractions and the separation distance/number of slices per lane for the 1D SDS gels. Increasing the number of fractions will increase the resolution and usually result in the identification of higher numbers of proteins and peptides (depth of coverage) at the expense of longer analysis time per proteome. The method presented here is aimed at maximizing the depth of coverage while keeping the sample size as low as possible.

3.1. Immunoaffinity Removal of Major Plasma Proteins

The first protein fractionation step is the depletion of major plasma proteins. We have examined a number of commercially available dye-based and immunoaffinity methodologies for depleting abundant proteins from human plasma. We found that the immunodepletion methods are more efficient at depleting targeted proteins, typically with greater than 98% removal of most targets and with minimal removal of untargeted minor proteins (11). Since major plasma proteins prevent the identification of lower-abundance proteins, immunoaffinity columns capable of efficiently depleting the most number of major plasma proteins would facilitate the greatest in-depth analysis of the plasma proteome. Aside from human plasma, immunodepletion methods are also available for a few other species, especially mouse and rat, due to their widespread use as animal models for human diseases, toxicology, and drug testing, as well as in protein biomarker discovery. The protocols for immunodepletion of major human and mouse proteins are very similar, although vendor-specific buffers for binding and elution of proteins are usually required. We have previously published protocols for immunodepletion of human plasma using the Sigma ProteoPrep 20®, which removes 20 major plasma proteins (12). In this protocol, we describe the use of the Agilent MARS-Mouse 3® column for depletion of three major proteins from mouse plasma as an example of major protein depletion. The depletion step is important for comprehensive analysis of the plasma proteome because these major proteins constitute approximately 80% of plasma proteins (1, 2). Hence, after depletion, far greater volumes of plasma can be separated by downstream analytical methods:

1. Transfer the MARS column to room temperature and prepare fresh daily the appropriate amount of Buffer A (see Note 2) and Buffer B. The final volumes needed for each depletion is approximately 50 mL of Buffer A and 25 mL of Buffer B. Filter the buffers through a 0.22-μm filter unit and degas for 5 min at 25°C using vacuum filtration with stirring.
2. Flush the FPLC system, including the column inlet line, with Buffer A for 15 min at 1 mL/min. Reduce the flow to 0.5 mL/min, remove the top plug of the column, and connect the inlet line to the top of the column, taking care not to introduce air into the column. Remove the bottom plug of the

column and connect the detector inlet solvent line. Continue to equilibrate with Buffer A for 5 min at 1 mL/min.

3. Prior to injection of the first sample of the day, run a blank to remove any residual protein from previous depletions and re-equilibrate the column using Buffer A.
4. Dilute the required volume of mouse plasma five times with Buffer A. Filter the diluted plasma with pre-rinsed, 0.22-μm microcentrifuge filters to remove particulates. Keep the filtered sample on ice until ready for use (see Note 3).
5. After the column has fully equilibrated, inject 375–500 μL of diluted plasma at 0.5 mL/min.
6. Elute the unbound proteins (depleted plasma) from the column with Buffer A at 0.5 mL/min for 10 min. Collect the flow-through fractions and pool the unbound fractions containing protein as it is eluted from the column. Keep the pooled sample on ice.
7. Elute the bound proteins with Buffer B at 1 mL/min for 14 min. Collect and pool the bound proteins and store on ice (see Note 4). Re-equilibrate the column with Buffer A for 22 min at 1 mL/min.
8. Repeat steps 5–7 until all samples are processed.
9. Disconnect and store the column at 4°C in Buffer A.
10. Replace Buffers A and B with Milli-Q® water and flush the FPLC system, including the detector, for 30 min at 1 mL/min, followed by 50% 2-propanol for 30 min at 1 mL/min.
11. Pre-rinse two or more 5-kDa MWCO spin separators by centrifuging 2 mL of Milli-Q® water through them, using the J6-B centrifuge at 1,500 × *g* and 4°C for 10 min.
12. Pool all unbound fractions and concentrate to 200 μL using the pre-rinsed spin separator.
13. Desalt the concentrated unbound fraction by adding 1.8 mL of 10 mM sodium phosphate buffer, pH 7.4, and again concentrating to 200 μL. Repeat at least four times to reduce ionic strength and replace the MARS Buffer A.
14. Rinse the concentrator unit with 50 μL of 10 mM Sodium phosphate, pH 7.4, and combine with the concentrated unbound fraction (see Note 5).
15. Run proportional amounts of unbound and bound fractions on a SDS-polyacrylamide gel to evaluate the efficiency of protein depletion. Examples of the results produced for mouse plasma, as described in this protocol, as well as for human plasma using the ProteoPrep® 20 column, are shown in Fig. 1.

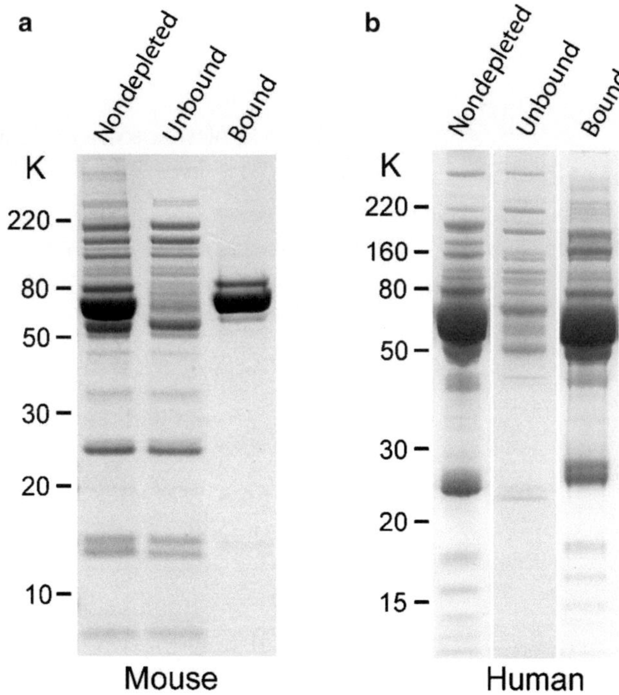

Fig. 1. Comparison of major protein depletion from mouse and human plasma. (**a**) Depletion of three major mouse proteins (albumin, IgG, and transferrin) using the Agilent MARS Mouse 3 LC-Column®. (**b**) Depletion of 20 major human proteins using Sigma ProteoPrep® 20 FPLC system. In both cases, depletion of major proteins enhanced detection of lower-abundant plasma proteins. Samples were separated on a NuPAGE® Bis–Tris gel and stained with colloidal Coomassie® blue.

3.2. MicroSol IEF Fractionation of Unbound Proteins

Solution IEF is used as the second fractionation step to further reduce the complexity of the plasma proteome. Solution IEF separation devices that rely on immobiline-buffered membranes for protein separation are capable of very high-resolution separations because membrane partitions can be selectively made at precise pH values, and proteins with pI values differing by as little as 0.01 pH units can be separated (13, 14). We developed a convenient, multi-chamber solution IEF device that was subsequently commercialized as Invitrogen's ZOOM® IEF Fractionator (14–17).

Solution IEF is performed using the ZOOM® IEF Fractionator, which contains seven 700-μL separation chambers and can, therefore, provide a maximum of seven pI fractions when eight immobiline/acrylamide partition disks are used. There is, however, great flexibility in its assembly, offering different configurations of immobiline/acrylamide partition disks and numbers of fractions for analyzing diverse types of samples and for different research needs (14, 17). This protocol describes a four-separation chamber configuration that we routinely use for comprehensive plasma proteome analysis:

1. Add sufficient 3 M Tris–HCl, pH 8.5, and dry urea to 3 mg of depleted plasma to yield final concentrations of 8 M urea, 20 mM Tris–HCl, pH 8.5, in a 500-μL final volume (see Note 6).
2. Add 3 M DTT to a final concentration of 20 mM to reduce disulfide bonds. Mix, blanket with argon, and incubate at 25°C for 30 min with gentle agitation.
3. Add DMA to a final concentration of 50 mM to alkylate cysteine residues. Mix, blanket with argon, and incubate at 25°C for 30 min with gentle agitation (see Note 7).
4. Terminate the reaction by adding DTT to a final concentration of 64.8 mM (1% DTT), taking into account that it already contains 20 mM DTT. Mix well. Incubate at 25°C for 15 min (see Note 8).
5. Add dry urea, thiourea, CHAPS, and ZOOM® focusing buffers, pH 3–7 and 7–12, to a final concentration of 8 M urea, 2 M thiourea, 4% (w/v) CHAPS, and 1% (v/v) each of ZOOM focusing buffers, pH 3–7 and 7–12 (the same final concentration as in the MicroSol sample buffer). If necessary, re-adjust DTT to 64.8 mM final concentration. The final sample volume should be less than 2,680 μL (step 8). Mix well.
6. Clean and assemble the ZOOM® IEF Fractionator according to the manufacturer's instructions. Assemble a four-chamber separation device with adjacent chambers separated by membrane disks in the following order: pH 3.0, 4.6, 5.4, 6.2, and 12.0 (see Note 9). The first two and the last chambers are blank chambers without a disk separating them from the anode and cathode chambers.
7. Load approximately 17.5 mL anode buffer (pH 3.0) and cathode buffer (pH 12.0) into each electrode reservoir of the chamber. Note that chambers 1, 2, and 7 will also contain the anode or cathode electrode buffer when the four-separation-chamber configuration is used.
8. Dilute the sample to a final volume of 2,680 μL (670 μL for each of the four chambers) with MicroSol sample buffer. Remove any undissolved particulate material by centrifugation at 16,000 × *g* for 10 min at 25°C. Load 670 μL of the diluted sample in all four chambers, taking care not to introduce bubbles into the chambers (see Note 10).
9. Insert sample chamber port plugs into all sample chambers. Place the lid on the assembled fractionator and focus the sample by applying an electrical field. Set the maximum limits of the power supply to 750 V, 1 mA, and 1 W. The separation is initially limited by current, and as the conductivity falls, the

voltage rises until 750 V is achieved. Continue the separation until a stable low current is reached (see Note 11).

10. After focusing is completed and prior to disassembling the device, remove the fractionated samples in chambers 3–6 through the fill ports using a gel-loading pipette (see Note 12). Transfer the contents into individual microfuge tubes. Rinse each chamber with 100 μL MicroSol sample buffer and combine with the corresponding fractions.
11. Disassemble the device. Place each of the membrane disks into individual wells of a 24-well flat-bottom plate, each containing 100 μL of MicroSol sample buffer.
12. Incubate at 25°C with shaking for 30 min.
13. Transfer the membrane disk extracts to individual microfuge tubes. Add another 100 μL of MicroSol sample buffer to the wells containing the membrane disks. Repeat step 12. Combine the second membrane disk extract with the first extract (see Note 13).
14. Run each fraction and membrane disk extract on a SDS polyacrylamide gel to evaluate the ZOOM® IEF separation. An example of the results produced is shown in Fig. 2.

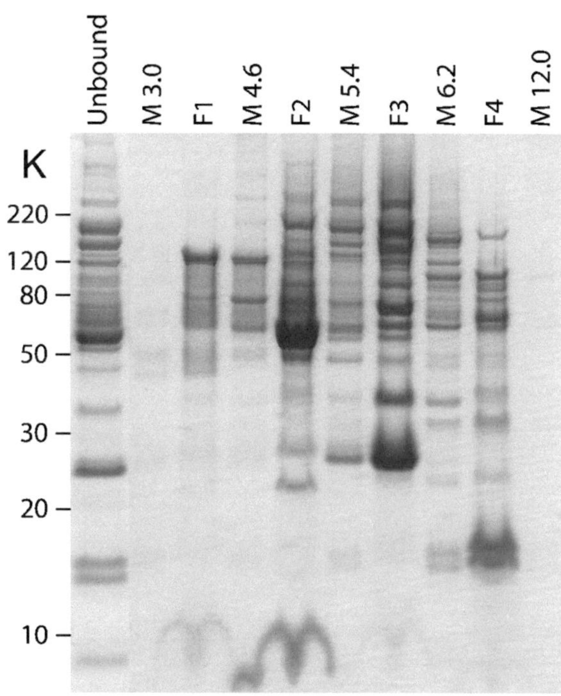

Fig. 2. MicroSol IEF fractionation of a depleted mouse plasma sample. Depleted plasma proteins (unbound) were separated by MicroSol IEF using a four-chamber configuration. The membrane extracts (M) and the pH values are indicated. The four MicroSol IEF fractions are indicated by F1–F4. Samples were run on a 10% NuPAGE® Bis–Tris gel with MES buffer and stained with colloidal blue. A good separation is evident from the presence of unique protein bands in each of the MicroSol IEF fractions.

3.3. Separation of MicroSol IEF Fractions by 1D SDS-PAGE

The third fractionation step utilizes 1D SDS-PAGE to separate proteins in each MicroSol IEF fraction and membrane disk extract efficiently according to their molecular weights. SDS-PAGE not only offers reproducible separation, but also provides a convenient method for sample clean-up prior to trypsin digestion. This protocol uses commercial precast gels for convenient and reliable protein separation. An important parameter to consider for SDS-PAGE is the length of electrophoretic separation, as it will affect the number of samples for downstream MS analysis and gel volume per trypsin digest. Generally, complex samples are electrophoresed for a longer distance to obtain better resolution, and less complex samples can be run for a short distance without drastically affecting the depth of analysis. In all cases, a maximum amount of proteins that will not cause excessive band distortion should be loaded in each gel lane to facilitate in-depth protein identification:

1. Precipitate desired amounts of each MicroSol IEF fraction and membrane disk extract with ethanol. Neutralize the F1 (pH 3.0–4.6) fraction to pH 7 with 1 N NaOH prior to ethanol precipitation. Add nine volumes of 200-proof ethanol (–20°C) to the samples.
2. Incubate the samples overnight on ice at 4°C. Centrifuge the samples for 30 min at 4°C. Carefully remove the supernatant without disturbing the pellets, and air-dry pellets briefly in a fume hood. Dissolve the pellets with the appropriate volume of SDS-solubilizing buffer (see Note 14).
3. Use separate precast gels for each fraction. Measure a distance of 1 or 3 cm from the bottom of the wells and mark the position on the gel cassette with a permanent marker. Use the 1-cm-marked gels for membrane disk extracts and the 3-cm-marked gels for F1–F4 samples (see Note 15).
4. Clean the gel unit and assemble as per the manufacturer's instructions. Fill the gel tank with 1× MES running buffer (see Note 16).
5. Heat the samples and protein standard at 90°C for 2 min. Load at least three lanes for each sample.
6. Run gels at 200 V constant. Terminate the electrophoresis when the bromophenol blue dye front reaches the 1- or 3-cm line.
7. Remove the gel from the unit chamber. Discard one of the plastic plates to expose the gel. Mark the dye front with a small amount of black India ink using a 1-cc syringe needle.
8. Fix and stain all gels using Colloidal blue stain according to the manufacturer's instructions. Store destained gels in clean Ziploc® plastic bags at 4°C until ready for trypsin digestion. Representative results are shown in Fig. 3.

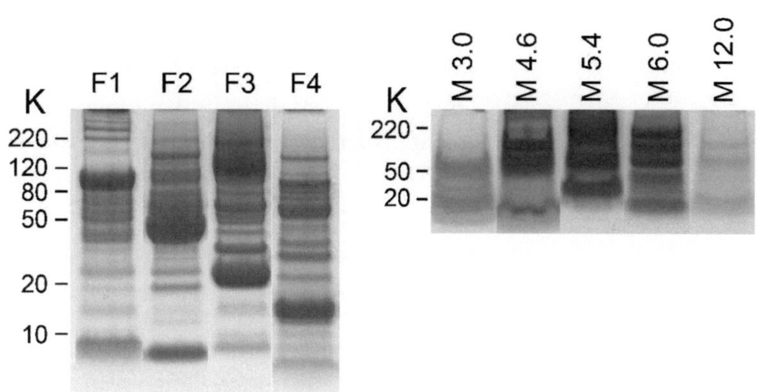

Fig. 3. Composite gels showing optimized loading of MicroSol IEF samples of fractionated mouse plasma for in-gel trypsin digestion. Samples were concentrated by ethanol precipitation and maximal amount of sample (not producing visible streaks) was loaded on a 12% NuPAGE® Bis–Tris gel. Each MicroSol IEF fraction and membrane extract were separately electrophoresed for 3 and 1 cm from the bottom of the sample well, respectively. All gels were stained with colloidal Coomassie® blue.

3.4. In-Gel Trypsin Digestion of Proteins

Due to the large number of digests that need to be performed, we have developed a high-throughput protocol that utilizes a 96-well plate format for in-gel digestion. Reagents are rapidly dispensed using an eight-channel pipette from disposable reservoirs, and gel slices are processed in 96-well plates with pierced bottoms that allow rapid centrifugal removal of reagents. This protocol allows a single person to perform up to 384 tryptic in-gel digests (4 × 96-well plates) easily in parallel:

1. Perform all procedures in a PCR hood to minimize airborne contamination. Clean all 96-well plates and glass plates by washing with 50% methanol containing 0.1% trifluoroacetic acid, followed by rinsing with methanol. Air-dry in the PCR hood.
2. Stack a 96-well pierced plate on top of a V-bottomed 96-well collecting plate.
3. Place the gel on a clean 20 × 20-cm glass plate. Cover the gel with just enough Milli-Q® water to prevent dehydration during the slicing process. Cut the gel lane using the gel cutting device into 1 × 4 mm slices (see Note 17). Transfer the gel slices into separate wells on the 96-well pierced plate using a microforcep. Combine gel slices from three adjacent replicate lanes (same molecular weight range) into the same well on the pierced plate (see Note 18).
4. Add 100 μL of destain solution to each gel-containing well using an eight-channel pipette. Place a cover on top of the pierced plate and incubate at 37°C for 30 min with shaking.
5. Remove destain solution from the pierced plate into the collecting plate by spinning for 1 min in the SpeedVac® concentrator.

Discard destain solution from the collecting plate and put it back under the pierced plate.

6. Dry the gel slices under vacuum in a SpeedVac® at 37°C for approximately 30 min. Replace the collecting plate with a clean 96-well plate (trypsin collecting plate) for trypsin digestion.
7. Add 40 μL of trypsin solution to each gel-containing well (see Note 19). Place an 8 × 12-cm glass plate on top of the pierced plate to prevent sample loss by evaporation. To further prevent evaporation of samples, add 80 μL of water to the wells of another 96-well collecting plate (humidifier plate) and place it under the trypsin collecting plate. Assemble the three plates, together with the glass plate and cover, in a Ziploc® bag to create a humidified chamber. Incubate the plate for 16 h at 37°C with no shaking.
8. Collect trypsin digests into the trypsin collecting plate with 1-min spin in the SpeedVac® concentrator. Add 30 μL trypsin wash buffer to each gel-containing well in the pierced plate and incubate with no shaking at 37°C for 30 min (see Note 20). Combine this tryptic supernatant with the first tryptic supernatant on the trypsin collecting plate with 1-min spin in the SpeedVac® concentrator.
9. Transfer the samples into cleaned autosampler vial and load directly into the LC–MS/MS system, or freeze the samples at −20°C until needed.

3.5. Reversed-Phase Separation of Tryptic Peptides and Tandem MS Analysis

Tremendous progress has been made in the field of protein/peptide MS in the past two decades, resulting in a wide variety of commercial mass spectrometers with different types of ionization methods and mass analyzers. Linear ion trap instruments (e.g., Thermo Fisher's LTQ series of instruments) – which feature high-duty cycle, tandem MS (MS/MS) capability, fast scan rates, and high sensitivity – are best suited for comprehensive, in-depth proteomic studies of complex samples (18, 19). Recently, hybrid mass spectrometers have been built that combine the advantages of multiple mass analyzers. One such instrument is the LTQ Orbitrap, which combines the speed and sensitivity of the LTQ with the high resolution and mass accuracy of the Orbitrap mass analyzer (20, 21).

Despite progress in MS instrumentation, successful MS analysis of complex samples requires front-end separations to simplify the extremely complex peptide mixtures prior to MS analysis. This is especially critical for the detection of low-abundance peptides that would otherwise be obscured by higher-abundance signals. The most common front-end separation technique, which is also used here as the fourth separation mode, is reversed-phase, high-performance liquid chromatography (RP-HPLC). Peptides are continuously separated based on their hydrophobicity by

RP-HPLC, and the effluent is directly analyzed using tandem MS analysis (LC–MS/MS). The introduction of UPLC systems has enabled the use of longer reversed-phase columns packed with small particle size (<2 μm) resins that result in improved resolution, peak capacity, and sensitivity of peptide separation (8, 22):

1. Tune and calibrate the LTQ Orbitrap according to the manufacturer's instructions (see Note 21).
2. Set the autosampler to pick up 4–8 μL of sample (see Note 22). Inject onto the trap column at 5 μL/min for 4 min with 100% Solvent A.
3. Set the binary solvent manager (UPLC pump) to deliver the following acetonitrile gradient at 0.2 μL/min for peptide separations: 3–28% B over 42 min, 28–50% B over 26 min, 50–80% B over 5 min, 80% B for 4 min, 3% B over 1 min, and hold for an additional 10 min (see Note 23).
4. To minimize carry-over between samples, run a quick blank gradient at 0.4 μL/min with the following gradient: 3–80% B over 11 min, hold at 80% B for 5 min, back to 3% B in 0.5 min, and hold for an additional 13.5 min.
5. Set the mass spectrometer to acquire MS/MS in data-dependent mode. Perform full MS scan (m/z 400–2,000) in the Orbitrap with 60,000 resolution at 400 m/z. Trigger MS/MS in the linear ion trap on the six most intense ions exceeding a minimum threshold of 1,000 (see Note 24).
6. Start flow at 0.5 μL/min with 3% Solvent B through the trap and analytical column for 15 min. Ensure that there is no leakage or blockage by monitoring the system pressure.
7. Place the sample vials on the chilled 48-well tray in the sample manager (autosampler) and start the MS run (see Note 25). An example of a typical LC–MS/MS run is shown in Fig. 4.
8. Analyze MS/MS results using software packages such as Sequest (23) or Mascot (Matrix Science, Boston, MA), and combine the results for all LC–MS/MS runs for the proteome using programs such as DTASelect (24) and Scaffold (Proteome Software, Portland, OR).

4. Notes

1. Blood samples should be collected in a consistent manner with a well-defined protocol in order to avoid introducing bias related to sample collection and handling. Samples should be stored in aliquots, preferably in liquid nitrogen or at –80°C, to prevent proteolysis or modifications. A recent

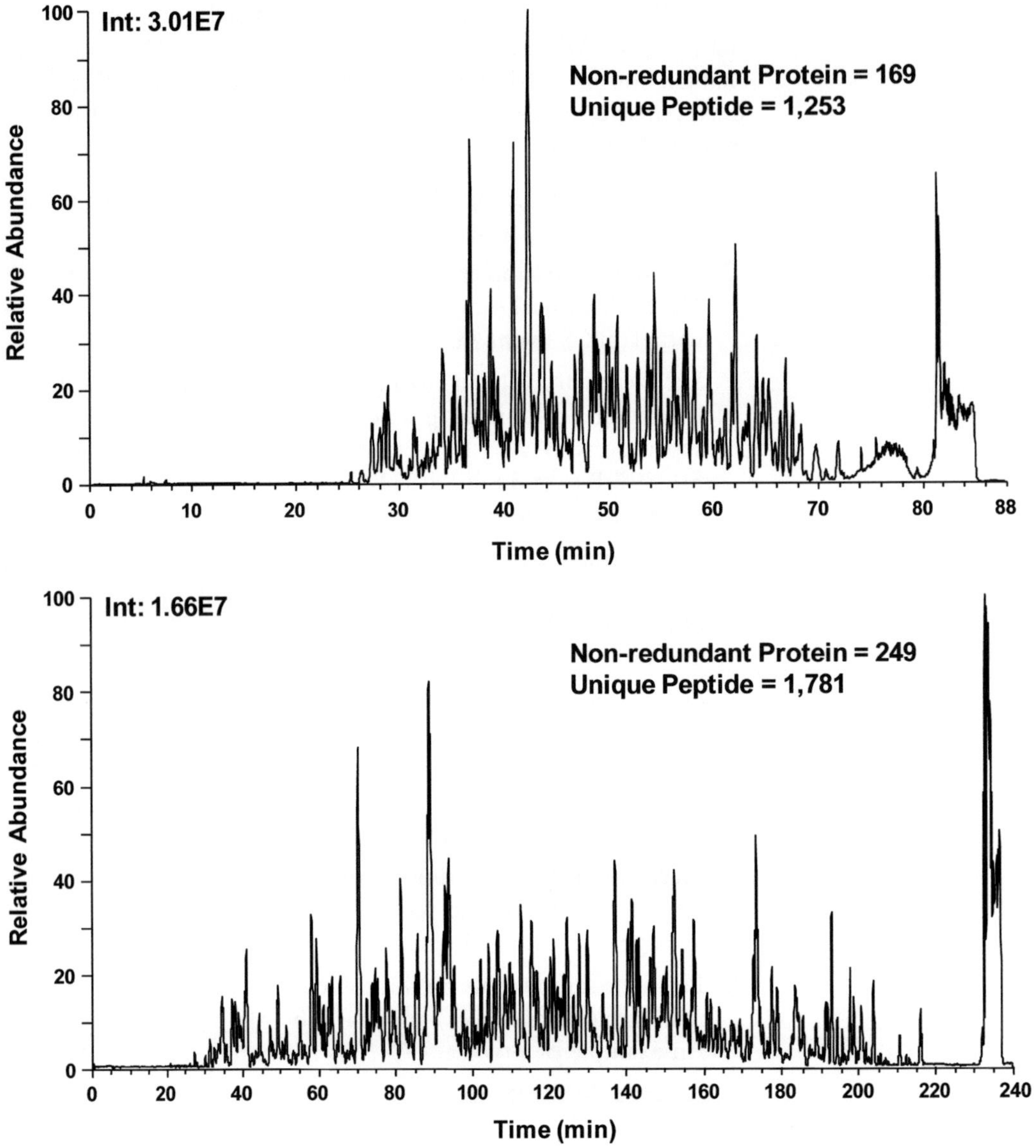

Fig. 4. Increased protein/peptide identification with longer LC–MS/MS analysis. Tryptic digests from two different gel slices of a depleted plasma proteome were combined and analyzed by LC–MS/MS. The MS method is identical for both analyses, but two different LC run times (88 and 240 min) were used. Base peak chromatograms indicate that the 88-min run (*top*) had higher ion intensities (Int) than the 240-min run (*bottom*), as expected. The 240-min run identified substantially more unique proteins and peptides than the 88-min run, but at the expense of decreased sample throughput. MS/MS spectra were searched using Sequest (23) and then processed using DTASelect (24) and custom software to obtain a comprehensive catalog of the plasma proteome.

study suggests that platelet-depleted plasma may be better than serum, which contains a large number of peptides that are thought to be generated as the result of platelets releasing peptides or from its blood-clotting process (25).

2. Buffer A is prepared fresh with the addition of protease inhibitors to minimize protein degradation.
3. The column can deplete approximately 75–100 μL of mouse plasma, depending upon the concentration of the samples and the actual column capacity. Typically, we obtained about 1 mg of unbound proteins from 100 μL of mouse plasma. For comprehensive plasma proteome analysis using this 4D method, we generally performed sufficient depletions to have at least 3 mg of unbound proteins for the next fractionation step. Protein concentrations can be determined using the BCA protein assay.
4. The bound fractions containing the major plasma proteins are usually not analyzed but are frozen for potential future analysis. Prior to storage, neutralize the bound fraction by adding 1 N NaOH to pH 7.0.
5. Set aside 1–3 mg of unbound fraction for subsequent analysis. Store the remainder in aliquots at –80°C.
6. The maximum amount of sample that can be separated effectively by MicroSol IEF varies slightly depending upon sample type, loading position, and the number of sample chambers used (14, 17). For most studies, a sample load capacity in the 1–3 mg range is recommended and should be adequate for downstream analysis. Vortex gently to ensure complete protein solubilization and denaturation. This will facilitate reduction and alkylation of all disulfide bonds, which are critical for efficient and reproducible separation of proteins by solution IEF. Since urea is used for sample denaturation, avoid excessive heating of samples to minimize unwanted protein carbamylation.
7. Do not use iodoacetamide for alkylation, as it will result in high current and interfere with MicroSol IEF separation.
8. This step not only quenches the alkylation reaction but also ensures that the sample contains a high concentration of DTT to keep any residual unmodified cysteines reduced during subsequent ZOOM® IEF fractionation.
9. The number of chambers used should be determined based on experimental needs. A seven-chamber approach, especially using custom pH disks (14), will increase the resolution of the separation but will substantially increase the total separation time. A frequent alternative for higher-resolution separation is using a five-chamber separation mode by inserting an additional chamber with a pH 7.0 disk between the pH 6.2 and pH 12.0 disks. Increase the maximum voltage to 1,000 V for five-chamber separations.
10. In general, a sample is loaded into all separation chambers to allow dilution to the largest possible volume. However, with

some samples (e.g., non-depleted plasma), restricting the loading to the middle chamber might result in better focusing. This presumably is because some proteins that are unstable at pH extremes may precipitate before they have sufficient time to move out of the highest and lowest pH chambers. Also, some proteins loaded into the most basic and acidic sample chambers must migrate the longest distance to reach the appropriate chamber at equilibrium.

11. The separation should be terminated as soon as a stable low current (usually around 0.3 mA) is achieved, because excessive focusing times can result in poorer separation. The total focusing time for 3 mg of plasma sample is approximately 3 h using the conditions described here. Samples containing higher protein loads, high salt concentrations, or higher ampholyte concentrations will usually require much longer separation times, and in some cases, it may not be feasible to reach 750 V or a minimum current below 1 mA.
12. After the MicroSol IEF run, the volumes in sample chambers may be reduced slightly due to electro-osmosis. Reduction of any fraction to less than 50% of its original volume indicates that the chamber was poorly sealed and the quality of the entire separation will probably be compromised.
13. Proteins with pI values equal to the membrane pH value will be trapped in the membrane disks and need to be extracted for optimal proteome coverage. Because the membrane disks are in contact with two chambers, the membrane extracts will also contain some proteins from the adjacent chambers (Fig. 2). In this protocol, all membrane extracts are analyzed individually, albeit at a lower gel resolution. Alternatively, the membrane extracts could be combined with an adjacent solution fraction to reduce the number of fractions for downstream analysis. In this case, the membrane disk extracts should be combined to an adjacent solution fraction that displays the protein profile most similar to that determined from the SDS-PAGE analysis.
14. Concentrate enough of each MicroSol IEF fraction for at least three lanes using a maximum load. We typically combine gel slices of the same molecular weight range from three adjacent replicate lanes to generate larger amounts of the tryptic digest. Generally, fractions with less protein (F1 and F4) can be concentrated up to 20-fold or more, and fractions with more proteins (F2 and F3) can be concentrated up to approximately tenfold. It is recommended to determine the maximum sample load by running an aliquot of the concentrated samples on a SDS-gel prior to running the preparative multiple lane gel.

15. Using a separate gel for each fraction will avoid contamination from other fractions, especially when maximum volumes are loaded and slight well overflow may occur. Since the sample complexity of membrane extracts is low, running the gel for 1 cm will compress all the protein bands, reducing the size of sparsely populated regions in the gel, and, therefore, reducing gel volumes (improved peptide recovery) and/or fractions for MS analysis (increases throughput). Running the more complex MicroSol IEF fractions for 3 cm typically yields adequate depth of analysis and is a good compromise between increased gel resolution and decreased throughput of MS analysis.
16. One of the major environmental contaminants identified by MS analysis is keratin. Make sure that the gel units are clean and protect them from dust. Wear gloves and wash gloves with Milli-Q® water before handling the gels.
17. Using 1-mm slices, a 1- and 3-cm gel lane will result in 10 and 30 gel slices, respectively.
18. Multiple replicate lanes from each MicroSol IEF fraction are cut and adjacent gel slices are combined to generate a larger amount of tryptic digest for potential additional LC–MS/MS runs. Corresponding gel slices from up to three or four replicate lanes can be digested per well. We prefer to use the custom gel-cutting tool because the width of the razor blade array allows up to four gel lanes to be cut simultaneously, thereby minimizing lane-to-lane variations in the locations of the gel slices. In addition, the sharp razor blades cut into the gel easily without crushing it.
19. The volume of trypsin solution used is dependent upon the number of gel slices. Use 40 μL for three gel slices. Increase the volume to 50 μL for digesting four gel slices, and reduce the volume to 20 μL for digesting one or two gel slices.
20. Add 30 μL of trypsin wash buffer for three gel slices. Increase volume to 40 μL for digesting four gel slices, and reduce volume to 15 μL for digesting one or two gel slices.
21. The major benefit of the Orbitrap mass analyzer is the ability to measure peptide masses accurately with less than 5 ppm error. High mass accuracy greatly increases the confidence of peptide identifications. To maintain this level of accuracy, the mass spectrometer usually needs to be calibrated every 2–3 days. The LTQ Orbitrap can also be run with "Lock Mass" on, which allows for real-time recalibration using polydimethylcyclosiloxane ions present in ambient air (26). However, it is recommended that the Orbitrap mass calibration still be performed regularly even if "Lock Mass" is used during data acquisition.

22. The injection volume depends upon the amount of protein in the gel slices that were digested. We typically inject 4 μL of digests from slices with very heavy protein loads, and 8 μL for all other gel slices. Generally, nano-C18 columns have a loading capacity of approximately 1–2 μg of tryptic peptides. If protein/peptide quantitation is critical, it is best to use a slightly lower load (~0.5 μg) to ensure that the signals are within the linear range of quantitation.
23. We find that this 88-min HPLC method is generally sufficient to handle the peptide complexity generated from this protocol. For highly complex peptide samples (e.g., a combination of multiple gel digests or non-MicroSol IEF fractionated samples), a longer gradient will certainly result in greater depth of peptide identifications. An example is shown in Fig. 4.
24. Other LTQ Orbitrap parameters that we typically use are as follows. Preview mode for FTMS master scans is enabled to allow parallel collection of MS/MS data in the linear ion trap, while the Orbitrap analyzes the full MS scan. Ions are isolated for MS/MS using an isolation width of 2.5 Da. Dynamic exclusion is enabled to prevent repeated analysis of the same ion (m/z within 5 ppm) with a duration of 45 s. The monoisotopic precursor selection is enabled to ensure correct reporting of the peptide monoisotopic mass, and charge state screen is used to exclude singly charged ions (mainly caused by chemical noise) from MS/MS analysis.
25. Always run a reference digest prior to running actual samples to ensure that the LC–MS/MS system is operating optimally.

Acknowledgments

This work was supported in part by National Institutes of Health Grants CA120393 and CA131582, and institutional grants to the Wistar Institute including an NCI Cancer Core Grant (CA10815) and the Commonwealth Universal Research Enhancement Program, Pennsylvania Department of Health.

References

1. Anderson, N. L., and Anderson, N. G. (2002) The human plasma proteome: history, character, and diagnostic prospects. *Mol Cell Proteomics* **1**, 845–867.
2. Tirumalai, R. S., Chan, K. C., Prieto, D. A., et al. (2003) Characterization of the low molecular weight human serum proteome. *Mol Cell Proteomics* **2**, 1096–1103.
3. Rifai, N., Gillette, M. A. and Carr, S. A. (2006) Protein biomarker discovery and validation: the long and uncertain path to clinical utility. *Nat Biotechnol* **24**, 971–983.

4. Pieper, R., Su, Q., Gatlin, C. L., et al. (2003) Multi-component immunoaffinity subtraction chromatography: an innovative step towards a comprehensive survey of the human plasma proteome, *Proteomics* **3**, 422–432.
5. Tang, H. Y., Ali-Khan, N., Echan, L. A. et al. (2005) A novel four-dimensional strategy combining protein and peptide separation methods enables detection of low-abundance proteins in human plasma and serum proteomes, *Proteomics* **5**, 3329–3342.
6. Liu, T., Qian, W. J., Gritsenko, M. A. et al. (2006) High dynamic range characterization of the trauma patient plasma proteome, *Mol Cell Proteomics* **5**, 1899–1913.
7. Washburn, M. P., Wolters, D., and Yates, J. R., 3rd. (2001) Large-scale analysis of the yeast proteome by multidimensional protein identification technology, *Nat Biotechnol* **19**, 242–247.
8. Shen, Y., Jacobs, J. M., Camp, D. G. et al. (2004) Ultra-high-efficiency strong cation exchange LC/RPLC/MS/MS for high dynamic range characterization of the human plasma proteome, *Anal Chem* **76**, 1134–1144.
9. Beer, L. A., Tang, H. Y., Barnhart, K. T., and Speicher, D. W. (2011) Plasma biomarker discovery using 3-D protein profiling coupled with label-free quantitation, Ed. Simpson RJ, Greening DW, *Serum/Plasma Proteomics, Methods Mol Biol* **728**, Humana Press.
10. Omenn, G. S., States, D. J., Adamski, M. et al. (2005) Overview of the HUPO Plasma Proteome Project: results from the pilot phase with 35 collaborating laboratories and multiple analytical groups, generating a core dataset of 3020 proteins and a publicly-available database, *Proteomics* **5**, 3226–3245.
11. Echan, L. A., Tang, H. Y., Ali-Khan, N., et al. (2005) Depletion of multiple high-abundance proteins improves protein profiling capacities of human serum and plasma, *Proteomics* **5**, 3292–3303.
12. Echan, L.A., and Speicher, D.W. (2009) Immunoaffinity depletion of high abundance plasma and serum proteins, in *The Protein Protocols Handbook*, J.M. Walker, Editor. Humana Press, New York, 139–153.
13. Righetti, P. G., Wenisch, E., Jungbauer, A., et al. (1990) Preparative purification of human monoclonal antibody isoforms in a multi-compartment electrolyser with immobiline membranes, *J Chromatogr* **500**, 681–696.
14. Tang, H. Y., and Speicher, D. W. (2005) Complex proteome pre-fractionation using microscale solution isoelectrofocusing, *Expert Rev Proteomics* **2**, 295–306.
15. Zuo, X., and Speicher, D. W. (2000) A method for global analysis of complex proteomes using sample pre-fractionation by solution isoelectrofocusing prior to two-dimensional electrophoresis, *Anal Biochem* **284**, 266–278.
16. Zuo, X., Echan, L., Hembach, P. et al. (2001) Towards global analysis of mammalian proteomes using sample pre-fractionation prior to narrow pH range two-dimensional gels and using one-dimensional gels for insoluble and large proteins, *Electrophoresis* **22**, 1603–1615.
17. Joo, W. A., and Speicher, D. (2009) Pre-fractionation using microscale solution IEF, *Methods Mol Biol* **519**, 291–304.
18. Schwartz, J. C., Senko, M. W., and Syka, J. E. (2002) A two-dimensional quadrupole ion trap mass spectrometer, *J Am Soc Mass Spectrom* **13**, 659–669.
19. Douglas, D. J., Frank, A. J., and Mao, D. (2005) Linear ion traps in mass spectrometry, *Mass Spectrom Rev* **24**, 1–29.
20. Hu, Q., Noll, R. J., Li, H., et al. (2005) The Orbitrap: a new mass spectrometer, *J Mass Spectrom* **40**, 430–443.
21. Makarov, A., Denisov, E., Kholomeev, A., et al. (2006) Performance evaluation of a hybrid linear ion trap/orbitrap mass spectrometer, *Anal Chem* **78**, 2113–2120.
22. Shen, Y., Zhang, R., Moore, R. J. et al. (2005) Automated 20 kpsi RPLC-MS and MS/MS with chromatographic peak capacities of 1000-1500 and capabilities in proteomics and metabolomics, *Anal Chem* 77, 3090–3100.
23. Eng, J. K., McCormack, A. L., and Yates III, J. R. (1994) An approach to correlate tandem mass spectral data of peptides with amino acid sequences in a protein database, *J Am Soc Mass Spectrom* **5**, 976–989.
24. Tabb, D. L., McDonald, W. H., and Yates, J. R., 3rd. (2002) DTASelect and Contrast: tools for assembling and comparing protein identifications from shotgun proteomics, *J Proteome Res* **1**, 21–26.
25. Rai, A. J., Gelfand, C. A., Haywood, B. C. et al. (2005) HUPO Plasma Proteome Project specimen collection and handling: towards the standardization of parameters for plasma proteome samples, *Proteomics* **5**, 3262–3277.
26. Olsen, J. V., de Godoy, L. M., Li, G. et al. (2005) Parts per million mass accuracy on an Orbitrap mass spectrometer via lock mass injection into a C-trap, *Mol Cell Proteomics* **4**, 2010–2021.

Chapter 4

Intact-Protein Analysis System for Discovery of Serum-Based Disease Biomarkers

Hong Wang and Samir Hanash

Abstract

Profiling of serum and plasma proteins has substantial relevance to the discovery of circulating disease biomarkers. However, the extreme complexity and vast dynamic range of protein abundance in serum and plasma present a formidable challenge for protein analysis. Thus, integration of multiple technologies is required to achieve high-resolution and high-sensitivity proteomic analysis of serum or plasma. In this chapter, we describe an orthogonal multidimensional intact-protein analysis system (IPAS) (Wang et al., Mol Cell Proteomics 4:618–625, 2005) coupled with protein tagging (Faca et al., J Proteome Res 5:2009–2018, 2006) to profile the serum and plasma proteomes quantitatively, which we have applied in our biomarker discovery studies (Katayama et al., Genome Med 1:47, 2009; Faca et al., PLoS Med 5:e123, 2008; Zhang et al. Genome Biol 9:R93, 2008).

Key words: Multidimensional protein fractionation, Serum, plasma, Quantitative proteomics, Disease biomarkers

1. Introduction

Comprehensive profiling of the serum and plasma proteomes has substantial relevance to the identification of circulating protein biomarkers that may have predictive value and utility for disease diagnosis and management. However, the complex spectrum of serum and plasma proteins spans a dynamic range of protein abundances of at least 12 orders of magnitude (1). Moreover, proteins often occur in multiple isoforms due to posttranslational modifications, cleavages, or splice variants. As a result, decomplexing biological fluids through fractionation or enrichment is beneficial for disease biomarker discovery. As an orthogonal separation technique, 2D-PAGE (2–4) was initially relied upon for quantitative separation of proteins. However, intrinsic limitations

Richard J. Simpson and David W. Greening (eds.), *Serum/Plasma Proteomics: Methods and Protocols*, Methods in Molecular Biology, vol. 728, DOI 10.1007/978-1-61779-068-3_4,

of 2D-PAGE, such as the dynamic range, inability to resolve small and large proteins, and the difficulty for online evaluation coupled with MS analysis, do not allow for sufficient capacity for analysis of low-abundance proteins in such complex proteomes as serum and plasma. Furthermore, extraction of proteins from gels for their identification following digestion or their recovery for other applications is difficult and associated with substantial losses. Therefore, alternative strategies using liquid-phase-based intact-protein fractionation approaches are increasingly relied upon (5–10). An advantage of separation of intact proteins prior to MS analysis is the ability to assess protein variation associated with posttranslational modifications, cleaved forms, or splice variants, given the altered chromatographic properties of such variants.

In this chapter, we describe a liquid-phase-based orthogonal multidimensional intact-protein analysis system (IPAS) that we have developed for disease biomarker discovery. We incorporate intact-protein isotopic labeling into IPAS to profile protein differences quantitatively in samples being compared. The strength of the intact-protein-based multidimensional separation system is in its ability to resolve related proteins, such as differentially modified forms, and to reduce the complexity of samples through fractionation prior to MS analysis. The IPAS platform is illustrated in Fig. 1.

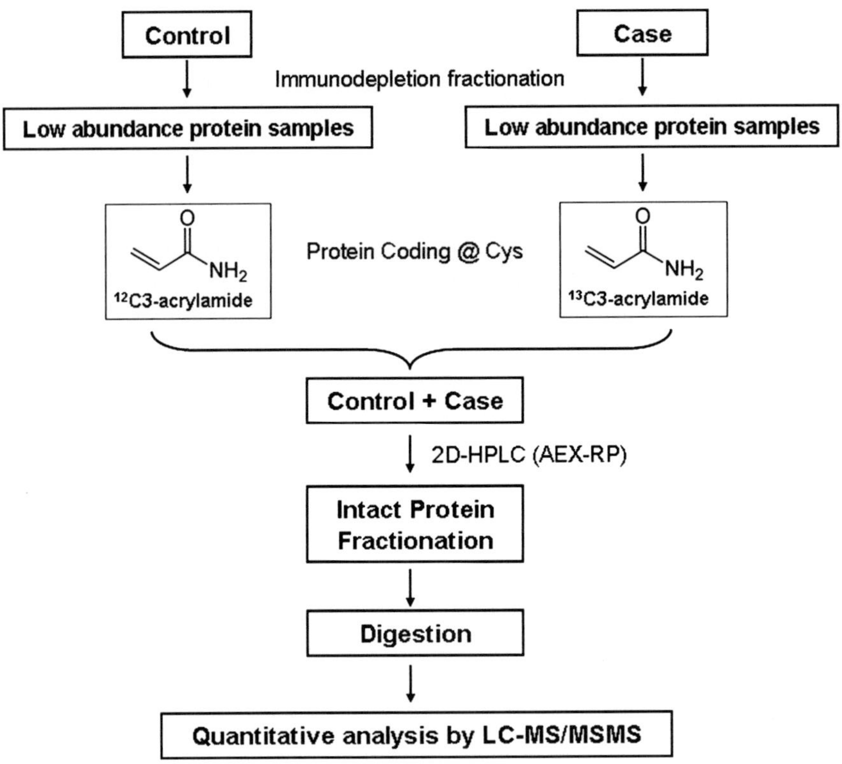

Fig. 1. Platform of intact-protein multidimensional HPLC fractionation, coupled with isotope labeling and nano-LC–ESI MS/MSMS for quantitative serum/plasma proteome analysis.

First, serum/plasma samples from control and case are processed through immunodepletion chromatography to remove the high-abundance proteins. Then, the remaining protein mass in each of the samples to be compared is concentrated and isotopically labeled with acrylamide, i.e., the control sample is labeled with one isotope, e.g., $^{12}C3$-acrylamide (light), and the case sample is labeled with another $^{13}C3$-acrylamide (heavy). A mixture consisting of each of the intact-protein-labeled samples is subjected to an orthogonal 2D-HPLC separation with anion-exchange chromatography (AEX) as the first dimension that separates proteins based on charge, and reversed-phase (RP) chromatography as the second-dimension separation that further separates proteins based on hydrophobicity. Collected protein fractions from 2D-HPLC are lyophilized by freeze-drying, followed by in-solution tryptic digestion. The digests are subjected to LC–MS/MS analysis. Relative quantification of identified proteins is achieved by comparing the signal intensities of identified peptides from the control and case samples in MS scan events. We have applied our current liquid-phase-based orthogonal multidimensional IPAS, coupled with the intact-protein isotopic labeling approach and nano-LC ESI MS, to studies of serum/plasma from several cancer types (11–14). Thus, with the multidimensional intact-protein fractionation approach, sample complexity is greatly reduced prior to MS analysis. In general, we have identified ~1,500 proteins with high confidence and obtained relative quantitative data for some 40% of identified proteins that span across seven orders of magnitude in protein concentration in a given experiment.

2. Materials and Instrumentation

2.1. Samples

Samples (control or case) are from individual subjects or pooled donors. The protocol described here is designed for a total of 600 μL of serum/plasma, i.e., 300 μL of serum/plasma from control and 300 μL of serum/plasma from case samples.

2.2. Intact-Protein Fractionation

2.2.1. Automated 1D-HPLC System

1D-HPLC system (Shimadzu Corporation, Columbia, MD). For immunodepletion chromatography, the system comprises the following:

1. Pump system: Two LC-10Ai pumps.
2. Degasser: DGU-20A3 Prominence Degasser.
3. Column oven: CTO-20 Prominence Column Oven.
4. Detector: SPD-20 UV/VIS Prominence Detector.
5. Collector: FRC-10A Fraction Collector.
6. Controller: SCL-10A VP System Controller.
7. Workstation: EZStart 7.4 SP1.

2.2.2. Automated Online 2D-HPLC System

Automated online 2D-HPLC system (Shimadzu Corporation, Columbia, MD).

For isotopically labeled intact-protein 2D-HPLC fractionation (see Fig. 2), the system comprises the following:

First dimension: AEX.

Second dimension: RP chromatography.

1. Auto sampler: SIL-10AP.
2. Pump system: One LC-10Ai pump for the first-dimension HPLC, Two LC-20AD pumps for the second-dimension HPLC.
3. Degasser: DGU-20A5 Prominence degasser.
4. Column oven: CTO-20AC Prominence column oven.
5. Detector: SPD-20 UV/VIS Prominence Detector.
6. Collector: Four FRC-10A Fraction Collectors.
7. Controller: A SCL-10A VP system controller.
8. Workstation: Class-VP 7.4 (Shimadzu Client/Server).

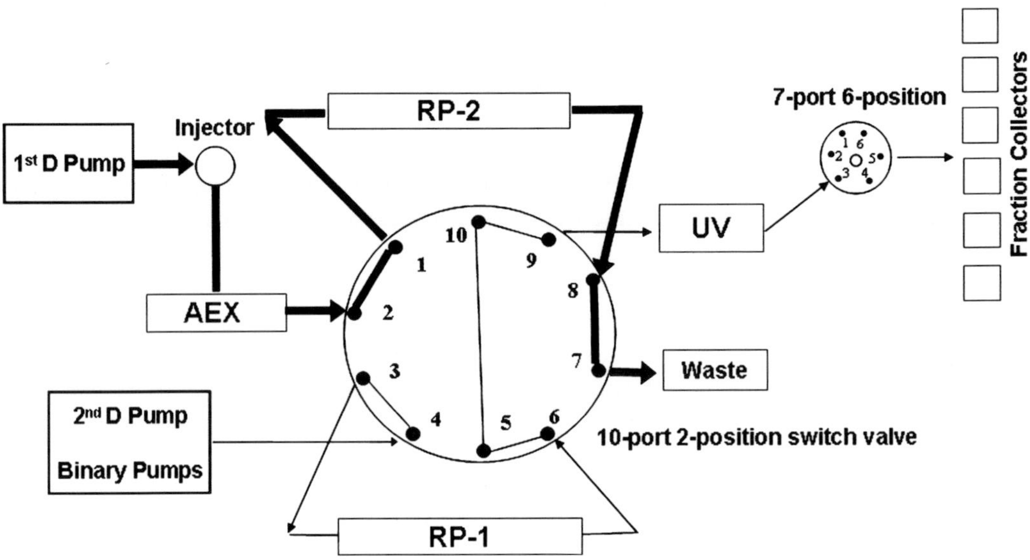

Fig. 2. *Work-flow diagram of automatic online 2D-HPLC system.* (a) The first-dimension pump loads the sample through the 10-mL sample loop onto the anion-exchange column with step elution Step 1. (b) Through the 10-port 2-position valve, the flow-through proteins (unable to be absorbed within anion-exchange column) will be trapped within one of the two RP columns, here the RP-2 column. (c) The second-dimension pump starts the gradient elution within the other RP column. Here is the RP-1 column that is directed to the Fraction Collector through the 7-port 6-position valve. (d) At the end of the RP-1 gradient elution, the 10-port 2-position valve will switch to another position. (e) The first-dimension pump will start the step elution Step 2. (f) Through the 10-port 2-position valve, proteins eluted from the anion-exchange column will be trapped within the RP-1 column. (g) The second-dimension pump starts the gradient elution within the RP-2 that is directed to the Fraction Collector through the 7-port 6-position valve and starts to collect protein fractions. (h) The 10-port 2-position valve automatically switches the position at the end of RP gradient elution, and the first-dimension pump will start the next step elution. The 2D-HPLC system will repeat the online two-dimension separation until it finishes the last step elution of anion-exchange chromatography.

2.2.3. Immunodepletion Chromatography

1. Hu-6 HC column (10×100 mm; Agilent Technologies, Wilmington, DE).
2. Ms-3 HC column (10×100 mm; Agilent Technologies, Wilmington, DE).
3. *Buffer A* (5185-5987), equilibration/loading/washing (Agilent Technologies, Wilmington, DE).
4. *Buffer B* (5185-5988), for elution (Agilent Technologies, Wilmington, DE).
5. 0.22-µm RC syringe filter (Fisher Scientific).

2.2.4. Anion-Exchange Chromatography

1. SAX/10 column, 7.5 mm ID×150 mm L (Column Technology Inc., Fremont, CA).
2. *AEX Mobile Phase A*: 20 mM Tris–HCl, 6% isopropanol, 4 M urea, pH 8.5 (pH is adjusted by adding 1 N HCl).
3. *AEX Mobile Phase B*: 20 mM Tris–HCl, 6% isopropanol, 4 M urea, and 1 M NaCl, pH 8.5 (pH is adjusted by adding 1 N HCl).
4. Trizma base (T1503, Sigma, St. Louis, MO).
5. Hydrochloric acid, trace metal grade (A508-500, Fisher Scientific).
6. Sodium chloride (S271-3, Fisher Scientific).
7. Urea (U15-3, Fisher Scientific).
8. Isopropanol, HPLC grade (A451-1, Fisher Scientific).

2.2.5. Reversed-Phase Chromatography

1. RP/5D column, 4.6 mm ID×150 mm L (Column Technology Inc., Fremont, CA).
2. *RP Mobile Phase A*: 95% HPLC-grade water, 5% acetonitrile+0.1% trifluoroacetic acid (TFA).
3. *RP Mobile Phase B*: 90% acetonitrile, 10% HPLC-grade water+0.1% TFA.
4. Water, HPLC grade (W5-1, Fisher Scientific).
5. Acetonitrile, HPLC grade (A998-1, Fisher Scientific).
6. TFA (3-3077, Supelco, Bellefonte, PA).

2.3. Acrylamide Isotope Labeling

1. Centricon YM-3 devices (Millipore).
2. *Labeling Buffer*: 8 M urea, 100 mM Tris–HCl, and 0.5% octyl-beta-D-glucopyranoside (OG), pH 8.5 (pH is adjusted by adding 1 N HCl).
3. Dithiothreitol (DTT, ultrapure, USB).
4. Light acrylamide: 1,2,3-^{12}C3-acrylamide (>99.5% purity, Fluka).
5. Heavy acrylamide: 1,2,3-^{13}C3-acrylamide (>98% purity, Cambridge Isotope Laboratories, Andover, MA).

2.4. Protein Fractionation Processing

FreeZone Plus 12 Liter Cascade Console Freeze Dry Systems (Labconco Corporation, Kansas City, MO).

2.5. Nano-LC–ESI MS/MS Analysis of 2D-HPLC Fractionated Samples

2.5.1. In-Solution Protein Digestion

1. Sequence-grade modified trypsin (Porcine, V511A/20 μg, Promega, Madison, WI).
2. *Digestion buffer*: aqueous 50 mM ammonium bicarbonate containing 4% acetonitrile in HPLC-grade water (v/v).
3. Ammonium bicarbonate (A643-500, Fisher Scientific).
4. Acetonitrile, Optima LC/MS grade (A955-1, Fisher Scientific).
5. Urea (U15-3, Fisher Scientific).
6. Formic acid, 99+% (28905, Pierce).

2.5.2. Nano-LC–ESI Mass Spectrometry Analysis

1. Nano-ESI LTQ-FT or Orbitrap nano-ESI mass spectrometer (Thermo Scientific).
2. Eksigent nano-LC-1D plus (Dublin, CA).
3. Capillary column: 25-cm Picofrit 75 μm ID (New Objectives) in-house packed with Magic C18 (100 Å pore size/5-μm particle size, Michrom Bioresources, Inc.).
4. Trap column: Symmetry C_{18} 180 μm × 20 mm, particle size 5 μm (Waters Corporation).
5. *Nano-LC Mobile phase A*: 0.1% formic acid in HPLC-grade water (v/v).
6. *Nano-LC Mobile phase B*: 0.1% formic acid in acetonitrile (v/v).
7. Water, HPLC grade (W5-1, Fisher Scientific).
8. Acetonitrile, Optima LC/MS grade (A955-1, Fisher Scientific).
9. Formic acid, 99+% (28905, Pierce).

2.5.3. Data Analysis

1. All the LC–MS/MS data management are performed using the Computational Proteomics Analysis System (15). https://proteomics.fhcrc.org/CPAS/project/home/begin.view?.
2. Search Algorithm: X!Tandem (16).
3. Scoring: PeptideProphet (16) and ProteinProphet (17).
4. Quantitation: *Q3* (18).

3. Methods

3.1. Immunodepletion Chromatography (Performed at Room Temperature)

1. Filter serum/plasma samples with a 0.22-μm RC syringe filter.
2. Equilibrate the Hu-6 HC immunodepletion column with *Buffer A* (Agilent) at 3 mL/min for 10 min.

Table 1
Immunodepletion chromatography program (UV = 280 nm)

Time (min)	Buffer B (%)	Total flow rate (mL/min)	Fraction collector
0.00	0	0.5	
6.30	0	0.5	Start to collect the flow-through fraction
25.00	0	0.5	
25.10	100	0.5	
26.30	100	0.5	Stop to collect the flow-through fraction
26.50	100	0.5	
27.00	100	3.0	
29.00	100	3.0	Start to collect the bound fraction
33.00	100	3.0	Stop collecting bound fraction
35.00	100	3.0	
35.10	0	3.0	
45.00	0	3.0	

3. Wash the 2-mL sample loop (PEEK) with 5 mL *Buffer A* (Agilent).
4. Load 300 μL of the serum/plasma sample onto 2-mL sample loop. Start the step elution at 0.5 mL/min (the step-elution program is listed in Table 1).
5. Collect the flow-through fraction (i.e., the low-abundance protein fraction) for 20 min and immediately store the fraction at –80°C until use.
6. Elute the bound fraction (i.e., the high-abundance protein fraction) and regenerate the column with *Buffer B* at 3 mL/min for 8 min (collect if wanted).
7. Re-equilibrate the system between sample injections.
8. The typical immunodepletion chromatogram is shown in Fig. 3.

3.2. Intact-Protein Isotopic Labeling (Performed at Room Temperature)

1. Concentrate the control and case immunodepleted serum/plasma samples separately (only the flow-through low-abundance protein fraction, see Fig. 3) with the Amicon YM-3 device until volume of <100 μL is reached.
2. Dilute each of the concentrated samples up to 1 mL with *Labeling buffer*: 8 M urea, 100 mM Tris–HCl, and 0.5% OG, pH 8.5.
3. Measure protein concentration using Bradford method (19) for case and control samples, respectively (see Note 1).
4. Reduce protein disulfide bonds for 2 h at room temperature by adding 0.66 mg DTT per milligram of protein (see Note 2).
5. Alkylate cysteines with acrylamide immediately after the reduction step by adding 7.1 mg/mg of total protein of ^{12}C3-acrylamide (*light*) to the control sample and 7.4 mg/mg of total protein of ^{13}C3-acrylamide (*heavy*) to the case sample. The reaction is carried out for 1 h in the dark at room temperature (see Note 3).
6. Mix the acrylamide-labeled control and case samples, dilute the mixture to 9 mL with AEX mobile phase A, and submit it immediately to 2D-HPLC intact-protein separation (see Note 4).

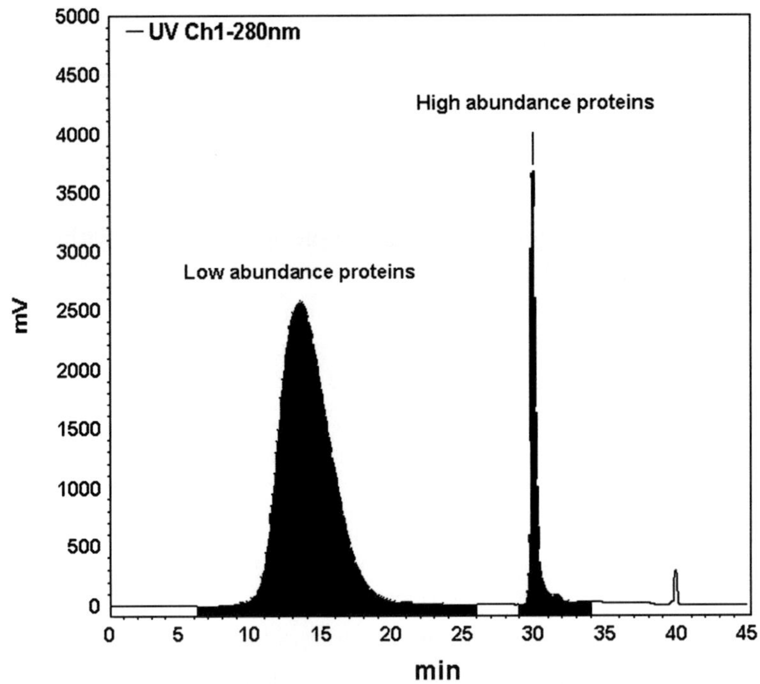

Fig. 3. Immunodepletion Chromatogram. The control and case samples are immunodepleted, respectively. The low-abundance protein fraction is collected and processed for further studies. A total of 300 μL of plasma/or serum from either control sample or case sample was loaded onto the immunodepletion column for immunodepletion chromatography.

3.3. Protein 2D-HPLC Fractionation (Performed at Room Temperature)

3.3.1. The First-Dimensional AEX

Step-elution program is listed in Table 2:

1. Equilibrate the anion-exchange column with AEX mobile phase A for 60 min at 1.0 mL/min.
2. Wash the 10-mL sample loop (PEEK) with 20 mL AEX mobile phase A.
3. Filter the mixed isotope-labeled sample (diluted with AEX mobile phase A) with a 0.22-μm RC syringe filter.
4. Load the filtered sample onto the 10-mL sample loop through the injector.
5. Start the run (the online 2D-HPLC system diagram is shown in Fig. 2).

3.3.2. The Second-Dimensional RP Chromatography

Gradient-elution program is listed in Table 3:

1. Equilibrate the two RP columns with 95% of RP *Mobile Phase A* (95% HPLC-grade water, 5% (v/v) acetonitrile + 0.1% TFA) for 15 min at 2.4 mL/min.
2. Protein eluted from the first-dimensional anion exchange column is loaded onto one of the two RP columns (the online 2D-HPLC diagram is shown in Fig. 2).
3. There is a total of eight-step elution's from the first-dimensional AEX (Table 2), and each of the eight eluted fractions is alternatively online loaded onto one of the two RP columns for sequential second-dimensional RP chromatography separation (see Fig. 2).
4. Desalt the RP column with 95% RP *Mobile Phase A* for 10 min before starting the gradient elution. The gradient-elution program for second-dimension RP chromatography is listed in Table 3 (see Note 5).

Table 2
Anion-exchange chromatography, flow rate 0.8 mL/min

Step elution	AEX mobile phase B (%)
1 (AEX01)	0
2 (AEX02)	5
3 (AEX03)	10
4 (AEX04)	15
5 (AEX05)	20
6 (AEX06)	30
7 (AEX07)	50
8 (AEX08)	100

Table 3
Reversed-phase chromatography, flow rate: 2.4 mL/min, UV = 280 nm

Time (min)	Solvent B (%)	Fraction collector
0	5	
10	5	Start to collect the fraction, 3 fractions/min, 800 μL/fraction
12	20	
30	65	
32	65	
33	90	
35	90	
37	5	
40	5	Stop collecting fractions
46	Stop	

5. Apply RP separation to each of the eight anion-exchange step elutions. Typical online 2D-HPLC chromatograms are shown in Fig. 4.
6. Collect the protein fractions from the RP column (3 fractions/min, 800 μL/per fraction) and keep the protein fractions at –80°C until use.
7. Lyophilize the protein fractions using the Freeze-Dry Systems. It usually takes 48 h to dry the protein fractions completely.

3.4. Nano-LC–ESI Mass Spectrometry Analysis

3.4.1. In-Solution Protein Digestion

1. Individual lyophilized protein fraction is suspended with 200 ng of trypsin in the *Digestion Buffer*. The protein digestion is carried out at 37°C overnight.
2. The digestion is quenched by adding 5 μL of 1.0% formic acid solution.
3. Based on the RP chromatography pattern, individual fractions (see Fig. 4) are combined into 12 pooled fractions for nano-LC–MS/MS analysis (the pooling sequence is listed in Table 4).

3.4.2. Nano-Flow Rate LC ESI Mass Spectrometry Analysis

Eksigent nano-LC-1D plus RP capillary HPLC system directly delivers nano-flow without using A splitter. This system is coupled with a nano-ESI LTQ-FT or LTQ-Orbitrap. The gradient elution program is listed in Table 5:

1. Load 20 μL of each pooled digested protein fraction into the 20-μL sample loop.

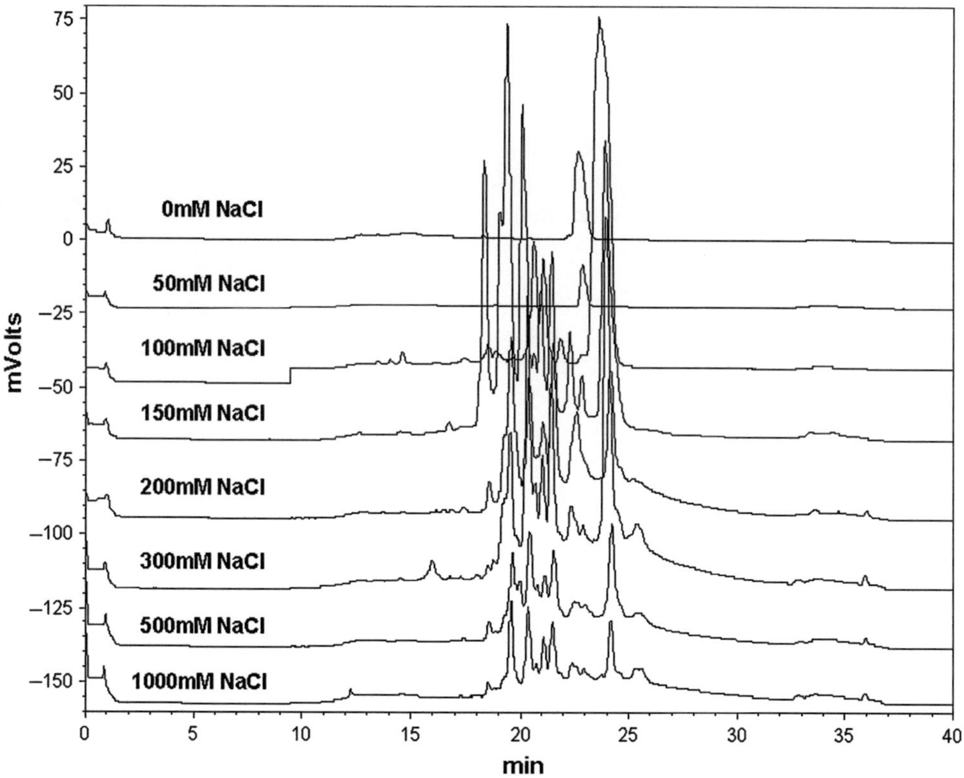

Fig. 4. 2D-HPLC Chromatogram (second-dimension RP chromatography). Protein fractions were collected between 10 and 40 min. A total of 84 fractions are collected. There are eight RP-HPLC chromatograms corresponding to the eight anion-exchange step elution (from 0 to 1,000 mM NaCl).

Table 4
Digested protein fraction pooling sequence for LC–MS/MS analysis

Pool no.	Fraction no.
1	From fraction RP #04 to #22
2	From fraction RP #23 to #24
3	From fraction RP #25 to #26
4	From fraction RP #27 to #28
5	From fraction RP #29 to #30
6	From fraction RP #31 to #32
7	From fraction RP #33 to #34
8	From fraction RP #35 to #36
9	From fraction RP #37 to #38
10	From fraction RP #39 to #40
11	From fraction RP #41 to #42
12	From fraction RP #43 to #72

Table 5
Nano-flow rate RPLC gradient elution, flow rate: 300 nL/min

Time (min)	Nano-LC mobile phase B (%)
0	3
2	7
92	35
93	50
102	50
103	95
108	95
109	3
130	3

2. After injection, the sample is trapped onto the trap column and desalted with 3% of nano-LC mobile phase B (0.1% formic acid in acetonitrile) for 10 min at 5 μL/min.
3. Start the nano-RPLC gradient elution (Table 5) and acquire spectra in a data-dependent mode in *m/z* range of 400–1,800. Select the five most abundant +2 or +3 ions in each MS spectrum for MS/MS analysis.
4. Nano-ESI LTQ-FT mass spectrometer parameters are capillary voltage of 2.0 kV, capillary temperature of 200°C, resolution of 100,000, FT target value of 800,000, and ion trap target value of 10,000.
5. Nano-ESI LTQ-Orbitrap mass spectrometer parameters are capillary voltage of 2.0 kV, capillary temperature of 200°C, resolution of 60,000, FT target value of 1,000,000, and ion trap target value of 10,000.

3.4.3. Data Analysis

The acquired LC–MS/MS data from nano-RPLC LTQ-FT or nano-RPLC LTQ-Orbitrap are automatically processed by the Computational Proteomics Analysis System (CPAS; see Subheading 2.5.3) (15, 16), https://proteomics.fhcrc.org/CPAS/project/home/begin.view?:

1. For protein database search, $^{12}C3$-acrylamide-labeled cysteine-containing peptide (light) is considered as the fixed modification and $^{13}C3$-acrylamide isotope-labeled cysteine-containing peptide (heavy) is detected using a delta mass of 3.01884 Da. The relative quantitation (ratio = heavy/light) information of

the paired cysteine-containing peptides is obtained by a script called "Q3" that was developed in-house to obtain the relative quantification for each pair of cysteine-containing peptides sequenced by MS/MS (18).

2. Apply the criteria for peptide sequence matching with a PeptideProphet score equal or greater than 0.2 (20). Then, the matched peptides are further grouped to protein queue using the ProteinProphet algorithm at an error rate of <5% (17). Furthermore, the sequenced peptides with PeptideProphet scores equal or greater than 0.75 are selected for quantitation (ratio). Based on the sequenced cysteine-containing peptides, their corresponding theoretical mono-isotopic mass and mass charge are obtained. The intensity of peaks in each MS scan is computed and each light peak (labeled with ^{12}C3-acrylamide) is paired with the corresponding heavy peak (labeled by ^{13}C3-acrylamide). For the final quantitation, a total of 20 MS scans before and after the MS/MS identification are summed. Proteins containing more than one pair of peptides that yield quantitation have all their peptide ratios averaged in that particular 2D-HPLC fraction (see Note 6).
3. Figure 5 shows a representative nano-flow rate LC–MS/MS analysis obtained by nano-ESI LTQ-FT, including the TIC chromatogram, MS spectrum, MS/MS sequence spectrum, and ^{13}C3/^{12}C3-acrylamide-labeled paired precursor ions for relative quantitation (ratio = heavy/light).
4. Figure 7 demonstrates the advantage of intact-protein-based multidimensional HPLC fractionation in studying serum/plasma samples in the pre-diagnosis of WHI-colon cancer (see Note 7).

4. Notes

1. As a result of immunodepletion, the total protein concentration is reduced by 90%, from approximately 70 to 7 mg/mL. These calculations are useful in evaluating the efficiency/performance of the immunodepletion column. The life time of the immunodepletion column is 200 injections, and efficiency of capture is between 90 and 95%.
2. Since the samples are in the *Labeling Buffer* that contains 8 M urea, the protein reduction reaction should be done at room temperature to avoid protein carbamylation due to the formation of isocyanic acid from urea at higher temperatures.
3. We suggest performing the labeling of control and case samples with ^{12}C3-acrylamide (*light*, control) and ^{13}C3-acrylamide

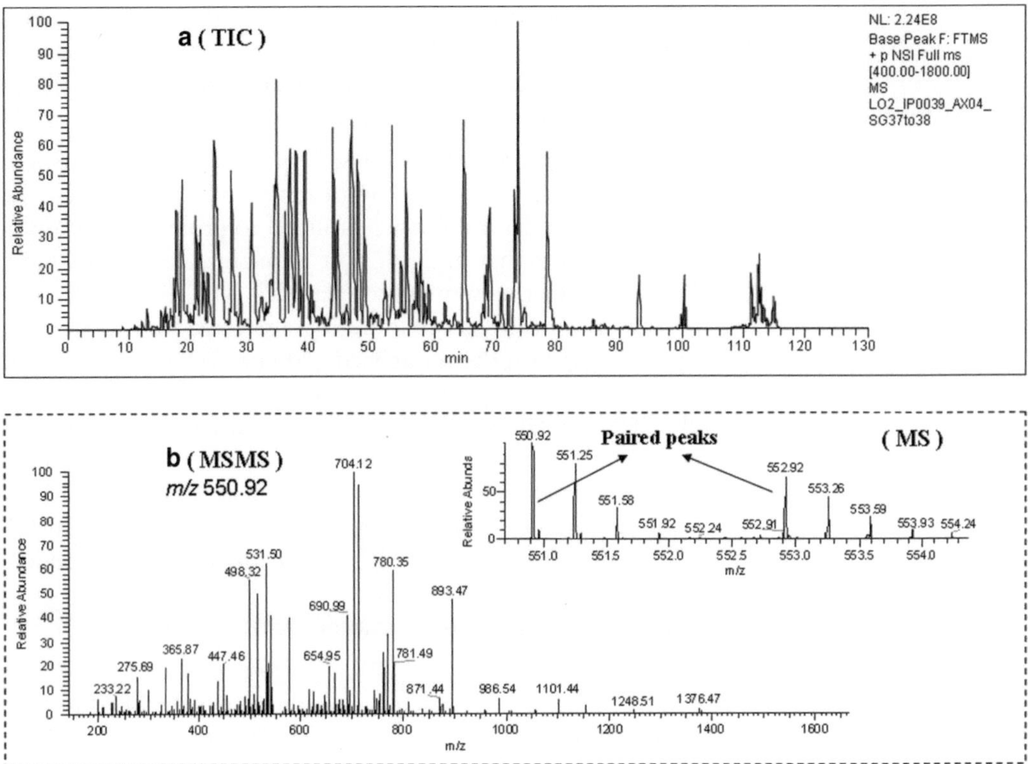

Fig. 5. Relative quantitation analysis of isotopic labeling of plasma protein by nano-LC–ESI LTQ-FT mass spectrometer. Two different plasma samples (control and cancer) were labeled with ^{12}C3-acrylamide (*light*) and ^{13}C3-acrylamide (*heavy*), respectively, after immunodepletion. The mixture of these samples is further fractionated using 2D-HPLC, followed by in-solution tryptic digestion for LC–MS/MSMS analysis. (**a**) The reconstructed total ion current (TIC) chromatogram corresponds to approximately 2.5 μg of protein on-column loading. (**b**) The MS/MS sequence spectrum of the precursor (*m/z* 550.92) and the quantifiable paired precursor ions from the MS scan.

(*heavy*, case). Also, the labeling reaction should be at room temperature as explained in Note 2.

4. When the labeling reaction is finished, add 1 mL of the AEX *Mobile phase A* to the control sample and the case sample, respectively. Then, mix these two samples and further dilute the mixed samples to 9 mL with AEX *Mobile phase A*. The size of the sample loop (PEEK) is 10 mL for the first-dimensional AEX.
5. Protein fractions eluted from the first-dimensional AEX are trapped onto the RP column with the anion-exchange buffer. It is important to desalt the protein fraction before initializing the second-dimensional RP chromatography to achieve higher resolution separation and prevent salt precipitation within the RP column in the presence of acetonitrile.
6. Acquired MS/MS data from LC-ESI mass spectrometry were automatically processed by the in-house built software

(https://proteomics.fhcrc.org/CPAS/project/home/begin.view). In brief, LC–MS and MS/MS data were first converted to mzXML format using ReAdW software (version 1.2) to generate the peak list for protein database search. X!Tandem search engine (version 2005.12.01) was configured with Cys labeled with ^{12}C3-acrylamide (71.03657@C) as the fixed modification and Cys labeled with ^{13}C3-acrylamide (74.04663@C) and Methionine (Met) oxidation (15.9949@M) as the variable modifications. Protein database used for this search is the International Protein Index (IPI) Human protein knowledgebase (version v3.68), and the number of protein entries in the database is 87,061. The minimum criterion for peptide matching was PeptideProphet score ≥0.2. Peptides that met with these criteria were grouped to protein sequences using the ProteinProphet algorithm at an error rate of ≤5%. In a given experiment, ~1,500 proteins are identified, with 40% of identified proteins yielding relative quantitation information. Figure 6 shows the distribution of identified proteins using the number of MS/MS events as a measure of protein

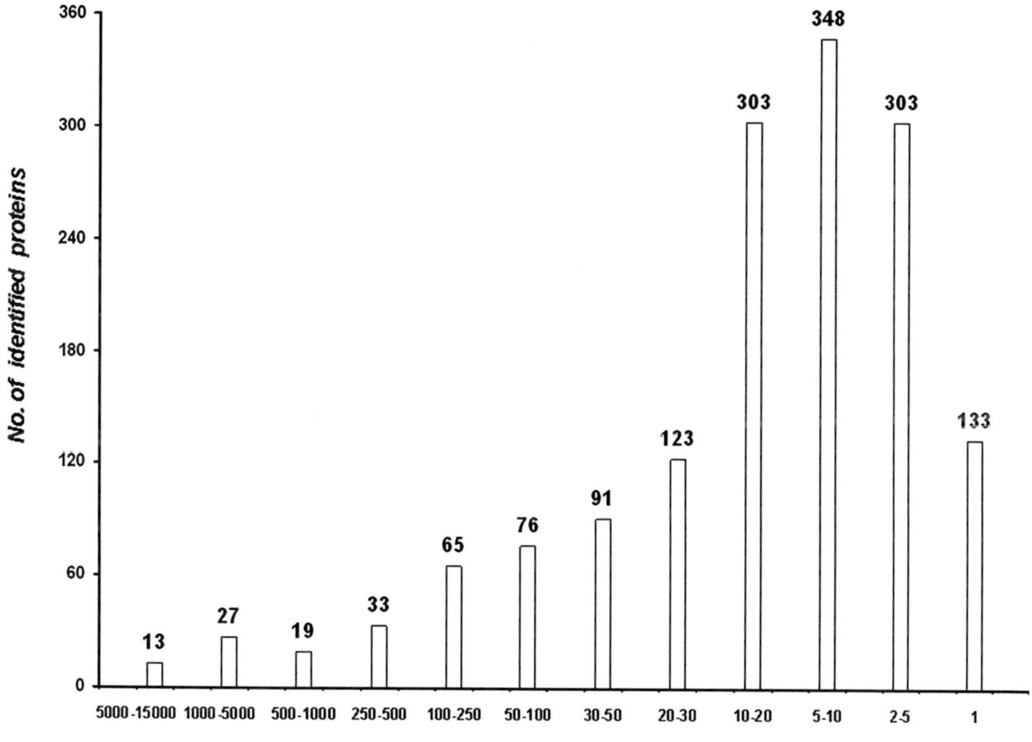

Fig. 6. Distribution of identified proteins using the number of MS/MS events as an indication of the relative abundance. This distribution illustrates that the majority (70%) of identified proteins are of low abundance, with ≤10 MS/MS events per protein. This indicates the efficiency of intact-protein multidimensional fractionation to identify low-abundance proteins in plasma.

abundance in a given study. From Fig. 6, it is clear that the majority (70%) of identified proteins occur at low abundance, with ≤10 MS/MS events per protein. This further illustrates the efficiency of intact protein multidimensional fractionation in reaching low-abundance proteins in serum/plasma.

7. Peroxiredoxin 6 (PRDX6) is a cytoplasmic protein involved in redox regulation of the cell, detoxifying hydrogen peroxide and various organic peroxides (21). Two different forms of PRDX6 are detected by using IPAS analysis in a study of pre-diagnosis of WHI-colon cancer. Quantitative data are obtained for the two different forms of PRDX6 (Fig. 7). Form (A) is the intact-protein form containing the AhpC/TSA domain and C-terminal domain. Form (A) is upregulated in cancer with a ratio of 1.60 (cancer vs. control). Form (B) is the cleavage form containing the AhpC/TSA domain only. There is no significant difference in levels between cancer and control with a ratio of 1.14 (cancer/control). It was reported that overexpression of PRDX6 leads to a more invasive phenotype and metastatic potential in human breast cancer, at least in part, through the regulation of the levels

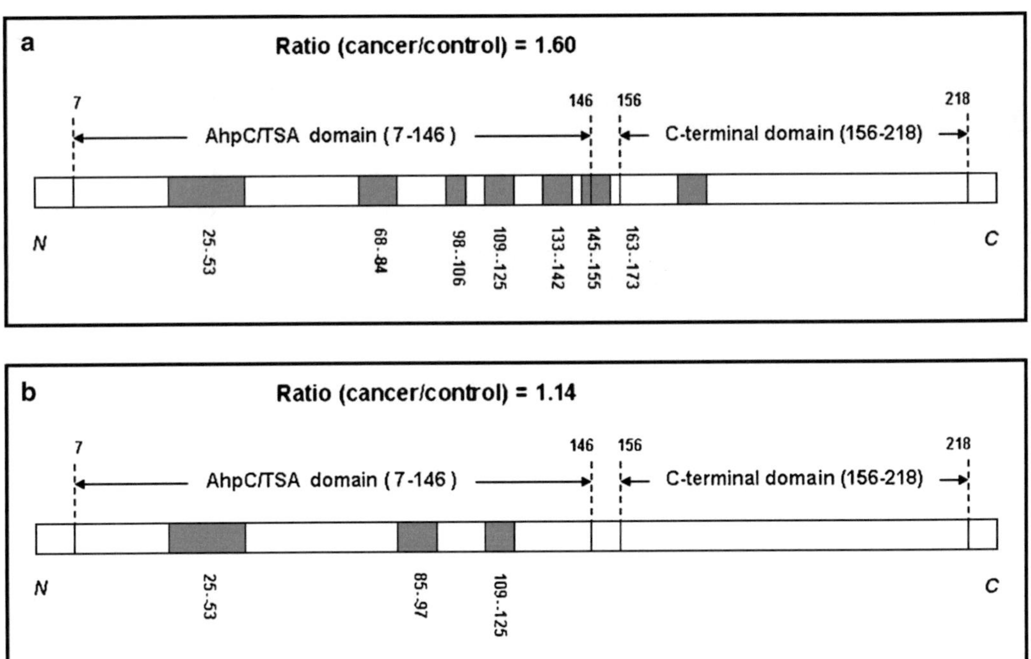

Fig. 7. Peroxiredoxin 6 (*PRDX6*) identified with different isoforms by 2D-HPLC and quantitatively measured by LC–MS. (**a**) Intact-protein isoform that is upregulated in cancer with a ratio of 1.60 (cancer/control). (**b**) Cleavage form resulted from C-terminal domain cleavage, where there is no significant change between cancer and control since the ratio is 1.14 (cancer/control).

of uPAR, Ets-1, MMP-9, RhoC, and TIMP-2 expression (22). It will be more interesting to investigate the PRDX6 serum level in other types of cancer, such as colon cancer.

References

1. Zhang Q., menon R., Deutsch E.W. et al. (2008). A mouse plasma peptide atlas as a resource for disease proteomics. Crenome Biol. **9**, R93.
2. Klose, J. (1975). Protein mapping by combined isoelectric focusing and electrophoresis of mouse tissues. A novel approach to testing for induced point mutations in mammals. *Hum. Genet.* **26**, 231–243.
3. O'Farrell, P. H. (1975). High resolution two-dimensional electrophoresis of proteins. *J. Biol. Chem.* **250**, 4007–4021.
4. Hanash, S. M. (2000). Biomedical applications of two-dimensional electrophoresis using immobilized pH gradients: Current status. *Electrophoresis.* **21**, 1202–1209.
5. Wang, H., Clouthier, S.G., Galchev, V. et al. (2005). Intact-protein-based high-resolution three-dimensional quantitative analysis system for proteome profiling of biological fluids. *Mol Cell Proteomics.* **4**, 618–625.
6. Stasyk, T., Huber, L. A. (2004). Zooming in: Fractionation strategies in proteomics. *Proteomics.* **4**, 3704–3716.
7. Betgovargez, E., Knudson, V., Simonian, M. H. (2005). Characterization of Proteins in the Human Serum Proteome. *J Biomol Tech.* **16**, 306–310.
8. Sheng, S., Chen, D., Van Eyk, J. E. (2006). Multidimensional liquid chromatography separation of intact proteins by chromatographic focusing and reversed phase of the human serum proteome: optimization and protein database. *Mol Cell Proteomics.* **5**, 26–34.
9. Ke, C. Y., Geng, X. D. (2008). A new chromatographic method for fast separation of active proteins. *Chinese Science Bulletin.* **53**, 1113–1116.
10. Dugo, P., Cacciola, F., Kumm, T., Dugo, G., Mondello, L. (2008). Comprehensive multidimensional liquid chromatography: theory and applications. *J Chromatogr A.* **1184,** 353–368.
11. Katayama, H., Paczesny, S., Prentice, R. et al. (2009). Application of serum proteomics to the Women's Health Initiative conjugated equine estrogens trial reveals a multitude of effects relevant to clinical findings. *Genome Med.* **1**, 47.
12. Faca, V.M., Song, K. S., Wang, H. et al. (2008). A mouse to human search for plasma proteome changes associated with pancreatic tumor development. *PLoS Med.* **5**, e123.
13. Pitteri, S.J., JeBailey, L., Faça, V. M. et al. (2009). Integrated proteomic analysis of human cancer cells and plasma from tumor bearing mice for ovarian cancer biomarker discovery. *PLoS One.* **4**, e7916.
14. Pitteri, S.J., Faca, V. M., Kelly-Spratt, K. S. et al. (2008). Plasma proteome profiling of a mouse model of breast cancer identifies a set of up-regulated proteins in common with human breast cancer cells. *J Proteome Res.* 7, 1481–1489.
15. Rauch, A., Bellew, M., Eng, J. et al. (2006). Computational Proteomics Analysis System (CPAS): an extensible, open-source analytic system for evaluating and publishing proteomic data and high throughput biological experiments. *J Proteome Res.* **5**, 112–121.
16. Maclean, B., Eng, J. K., Beavis, R. C., McIntosh, M. (2006). General framework for developing and evaluating database scoring algorithms using the TANDEM search engine. *Bioinformatics.* **22**, 2830–2832.
17. Nesvizhskii, A. I., Keller, A., Kolker, E. & Aebersold, R. (2003). A statistical model for identifying proteins by tandem mass spectrometry. *Anal Chem* **75**, 4646–4658.
18. Faca, V., Coram, M., Phanstiel, D. et al. (2006). Quantitative analysis of acrylamide labeled serum proteins by LC-MS/MS. *J Proteome Res.* **5**, 2009–2018.
19. Bradford, M. M. (1976). A rapid and sensitive method for the quantitation of microgram quantities of protein utilizing the principle of protein-dye binding. *Anal. Biochem.* **748**, 248–254.
20. Keller, A., Nesvizhskii, A. I., Kolker, E., Aebersold, R. (2002). Empirical statistical model to estimate the accuracy of peptide identifications made by MS/MS and database search. *Anal. Chem.* **74**, 5383–5392.
21. Kümin, A., Christine Huber, C., Thomas Rülicke, T., Wolf, E., Werner, E. (2006). Peroxiredoxin 6 Is a Potent Cytoprotective Enzyme in the Epidermis. *Am J Pathol.* **169**, 1194–1205.
22. Chang, X. Z., Li, D. Q., Hou, Y. F. et al. (2007). Identification of the functional role of peroxiredoxin 6 in the progression of breast cancer. *Breast Cancer Res.* **9**, R76.

Chapter 5

Model-Based Discovery of Circulating Biomarkers

Maryann S. Vogelsang, Kian Kani, Jonathan E. Katz, and Parag Mallick

Abstract

Proteomic-based biomarker discovery approaches broadly attempt to identify proteins whose basal abundance, or change in abundance in response to a perturbation (e.g., a therapeutic intervention) is able to discriminate between populations of patients. Up until recently, the majority of approaches for discovering circulating biomarkers have focused on directly profiling serum or plasma to identify such proteins. However, the complexity and dynamic range of protein abundance in serum and plasma create a significant challenge for proteomics methods. To overcome these barriers, diverse approaches to simplify or to fractionate serum and plasma have been developed. For some diseases, such as those related to specific organs, there may be useful marker proteins that originate in the organ. Here, we describe an approach for marker discovery that focuses on the profiling of either primary tissue or cell culture models thereof.

Key words: Model-based biomarker discovery, Serum plasma, Quantitative proteomics, Disease biomarkers

1. Introduction

Proteomics has become an important component of biological and clinical research. Numerous proteomics methods have been developed to identify and to quantify the proteins present in biological and clinical samples (1, 2). Differences between cell types or treatment groups have been used to identify cellular functions and pathways affected by disease or perturbations (3, 4), new components and changes in the composition of protein complexes and organelles (5–7), and putative disease biomarkers (8). Serologic markers typically attempt to differentiate classes of patients, such as diagnostic markers that attempt to differentiate patients with a disease from those without a disease, or prognostic markers that indicate which patients are likely to respond to a therapeutic intervention.

Richard J. Simpson and David W. Greening (eds.), *Serum/Plasma Proteomics: Methods and Protocols*, Methods in Molecular Biology, vol. 728, DOI 10.1007/978-1-61779-068-3_5, © Springer Science+Business Media, LLC 2011

The application of proteomics approaches to the discovery of relevant protein markers from clinical samples has been hampered by sample complexity and variability. To begin to broach this challenge, complex experimental protocols for enrichment, separation, and quantification have been developed for selective or comprehensive proteome analysis (as discussed elsewhere in this book). Though these approaches have been tremendously effective, it is still nontrivial to perform discovery proteomics in serum or plasma. An alternate approach argues that for a subset of biological and clinical questions such as those related to specific organs and their interaction with the circulatory system, a subset of proteins of interest may originate in the specific organ. For example, it is well known that following injury, sets of organ-derived proteins can be found in the circulation. Likewise, in cancer, proteins derived from the tumor are often found in the circulation. In our own studies, we have observed a notable number of tumor-derived proteins in the circulation (unpublished work) (9). In fact, profiling of both tumors and sera revealed that greater than 99% of the tumor-derived proteins identified by serum profiling were also identified by tumor profiling. Furthermore, we observed a notable bias in the serum proteome for cell surface and secreted proteins. Lastly, we have repeatedly observed a strong correlation between quantitative protein abundance changes in the organ, mapping directly to quantitative protein abundance changes in the peripheral circulation. Beyond the significant experimental ease of cell culture models over biofluids, it is also frequently easier to connect discovered markers to biochemical mechanism. Given these observations, here, we present approaches for serologic marker discovery that focuses on the profiling of either primary tissue or cell culture models when available.

Specifically, we begin with a modest adaptation of the liquid-phase-based orthogonal multidimensional intact-protein analysis system (IPAS) for whole cell lysate developed by the Hanash lab (10) for disease biomarker discovery as illustrated in Fig. 1. Additional detailed protocols from the Hanash lab are found in this book (11). Next, we discuss cell surface capture adapted from Faca and Hanash (2009) (12) and last glycocapture, adapted from Zhou et al (2007) (13) for analysis of the secretome. As with many label-based quantification studies, quantification is achieved by comparing the signal intensities of identified peptides from the control and case samples in MS scan events. For cell culture experiments, we typically incorporate SILAC as described below. For tissue analysis, acrylamide-based labeling has been highly successful. When analyzing SILAC-labeled whole cell lysate, we identify more than 5,000 proteins with high confidence and obtained relative quantitative data greater than 65% of identified proteins spanning 6–8 orders of magnitude of protein concentration in the primary substrate and

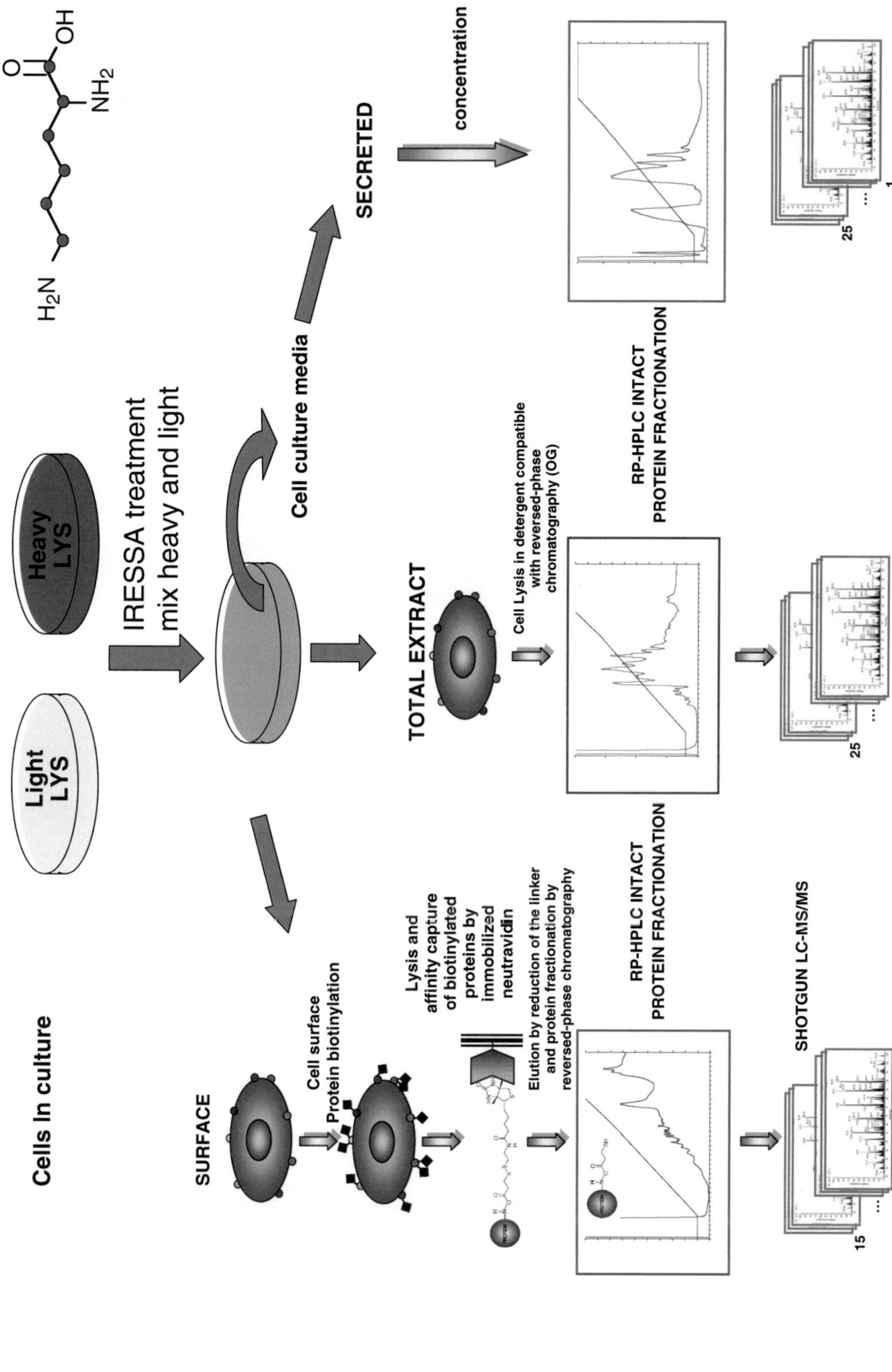

Fig. 1. Experimental design.

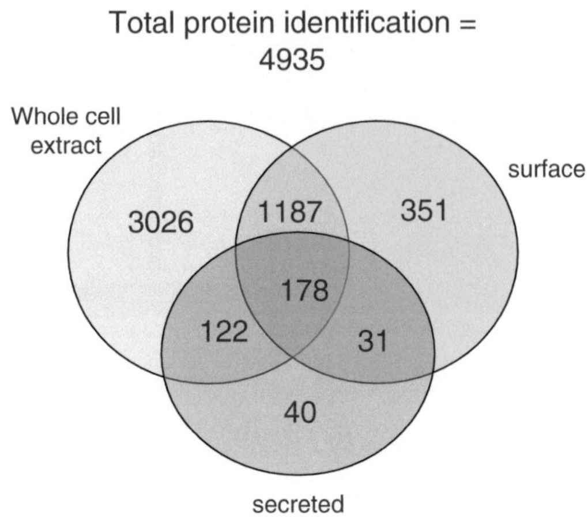

Fig. 2. Summary of proteins identified in A431 in vitro discovery experiment. Number of proteins identified (Peptide and ProteinProphet >0.9) per compartment. Roughly two-thirds of all proteins had SILAC ratio.

more so in the circulation (Fig. 2). By way of example, we use the A431 cell line from ATCC. However, the methods are applicable to other cell lines.

2. Section I Whole Cell Lysate Analysis

2.1. Materials

1. Complete Mini Protease Inhibitor, EDTA-free tablets, (Roche, #11836170). Use one tablet per 7 mL of PBS.
2. *N*-Octylglucoside (Roche, #10634425001).
3. Dithiothreitol, Mass spectrometry grade (DTT) (Sigma, #D5545).
4. Iodoacetamide, Mass spectrometry grade (IAA) (Sigma, #I1149).
5. Guanidine HCl, Mass spectrometry grade (GuHCl; Sigma, #G3272).
6. Ammonium bicarbonate (AMBIC; Sigma, #A-6141).
7. Needles, 1 mL 27½-gauge needle (B-D, #309623).
8. PBS pH 7.4, supplemented with 1 mM $CaCl_2$ and 1 mM $MgCl_2$ (Gibco, #10010-023).
9. Poros R1 column (4.6 mm × 100 mm, Applied Biosystems, #1-1012-26).
10. BETASIL C18 column (100 mm × 1 mm, 3 μm particle size, Thermo Scientific, #70103-101030).

11. Solvent A: 0.1% Formic acid in water (Honeywell Burdick & Jackson, #BJLC452-1).
12. Solvent B: 0.1% Formic acid in acetonitrile (Honeywell, Burdick & Jackson, #BJLC441-1).
13. Sequencing-grade modified trypsin (Promega, #V5111).
14. Digestion buffer: 0.25 M urea, 50 mM ammonium bicarbonate, and 4% acetonitrile (v/v).
15. BCA Protein Assay Kit (Pierce, #23227).
16. Bovine Serum Albumin Standard Ampules, 2 mg/mL (Pierce, #23209).
17. Quality-control material (see Note 1): Angiotensin II (Sigma A – 9525), Bombessin (Sigma B – 4272), Substance P (Sigma S – 6883), Neurotensin (Sigma N – 6383), and Alpha 1–6 (Sigma T – 2903). The peptides are reconstituted in water to a concentration of 1 mM, for stock solutions. For working standards, the peptides are combined and diluted to a final concentration of 1 μM of each peptide, in 3 M GuHCl, 0.1% formic acid in water (see Note 2).

3. Method

3.1. Reagents for WCL Analysis

1. Prepare 1% OG (w/v) in PBS. For single plate lysis, prepare 1–2 mL. When working with larger-scale experiments, 50 mL or more of buffer is prepared.
2. 6 M GuHCl, plus 1% OG (w/v). (Final concentration will be 3 M).
3. 100 mM Ammonium bicarbonate (AMBIC), pH 8.0.
4. 1 M Iodoacetamide (IAA) in PBS.
5. 1 M Dithiothreitol (DTT) in PBS.
6. 3 M GuHCl in 100 mM AMBIC, pH 8.

3.2. Culture and Isotopic Labeling of A431 Cells

1. Grow A431 in DMEM media (Cellgro) containing 1% dialyzed fetal bovine serum (FBS) (Cellgro) and an antibiotic of choice.
2. Not all of the following are required; choose those necessary for the desired SILAC experimental design: L-arginine-$^{13}C_6$ monohydrochloride (Cambridge Isotope Laboratories, #CLM-2265), L-arginine-$^{13}C_6$,$^{15}N_4$ hydrochloride (Sigma-Isotec, #608033), L-lysine-$^{13}C_6$,$^{15}N_2$ hydrochloride (Sigma-Isotec, #60804), L-methionine-^{13}C,$^{2}H_3$ (methyl-^{13}C,^{2}H) (Sigma-Isotec, #299154).
3. Change medium (using either light or heavy SILAC medium) every 2–3 days if cells are not ready for subculture.

4. After the first passage, begin expansion of light and heavy SILAC cell populations by culturing cells in larger dishes or a larger number of plates, as needed.
5. Grow cells in SILAC media for at least seven passages.
6. Grow plates to 80% confluence.

3.3. Cell Lysis Protocol

1. Start with 15-cm plates at 80% confluence.
2. Wash cells thrice with PBS.
3. Lyse the cells with PBS supplemented with 1% OG, and protease inhibitor cocktail.
4. We typically use between 800 μL and 1.5 mL per plate depending on cell type.
5. Place lysates on ice and disrupt cell membrane by needle treatment or sonication.
6. Centrifuge lysates at 4°C at 15,000 × g for 10 min.
7. Decant soluble fraction and centrifuge with a 0.2-μm filter device.
8. Determine protein concentration of lysate solutions by BCA or any equivalent assay (see Note 3).
9. We typically mix 1 mg of heavy lysate solution with 1 mg of light lysate solution.
10. Typical volumes should be around 1 mL.
11. We add DTT to a final concentration of 25 mM in 6 M GuHCl [1% OG (w/v)] to bring the final GuHCl concentration to 3 M.
12. Rotate at 65°C for 1 h.
13. Allow to cool to room temperature for 15 min.
14. Make a solution of 1 M IAA in PBS and add to a final concentration of 55 mM (see Note 4).
15. Typically, we end up at a final volume between 1.5 and 2 mL.
16. Incubate for 1 h, at room temperature in the dark.

3.4. Reversed-Phase Whole Protein Fractionation

1. Protein sample should be approximately 1 mg in about 1–2 mL.
2. Filter lysed cells through a 0.22-μm syringe filter, before loading onto column.
3. Load sample onto a POROS R1/10 column, typically 1 mL volume loaded per run.
4. After column equilibration:

   Gradient: Flow rate: 1 mL/min

   0 min: 5% solvent B

23 min: 50% solvent B

53 min: 80% solvent B

55 min: 95% solvent B

65 min: 100% solvent B

69 min: 5% solvent B

89 min: end run

5. Equilibrate the column for at least 10 min at 1 mL per min with solvent A before injections.
6. Inject the sample and keep the system at the initial solvent concentration of 5% solvent B until absorbance returns to baseline before starting the gradient.
7. Collect fractions of 90-min run.
8. Determine protein concentration of each fraction by BCA or any equivalent assay (see Note 3).
9. Pool the fractions into roughly 100 μg protein per fraction. In our case, typically 20 plus fractions are collected, digested, and then analyzed.
10. The fractions are then placed in a –80°C freezer for approximately 1 h.
11. The fractions are then dried down in a SpeedVac. They are centrifuged under vacuum until dryness (see Note 5).
12. After dryness is reached, the samples may be stored at –80°C.

3.5. Protein Digestion

1. Resuspend individual fractions in 100 μL of 3 M GuHCl in 100 mM ammonium bicarbonate, pH 8.0.
2. Shake for 10 min at room temperature.
3. Add 180–190 μL of ammonium bicarbonate, pH 8.0.
4. Add 1–4 μg of trypsin (from a 0.1 μg/μL trypsin suspension in 100 mM ammonium bicarbonate). The final cell protein extract to trypsin mass ratio is approximately 50:1.
5. Incubate at 37°C for at least 18 h.
6. Interrupt the digestion by adding 9.9 μL of glacial acetic acid.
7. Freeze samples at –80°C for at least 2 h.
8. Centrifuge sample under vacuum to dryness (see Note 5).
9. Digested sample may be stored at –80°C for no longer than 6 months.

3.6. Mass Spectrometry Analysis

1. Resuspend dried fractions into 40 μL of 95:5:0.1 (H_2O:ACN:FA).
2. Load 10 μL/fraction/run (see Note 6).
3. Inject 10 μL of sample onto BETASIL C18 column.

4. After injection, the sample is desalted directly on the column, for 20 min, prior to elution.
   Gradient: Flow rate: 65 μL/min

Time (min)	Pump	Start (%)
0.01	B	5
3	B	8
7	B	15
60	B	35
65	B	95
75	B	95
78	B	5
80	B	5
80.01	Controller	Stop

5. Acquire spectra in a data-dependent mode in *m/z* range of 400–1,800. Select the six most abundant +2 ion or greater for each MS spectrum for MS/MS analysis. Mass spectrometer parameters are as follows: electrospray ionization voltage of 3.75 kV, capillary temperature of 200°C, resolution of 60,000, and FT target value of 1,000,000.

   Scan event details:

1. ITMS + p norm o(400.0-1800.0)
   CV = 0.0V
2. FTMS + p norm res = 60000 o(400.0-1800.0)
   CV = 0.0V
3. ITMS + c norm Dep MS/MS Most intense ion from (2)
   Activation type: CID
   Min. signal required: 500.0
   Isolation width: 2.00
   Normalized coll. energy: 35.0
   Default charge state: 2
   Activation Q: 0.250
   Activation time: 30.000
   CV = 0.0V
4. ITMS + c norm Dep MS/MS second most intense ion from (2)
   Activation type: CID
   Min. signal required: 500.0
   Isolation width: 2.00
   Normalized coll. energy: 35.0
   Default charge state: 2

Activation Q: 0.250
Activation time: 30.000
CV=0.0V

5. ITMS+c norm Dep MS/MS third most intense ion from (2)
Activation type: CID
Min. signal required: 500.0
Isolation width: 2.00
Normalized coll. energy: 35.0
Default charge state: 2
Activation Q: 0.250
Activation time: 30.000
CV=0.0V
6. ITMS+c norm Dep MS/MS fourth most intense ion from (2)
Activation type: CID
Min. signal required: 500.0
Isolation width: 2.00
Normalized coll. energy: 35.0
Default charge state: 2
Activation Q: 0.250
Activation time: 30.000
CV=0.0V
7. ITMS+c norm Dep MS/MS fifth most intense ion from (2)
Activation type: CID
Min. signal required: 500.0
Isolation width: 2.00
Normalized coll. energy: 35.0
Default charge state: 2
Activation Q: 0.250
Activation time: 30.000
CV=0.0V
8. ITMS+c norm Dep MS/MS sixth most intense ion from (2)
Activation type: CID
Min. signal required: 500.0
Isolation width: 2.00
Normalized coll. energy: 35.0
Default charge state: 2
Activation Q: 0.250
Activation time: 30.000
CV=0.0V

3.7. Data Analysis

The acquired data is automatically processed by the Computational Proteomics Analysis System – CPAS (14). Tandem mass spectra are searched against the human IPI database (60,428 protein entries) with sequences for human and bovine trypsin added. Searches are typically performed with X!Tandem (2005.12.01) using a mass tolerance for precursor ions set to 20 ppm. The mass tolerance for fragment ions is set to 0.5 Da. A fixed modification of 6.020129 mass units was added to lysine residues for database searching to account for incorporation of the heavy lysine isotope. All identifications with a PeptideProphet (15) probability greater than 0.75 are submitted to ProteinProphet (16), and the subsequent protein identifications are filtered at a 1% error rate with tryptic fragments (one missed cleavage), with an allowance for fixed modification on $C = 57.021$ and variable amino acid modifications on $C = -17.027$, $E = -18.011$, $K = 6.020$, $M = 15.995$, and $Q = -17.027$.

The quantitative approach consisted of differential labeling of proteins in culture with lysine isotopes (heavy or light) (17). Quantitative information is extracted using a script designated "Q3" that was developed originally to obtain the relative quantification for each pair of peptides identified by MS/MS that contains acrylamide-labeled peptides (18). Only peptides with a minimum of 0.75 PeptideProphet score and mass deviation less than 20 ppm were considered in the analysis. All normalized peptide ratios for a specific protein were averaged to compute an overall protein ratio. A single sample *t*-test is used to assess confidence in the protein quantitation ratio (19). In addition, protein quantification values are manually verified to ensure that quantified peptides have clear elution profiles and isotope envelopes at the appropriate mass spacing.

4. Section II Cell Surface Analysis

The cell surface proteome is of substantial interest, especially in being rich with therapeutic targets. Gaining a better understanding of the surface membrane proteins and their role in cellular growth can help lead to the development of new drug targets. Here, we introduce a cell surface protein capture technique that employs biotinylation of proteins of interest.

4.1. Materials

1. Epidermoid carcinoma cell line A-431 (ATCC, #CRL-1555).
2. DMEM F-12 medium (Invitrogen, #11320-033).
3. SILAC DMEM media (Invitrogen, #MS10030).
4. Dialysed FBS (Invitrogen, #26400-044).
5. Trypsin (with 0.05% EDTA) (Cellgro, #25-052-Cl).
6. Dulbecco's PBS (D-PBS, Invitrogen, #14190-144).
7. EZ-Link sulfo-NHS-LC-biotin (Pierce, #21217).
8. Glycine (Sigma-Aldrich, #241261).
9. NP40 (Sigma-Aldrich, #100511-472).

10. HALT cocktail protease inhibitor (Pierce, #PI78430).
11. D-Biotin (US Biological, #B1750).
12. Pierce Monomeric Avidin Kit (Pierce, #20227).
13. Sodium Orthovanadate (Vanadate, New England Biolabs, #P0758L).
14. Centricon YM-3 columns (Millipore, #42404).

5. Methods

5.1. Reagents for Cell Surface Analysis

1. 50 mM Tris–HCl (pH 7.5)
2. 2 mM Biotin in PBS
3. 0.1 M glycine (pH 2.8)
4. Prepare SILAC media according to the manufacturer's instructions:
   (a) Prepare 1 L of complete medium as follows:
      - Replace 100 mL of basal medium with dialyzed FBS.
      - Add 10 mL of 100× L-glutamine, if basal medium does not contain glutamine.
      - Add 10 mL of 100× penicillin–streptomycin, if needed.
      - Add any additional growth factors required for your cell line.
      - Mix well and filter-sterilize the medium using a 0.22 μm filtration device.
      - Store the complete medium at 4°C protected from light until use.
5. 100 mL of NHS-SS-BIOTIN (at 1 mg/mL) in PBS, pH 8.0 at 4°C.

5.2. Culture and Isotopic Labeling of A431 Cells

1. Grow cells in SILAC media containing 1% dialyzed fetal bovine serum (FBS) (Cellgro) and penicillin–streptomycin (or antibiotic of choice).
2. Not all of the following are required; choose those necessary for the desired SILAC experimental design: L-arginine-$^{13}C_6$ mono-hydrochloride (Cambridge Isotope Laboratories, #CLM-2265), L-arginine-$^{13}C_6$,$^{15}N_4$ hydrochloride (Sigma-Isotec, #608033), L-lysine-$^{13}C_6$,$^{15}N_2$ hydrochloride (Sigma-Isotec, #60804), L-methionine-^{13}C,$^{2}H_3$ (methyl-^{13}C,^{2}H) (Sigma-Isotec, #299154).
3. Change medium (using either light or heavy SILAC medium) every 2–3 days if the cells are not ready for subculture.

4. After the first passage, begin expansion of light and heavy SILAC cell populations by culturing the cells in larger dishes or a larger number of plates, as needed.
5. Grow the cells in SILAC media for at least seven passages.
6. Grow plates to 80% confluence.
7. The cells are typically expanded to 50 cm × 15 cm plates of H1975 (Light) and 50 cm × 15 cm plates of H1975 (Heavy).
8. The cells are maintained in SILAC, 1% FBS until 100 plate expansion is reached (see Note 7).
9. Starting with 50 cm × 15 cm plates of H1975 and 50 cm × 15 cm plates of H1975.
10. Maintain in SILAC, 1% FBS.
11. Split the cells with 1 mL trypsin.
12. Spin the cells 1,000 × *g* for 10 min.
13. Aspirate the media.
14. Add PBS.
15. Spin the cells 1,000 × *g* for 10 min.
16. Count the cells and separately seed equivalent light- and heavy-labeled cells.
17. Aspirate the media.
18. Resuspend in SILAC media.

5.3. Gefitinib (Model Therapeutic) Treatment

1. When the cells reach 65–90% confluency, wash the cells with PBS.
2. Resuspend the cells in either SILAC media with DMSO or SILAC media with 100 nM Iressa.
3. Treat 25 heavy plates with DMSO, 25 heavy plates with Gefitinib, 25 light plates with DMSO, and 25 light plates with Gefitinib.
4. Follow cells through a 16 h time course (see Note 8).

5.4. Cell Surface Protein Biotinylation

1. At 1 h time points per 12 plates, aspirate cell media and wash cells with 15 mL of PBS three times on ice.
2. Add 5 mL of NHS-SS-BIOTIN in PBS pH 8.0 at 4°C to each plate.
3. Incubate for 20 min on ice.
4. Aspirate NHS-SS-BIOTIN solution.
5. To stop biotinylation, wash cells with 5 mL of 100 mM glycine, chilled at 4°C (made in 50 mM Tris–HCl).
6. Incubate cells in glycine–Tris–HCl for 5 min.
7. Aspirate the glycine–Tris–HCl.

8. Wash with ice-cold PBS twice.
9. Harvest the cells by scraping in PBS (supplemented with 1% NP40, HALT protease inhibitor, and Na Vanadate).
10. The cells may be flash-frozen in dry ice and stored at −80°C until further processing.
11. If continuing on, the cells may be further disrupted by a brief sonication cycle to initiate cell lysis.

5.5. Cell Lysis

1. Sonicate the cells (according to manufacturer's recommended settings).
2. Pipette up and down to suspend the cells.
3. Use lower power to prevent foaming and disrupt cells by sonicating on ice using five 1 s bursts.
4. Incubate cells for 30 min on ice, vortexing every 5 min for 5 s.
5. To improve solubilization efficiency, 2–3 additional sonication cycles may be added during incubation.
6. Centrifuge cell lysate at $20{,}000\times g$ for 20 min at 4°C.
7. Transfer clarified supernatant to a new tube.
8. Keep on ice until ready to load onto an avidin column.

5.6. Purification of Biotinylated Surface Proteins

1. Wash the avidin (2 mL) column with at least 8 mL of PBS.
2. Wash the column with 6 mL of 2 mM biotin in PBS. This blocks any nonreversible biotin binding sites on the column.
3. Remove excess biotin by washing the column with 12 mL of 0.1 M glycine (pH 2.8).
4. Wash the column extensively with at least five column volumes of PBS (10 mL).
5. Load clarified solubilized supernatant to the column.
6. Pass supernatant through the column at least three times.
7. Wash unbound protein by applying 8 mL of PBS containing 1% NP40.

   Elute bound biotinylated proteins from the column with 12 mL of 5 mM biotin in PBS containing 1% NP40. These may be collected as 6×2 mL fractions (see Note 9).
8. Load samples into Centricon YM-3 columns and spin at $5{,}000\times g$ until volume is approximately 500 μL–1 mL (see Note 10).

Following isolation and mass spectrometric and computational analyses, proceed as described (Section I, Subheadings 3.4–3.6) with the following modification: residue mass alteration must include the mass of the linked portion of the NHS-Sulfhydryl that remains post biotin cleavage.

6. Section III Glycocapture

As noted (Sections I–II), analysis of conditioned media and the secretome can provide significant insight into which proteins are likely to be found in the circulation. Here, we introduce the glycocapture technique for secretome analysis.

6.1. Materials

1. 10DG Desalting Columns (Econo-Pac; Bio-Rad Laboratories, #732-2010).
2. Vial 1 mL glass with polyethylene snap cap (Waters, #WAT025054C).
3. Affi-Prep Hydrazide resin (Bio-Rad, #156-001 6). Supplied as a 50% suspension in isopropanol.
4. Trizma base (Sigma-Adlrich, #T1503-500G).
5. Sodium dodecyl sulfate, SDS (Sigma-Aldrich, #436143-25G).
6. Urea (Fisher Scientific, #BP169-212).
7. Ethylenediaminetetraacetic acid disodium salt dihydrate, EDTA (Sigma-Aldrich, #E5134-50G).
8. Ammonium bicarbonate (AMBIC; Sigma, #A-6141).
9. Sodium acetate trihydrate (Fisher, #BP334-500).
10. Sodium chloride (Sigma-Aldrich, #S3014-5KG).
11. Hydrochloric acid, HCl (Fisher Scientific, #A144-500).
12. Acetic acid, glacial (Sigma-Aldrich, #A6283-500ML).
13. TCEP HCI (Pierce, #20490); used as a 200 mM stock solution, 5 mM final concentration in sample.
14. Sodium periodate (Pierce, #20504).
15. lodoacetamide, Mass spectrometry grade (IAA) (Sigma, #I1149).
16. Sequencing-grade modified Trypsin, (Promega, #V5113) 20 μg/tube, 100 μg total
17. *N*-glycosidaseF(PNGaseF,RocheDiagnostics,#11365185001); dissolve in double-distilled water to get a 1 U/μL solution.
18. Spin column (MoBiTec, #M1002).
19. 10-μm filter (MoBiTec, #M2210).
20. Spin columns with 10-μm filter (MoBiTec, #M1002S).
21. SepPak Ci 8 column (Waters, #WAT023590); C-18 Vydek (The Nest Group, #SNS SS18V).
22. Trifluoroacetic acid, sequencing grade (Pierce, #28904).
23. Methanol (Honeywell Burdick & Jackson, #BJLC230-1).

7. Methods

7.1. Reagents to Be Prepared

Following items must be prepared fresh each time:

1. Urea buffer: 8 M urea, 200 mM Tris pH 8.3, 0.1% SDS, 5 mM EDTA in water; each sample requires approximately 10 mL (see Note 11).
2. 50 mM Ammonium bicarbonate buffer in water, pH 7.8, 0.1% SDS.
3. 250 mM stock Iodoacetamide in water. Prepare this solution immediately prior to use and keep it in dark (light sensitive).

Following items may be made as stock solutions:

4. Oxidation buffer: 20 mM Sodium acetate in water, 150 mM Sodium chloride, pH 5.0. May be stored at room temperature (see Note 12).
5. Coupling buffer: 100 mM Sodium acetate in water, 1 M Sodium chloride, pH 4.5. May be stored at room temperature (see Note 12).
6. 80% (v/v) Acetonitrile in water. May be stored at room temperature.
7. 1.5 M Sodium chloride in water. May be stored at room temperature.
8. 0.4% (v/v) Acetic acid in water. May be stored at room temperature.

7.2. Glycocapture

1. Equilibrate desalting columns with 20 mL of Oxidation Buffer (OB). Add samples to the columns and drain to waste. Add OB to the columns to bring total volume to 3 mL. Add 120% sample volume of OB and collect flow-through.
2. Prepare 50 mM Sodium periodate solution in water. This solution is light sensitive. Protect from light, by wrapping tube in foil.
3. Add the periodate solution at one-fifth sample volume (10 mM final concentration).
4. Oxidize samples in darkness (typically wrapped in foil) at room temperature for 1 h on an end-over-end rotator.
5. Equilibrate the desalting columns with 20 mL of Coupling Buffer (CB). Add samples to the columns and drain to waste. Add CB to the columns to bring total volume to 3 mL. Add 120% sample volume of CB and collect flow-through in clean tubes.
6. Prepare 100 μL of hydrazide resin (200 μL of a 50% slurry in isopropanol) by washing resin with ten CV (column volumes)

water (or three times three CV water), followed by ten CV of CB. Remove washes by centrifuging at 1,000 × *g* for 2 min. After washing, resuspend the hydrazide resin in 50% slurry of CB. Add the oxidized samples to hydrazide resin slurry. Conjugate the glycoproteins to the hydrazide resin at room temperature with mixing for 10–24 h (see Note 13).

7. After coupling reaction is complete, spin down the resin and remove supernatant. Wash thrice with 1 mL of Urea buffer; remove by centrifuging at 1,000 × *g* for 2 min. Resuspend in 200 μL of Urea buffer after last wash.
8. Add 5 μL of TCEP solution (5 mM final concentration) and incubate at room temperature for 30 min. During the incubation, prepare the 250 mM iodoacetamide solution in water; add 8 μL (10 mM final concentration) of prepared solution to the hydrazide resin slurry. Incubate in darkness at room temperature for another 30 min.
9. Wash the resin thrice with 1 mL of Urea buffer; remove by centrifuging at 1,000 × *g* for 2 min. Resuspend in 150 μL of Urea buffer after last wash.
10. Incubate the resin for 30 min at 55°C to denature proteins. After incubation, wash thrice with 1 mL of Urea buffer; remove by centrifuging at 1,000 × *g* for 2 min.
11. Wash resin thrice with 1 mL of a diluted form of Urea buffer (prepare a 1:4 dilution in water).
12. Remove by centrifuging at 1,000 × *g* for 2 min. Resuspend the hydrazide resin in 150 μL of the 1:4 diluted Urea buffer and add 20 μg of trypsin solution. Check pH; it should be approximately pH 8.3. Incubate overnight at 37°C with gentle agitation.
13. Remove the trypsin-released peptides by washing the resin thrice, each time with 1 mL of 1.5 M NaCl, 80% Acetonitrile, methanol, and 50 mM Ammonium bicarbonate buffer, pH 7.8.
14. Remove supernatant by centrifuging at 1,000 × *g* for 2 min at each step.
15. Resuspend the resin with 150 μL of 50 mM Ammonium bicarbonate buffer, pH 7.8. Transfer the resin to a 0.65-mL tube.
16. *N*-linked glycopeptides are released from the resin by adding 2.5 μL of the 1 U/μL *N*-glycosidase F solution to the resin and by incubating at 37°C with mixing overnight.
17. Centrifuge at 1,000 × *g* for 2 min. Transfer supernatant to a clean glass vial.
18. Wash the resin twice with 200 μL of 80% acetonitrile (Be mindful to keep the washes for the next step). Centrifuge at 1,000 × *g* for 2 min.

19. Combine the washes with the initial supernatant in the glass vial, and dry the samples using a SpeedVac system (see Note 5).
20. Perform C18 cleanup on samples.

Following isolation and mass spectrometric and computational analysis, proceed as described above with the following modification: asparagine residue mass alteration must include the mass of the glycosylation anchor.

8. Notes

1. Quality control: As an internal control for properly running instruments and samples, we typically recommend quality-control runs every 4–5 samples using a known mixture. One simple mixture that can be effective is an equimolar mix of angiotensin II, bombesin, substance P, neurotensin, and alpha 1–6. For microspray experiments, 2 μL injections of 1 μM of each peptide as a final concentration in water are used. For nanospray experiments, 2 μL injections of 0.1 μM of each peptide as a final concentration in water are used.

   The following are the intensities of the above-mentioned peptides. These peptides represent the dominant H+ adducts for the five quality-control peptides (Fig. 3):

   Angiotensin II: 523.7747

   Neurotensin : 558.3107

   Substance P : 674.3715

   Bombesin : 810.4148

   Alpha 1–6 : 882.4628
2. Depending on the application in our lab, the five-peptide mixture is reconstituted in different final buffers. For application as standards between runs, the final concentration of each peptide is 1 μM, in 3 M GuHCl. For standard infusion mix, the concentration is 0.1 μM in 50:50:0.1% (acetonitrile–water–formic acid).
3. Due to our small sample size, we follow the microplate procedure in the Pierce BCA Assay kit instructions manual for kit #23277.
4. Make up 1 M DTT and add about 10 μL of 1 M to 1 mL of cell solution. This renders a solution approximately 10 mM DTT final.
5. The samples are dried down in a vacuum centrifuge, with a -110°C cooling trap. This step helps in two ways:first, we boil-off the solvents from the fractionation run; second, we have smaller volumes to work within the subsequent digestion steps.

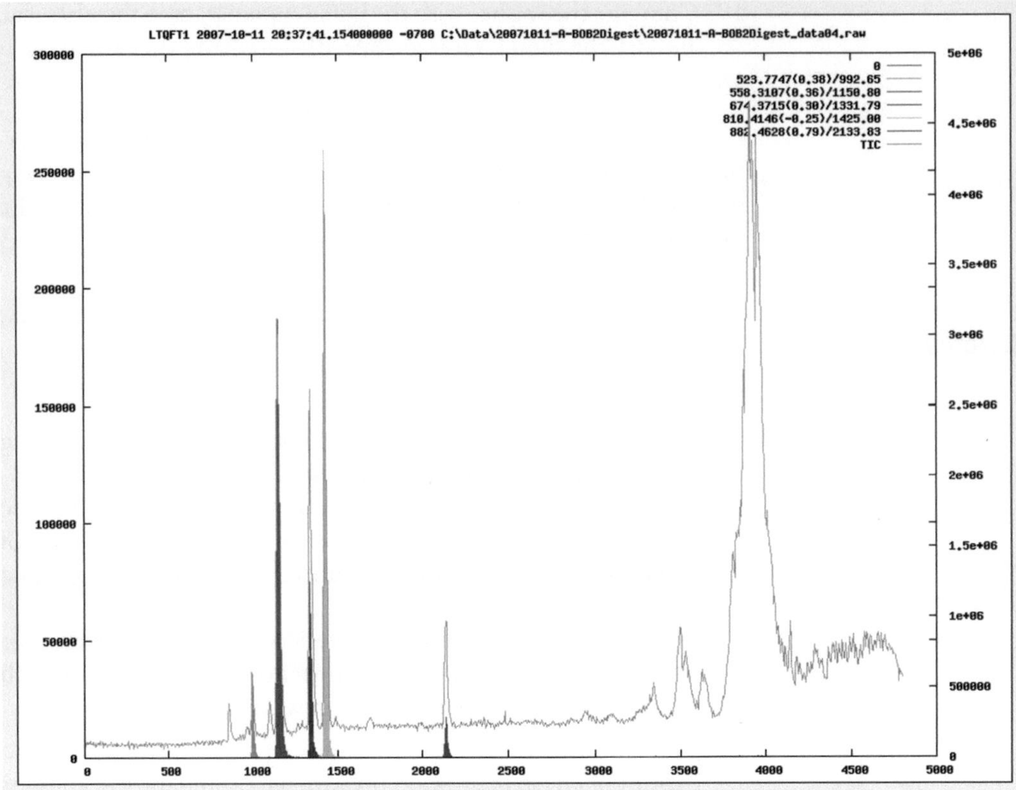

Fig. 3. An example of an automatically-generated picture created at the end of each of our 5-peptide quality control (QC) runs. Shown is the total ion chromatogram (scan # vs. intensity) on top of which is overlayed 5 extracted ion chromatograms for the m/z values (+/- 8 ppm) of the dominant charge states for each of the five peptides that make up our QC mixture. From a quick inspection the investigators can rapidly determine if the chromatography and sensitivity are within acceptable tolerances.

6. Our system utilizes two-column setup on a ten-port valve. This enables a sample to be analyzed on one column while the offline column is washed and having the next sample applied, as shown in Fig. 4.
7. In this example experiment, we scale up to 100 plates to (1) account for lower abundance of protein at the cell surface level. A significant level of protein is lost during the cell surface protein capture experiment, and to account for that loss and to ensure that we have enough protein for mass spectrometry analysis, we increase the number of plates we lyse. (2) 100 plates provide us with a quantity to work within the varying drug treatment conditions. As a side note, for nonenriched protein preparations, a typical plate yields approximately 500 μg to 2 mg of protein.
8. We divided our plates as follows:

   24 plates H (+ Gefitinib) vs. 24 plates L (– no treatment)

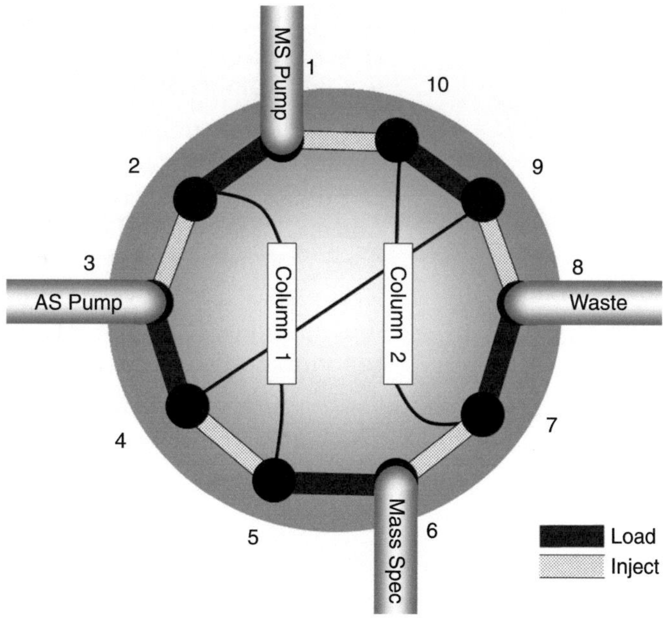

Fig. 4. Electrospray double column setup.

24 plates L (+ Gefitinib) vs. 24 plates H (– no treatment)

2 plates H (+ Gefitinib) vs. 2 plates L (– no treatment)

2 plates L (+ Gefitinib) vs. 2 plates H (– no treatment)

9. Column may be regenerated with 8 mL of 0.1 M glycine (pH 2.8).
10. Samples are concentrated using Centricon centrifugal filter device to concentrate samples to a workable volume without a concomitant increase in salt concentration as would occur in evaporative concentrator. Following concentration, the samples may be digested as described in Section I, Subheading 3.4, with few exceptions. Since this preparation renders the sample in solution, you need to add 3 M GuHCl in 100 mM AMBIC, to a final working concentration of GuHCl is 1 M.
11. First, prepare the 200 mM Tris buffer, pH 8.3. For a 1-L stock solution, add 24.24 g of Trizma base to approximately 400 mL of water and titrate to pH 8.3 with concentrated HCl, with stirring. Then, add remaining water to a final volume of 1 L. Alternatively, it is useful to prepare a 2× stock solution. Follow the above preparation, but use 48.48 g Trizma base in a final volume of 1 L. To complete a 50-mL preparation of Urea buffer, weigh 24.02 g of urea, 93 mg of

EDTA, and 50 mg of SDS and add these items to one appropriately sized container, then add 200 mM Tris buffer to a final volume of 50 mL.

12. To prepare 500 mL of oxidation buffer, add 1.36 g of sodium acetate to a beaker. Add about half the water (200–250 mL) and titrate the solution with glacial acetic acid until pH 5.0. Add 4.38 g of NaCl, stir, and check pH. Then, add remaining water to a final volume of 500 mL. 500 mL of coupling buffer is prepared as above, but the following are used: 6.8 g of sodium acetate (for 100 mM) and 29.22 g of NaCl (for a 1 M concentration). The coupling buffer is titrated to pH 4.5 with glacial acetic acid.
13. The binding capacity is ~1 mg pure beads per 10 mg bound protein.

References

1. Gerber, S. A., Rush, J., Stemman, O., et al. (2003) Absolute quantification of proteins and phosphoproteins from cell lysates by tandem MS, *Proc Natl Acad Sci U S A* ***100***, 6940–6945.
2. Ong, S. E., Foster, L. J., and Mann, M. (2003) Mass spectrometric-based approaches in quantitative proteomics, *Methods (San Diego, Calif* ***29***, 124–130.
3. Durr, E., Yu, J., Krasinska, K. M., Carver, L. A., et al. (2004) Direct proteomic mapping of the lung microvascular endothelial cell surface in vivo and in cell culture, *Nature biotechnology* ***22***, 985–992.
4. Wright, M. E., Eng, J., Sherman, J., et al. (2003) Identification of androgen-coregulated protein networks from the microsomes of human prostate cancer cells, *Genome biology* ***5***, R4.
5. Ranish, J. A., Hahn, S., Lu, Y., et al. (2004) Identification of TFB5, a new component of general transcription and DNA repair factor IIH, *Nature genetics* ***36***, 707–713.
6. Blagoev, B., Kratchmarova, I., Ong, S. E., et al. (2003) A proteomics strategy to elucidate functional protein-protein interactions applied to EGF signaling, *Nature biotechnology* ***21***, 315–318.
7. Andersen, J. S., Wilkinson, C. J., Mayor, T., et al. (2003) Proteomic characterization of the human centrosome by protein correlation profiling, *Nature* ***426***, 570–574.
8. Marko-Varga, G., Lindberg, H., Lofdahl, C. G., et al. (2005) Discovery of biomarker candidates within disease by protein profiling: principles and concepts, *J Proteome Res* ***4***, 1200–1212.
9. Kani, K., Faca, V. M., Hughes, L., et al. (2010) Proteomics reveals protein indicators of HER axis therapy response, *submitted.*
10. Wang, H., Clouthier, S. G., Galchev, V., et al. (2005) Intact-protein-based high-resolution three-dimensional quantitative analysis system for proteome profiling of biological fluids, *Mol Cell Proteomics* ***4***, 618–625.
11. Wang, H., and Hanash, S. (2011) Intact-protein analysis system for discovery of serum-based disease biomarkers, Ed. Simpson RJ, Greening DW, *Serum/Plasma Proteomics, Methods Mol Biol* ***728***, Humana Press.
12. Faca, V. M., and Hanash, S. M. (2009) In-depth proteomics to define the cell surface and secretome of ovarian cancer cells and processes of protein shedding, *Cancer Res* ***69***, 728–730.
13. Zhou, Y., Aebersold, R., and Zhang, H. (2007) Isolation of N-linked glycopeptides from plasma, *Anal Chem* ***79***, 5826–5837.
14. Rauch, A., Bellew, M., Eng, J., et al. (2006) Computational Proteomics Analysis System (CPAS): an extensible, open-source analytic system for evaluating and publishing proteomic data and high throughput biological experiments, *J Proteome Res* ***5***, 112–121.
15. Keller, A., Nesvizhskii, A. I., Kolker, E., and Aebersold, R. (2002) Empirical statistical model to estimate the accuracy of peptide identifications made by MS/MS and database search, *Anal Chem* ***74***, 5383–5392.

16. Nesvizhskii, A. I., Keller, A., Kolker, E., and Aebersold, R. (2003) A statistical model for identifying proteins by tandem mass spectrometry, *Anal Chem 75*, 4646–4658.

17. Ong, S. E., and Mann, M. (2006) A practical recipe for stable isotope labeling by amino acids in cell culture (SILAC), *Nat Protoc 1*, 2650–2660.

18. Faca, V., Coram, M., Phanstiel, D., et al. (2006) Quantitative analysis of acrylamide labeled serum proteins by LC-MS/MS, *J Proteome Res 5*, 2009–2018.

19. Faca, V. M., Ventura, A. P., Fitzgibbon, M. P., et al. (2008) Proteomic analysis of ovarian cancer cells reveals dynamic processes of protein secretion and shedding of extra-cellular domains, *PLoS ONE 3*, e2425.

Chapter 6

Low-Molecular Weight Plasma Proteome Analysis Using Centrifugal Ultrafiltration

David W. Greening and Richard J. Simpson

Abstract

The low-molecular weight fraction (LMF) of the human plasma proteome is an invaluable source of biological information, especially in the context of identifying plasma-based biomarkers of disease. This protocol outlines a standardized procedure for the rapid/reproducible LMF profiling of human plasma samples using centrifugal ultrafiltration fractionation, followed by 1D-SDS-PAGE separation and nano-LC–MS/MS. Ultrafiltration is a convective process that uses anisotropic semipermeable membranes to separate macromolecular species on the basis of size. We have optimized centrifugal ultrafiltration for plasma fractionation with respect to buffer and solvent composition, centrifugal force, duration and temperature to facilitate >95% recovery, and enrichment of low-M_r components from human plasma. Using this protocol, >260 unique peptides can be identified from a single plasma profiling experiment using 100 μL of plasma (Greening and Simpson, J Proteomics 73:637–648, 2010). The efficacy of this method is demonstrated by the identification, for the first time, of several plasma proteins (e.g., protein KIAA0649 (*Q9Y4D3*), rheumatoid factor D5, serine protease inhibitor A3, and transmembrane adapter protein PAG) previously not reported in extant high-confidence Human Proteome Organization Plasma Proteome Project datasets.

Key words: Blood, Plasma, Low-molecular weight, LMF, LMW, Ultrafiltration, Proteomics, HUPO

1. Introduction

Human plasma is one of the most informative and important proteomes from a clinical perspective. For example, characteristic changes in protein levels in plasma are indicative of many clinical conditions, including severe liver disease, hemolytic anemia and Down's syndrome, schizophrenia, Alzheimer's disease, amyotrophic lateral sclerosis, and Creutzfeldt–Jakob disease (1). Hence, characterization of plasma proteins (both in qualitative and quantitative

Richard J. Simpson and David W. Greening (eds.), *Serum/Plasma Proteomics: Methods and Protocols*, Methods in Molecular Biology, vol. 728, DOI 10.1007/978-1-61779-068-3_6, © Springer Science+Business Media, LLC 2011

terms) should provide a foundation for the discovery of candidate markers for disease diagnosis and development of new therapeutics. However, human plasma is limited by its dynamic range of protein abundances [ten orders of magnitude between the least abundant (1–5 pg/mL, e.g., interleukins and cytokines) and most abundant ($35–70\times10^9$ pg/mL, e.g., albumin and IgG (2))]. For example, albumin and immunoglobulin G constitute approximately 51–71 and 8–26% of the total protein content in human plasma, respectively (3). This complexity creates extensive difficulties in the use of many proteomic separation tools (e.g., free-flow electrophoresis, 10^5 (4)) for the identification of low-abundance species directly in plasma (overview: (2)). The strategies that have been most frequently used to overcome this issue of dynamic range are to fractionate the plasma proteome into smaller subsets and/or to deplete one or more of the major proteins. Immunoaffinity is an established method that addresses the dynamic range of plasma by specific depletion of high-abundance proteins (5). However, although the efficiency of immunodepletion ranges from 96 to 99%, the remaining concentration of albumin, for example, would still be ~50–1,000 μg/mL – a value ~10^4-fold higher than blood CEA levels (~5 ng/mL) and 5×10^6-fold higher than blood IL-6 levels (~10 pg/mL). Hence, MS-based detection of most already known biomarkers in blood requires the use of additional separation/enrichment technologies.

In 2005, Human Proteome Organization (HUPO) Plasma Proteome Project (PPP) generated a high-confidence core set of 889 serum and plasma proteins (6). Interestingly, the low-molecular weight (low-M_r, ≤25K) component of the blood proteome (considered a rich source of plasma biomarkers) was significantly under-represented (2.9% coverage (6)) in these studies. Known plasma polypeptides such as the defensins, and bioactive peptides such as glucagon, insulin, growth hormone, and neuropeptides are involved in a variety of biological functions. The low-molecular weight fraction (LMF) also contains proteolytic peptide fragments of several abundant proteins such as albumin, transthyretin, and the apolipoproteins (7, 8). The plasma or serum proteome has been the focus of recent attempts to identify low-abundance and low-M_r endogenous peptides which hold diagnostic and prognostic potential in cancer biology (9–11) (reviewed in ref. 12).

Centrifugal ultrafiltration has been the most widely used method to extract peptides and remove proteins with higher molecular weights from plasma/serum based on a size-exclusion filtration mechanism (13–17). Typically, membranes have a mean pore size between 10 and 500 Å (or 1.0 and 50 nm).

In the proteomic studies investigating the low-M_r region of plasma, issues with membrane selectivity, centrifugal conditions, buffers and solvents, filtrate heterogeneity, and contamination with abundant, high-M_r plasma proteins have limited the enrichment of the LMF at present (13, 14, 18, 19). Here, we report optimized conditions for the use of Sartorius Vivaspin® tangential centrifugal ultrafiltration membranes that influence transmembrane pressure and permeability (20).

2. Materials

Throughout the protocol, Milli-Q deionized water (HPLC grade, ≥18 MΩ) should be used for making up all aqueous solutions. All washing, lysis, and HPLC buffers should be prepared using clean glassware on the day analysis is to be performed.

2.1. Blood Collection, Plasma Preparation, and Storage (See Notes 1–4)

1. EDTA blood collection tubes (e.g., BD Vacutainer® #366450).
2. Polypropylene tubes (1.5, 15 mL).
3. Freezer (−80°C or lower).
4. Gloves, gown, and eye protection.
5. Pipettes.
6. Disposal container for contaminated tubes.
7. Centrifugation unit (either/or bench-top/swing bucket rotor – compatible with 1.5/15-mL tubes, programmable temperature setting; range 4–25°C).
8. Labels for blood sample tubes.
9. Alcohol (70% (v/v) aqueous ethanol) and swabs for cleaning venipuncture site.
10. Micro BCA protein assay kit, sufficient reagent to perform 480 standard tube assays or 3,200 microplate assays (#23235, Pierce, Rockford, IL).
11. Water bath or incubator set at 37°C.

2.2. Centrifugal Ultrafiltration

1. Centrifugation unit (bench-top series – compatible with 1.5-mL tubes, programmable temperature setting; range 4–25°C).
2. Centrifugation unit (swing bucket rotor – compatible with 15-mL tubes, programmable temperature setting; range 4–25°C).
3. Centrifugal ultrafiltration membranes – Vivaspin-2® MWCO of 20,000 (#VS02X1, cellulose triacetate (CTA), Sartorius Stedim Biotech, Aubagne, France) (see Notes 5 and 6). For selecting the correct NMWL of the filtration membrane device, refer Notes 7 and 8.
4. Acetonitrile, HPLC grade (Fisher, A998-1 or equivalent) (see Note 9).
5. Water, HPLC grade (Fisher, W5-1 or equivalent).

2.3. SDS-PAGE

1. Laemmli nonreducing sample buffer (0.2 M Tris–HCl, 40% (v/v) aqueous glycerol, 4% SDS, and trace bromophenol blue).
2. Heat block – up to 95°C (compatible with 1.5-mL centrifuge tubes).
3. NuPAGE® LDS sample buffer (Invitrogen), store at 4°C.
4. 1D-Gel apparatus (Invitrogen Novex Mini-Cell).
5. Precast SDS polyacrylamide 12-well, 1.5-mm gel (4–12% Bis–Tris precast gel, Invitrogen).
6. 20 × NuPAGE® MES SDS running buffer (Invitrogen): 50 mM MES, pH 7.2, 50 mM Tris–NaOH, 0.1% SDS, and 1 mM EDTA, pH 7.3, stored at room temperature (RT). Add 25 mL of 20 × running buffer to 475 mL water for preparing 1 × SDS running buffer.
7. Benchmark or Mark 12 protein standard mix, store at 4°C.

2.4. Protein Visualization

1. SilverSNAP® Stain Kit II (#24612, Pierce, Rockford, IL) gel stain, sufficient reagents to stain up to 20 SDS-PAGE minigels.
2. Fixing solution, 30% (v/v) aqueous ethanol containing 10% (v/v) aqueous acetic acid (>99.7%, Sigma–Aldrich, Saint Louis, MO).
3. Personal Densitometer SI (Molecular Dynamics).
4. Coomassie R-250 (#24615, Imperial Protein Stain, Pierce Biotechnology), 1 L, sufficient reagent for staining up to 50 minigels (see Note 10).
5. ImageQuant™ software (Molecular Dynamics).

2.5. In-Gel Digestion and Peptide Extraction

1. Gel cutter – 40 slices (or could use scalpel for gel lane excision).
2. 96-Well polypropylene plates (#AB-1058, ABgene™ Thermo-Fast, Thermo Fisher Scientific).
3. 100 mM Ammonium bicarbonate. Dissolve 0.79 g of ammonium bicarbonate in 100 mL of Milli-Q water to make 100 mM ammonium bicarbonate. Prepare fresh for every digest.
4. Dehydration buffer, 100% acetonitrile (>99.7%, Sigma–Aldrich, Saint Louis, MO).
5. Reduction buffer. Dissolve 15.4 mg of dithiothreitol (DDT, Clelands reagent) in 10 mL of 100 mM ammonium bicarbonate to make 10 mM DDT. A volume of 1 mL is adequate for ten samples (prepare fresh for each digest). Preweighed DDT can be stored at –20°C.

6. Alkylation buffer. Dissolve 90 mg of iodoacetic acid (IAA) in 10 mL of 100 mM ammonium bicarbonate to make 50 mM IAA. Preweighed IAA can be stored at –20°C.
7. Promega trypsin Gold, mass spectrometry grade (#V5280), in 50 mM acetic acid, diluted to 6 ng/μL in 100 mM ammonium bicarbonate, stored covered at –20°C.
8. Extraction buffer, 1% (v/v) aqueous formic acid/2% (v/v) aqueous acetonitrile in Milli-Q water. Stock solutions can be stored. A volume of 150 μL of buffer is required for each well.
9. SpeedVac, centrifugal lyophilization (Savant AES1010, Savant, USA).
10. Adhesive plate seals or polypropylene plate covers.

2.6. Nano-LC Analysis

1. HPLC solvents. Solvent A: 0.1% (v/v) aqueous formic acid (HPLC/Spectrograde). Solvent B: 60% (v/v) aqueous acetonitrile (ChromAR grade; Mallinkrodt) containing 0.1% formic acid (v/v; HPLC/Spectrograde). For Solvent A, mix 1 mL of neat formic acid (pipette) in 1 L of Milli-Q water in a glass-stoppered measuring cylinder, into a clean HPLC reservoir bottle. For Solvent B, add 600 mL of acetonitrile to 1-L glass-stoppered measuring cylinder, adjust the volume to 1 L with HPLC-grade water, add 0.9 mL of neat formic acid [final concentration 0.09% (v/v)], and mix thoroughly.
2. RP-capillary column, nanoACQUITY™ (C18) 150×1.0 mm I.D. (nanoACQUITY™-C18, 1.8 μm, Waters Corp, MA, USA).
3. Software: Chemstation 1200 series (Agilent Technologies).

2.7. MS/MS Analysis

1. Mass spectrometer with fast scan rate [e.g., Electrospray-Ion Trap (ESI-IT) tandem mass spectrometry (MS/MS) (LTQ-Orbitrap, Thermo Fisher Scientific, MA, USA)].

2.8. Data Processing and Analysis

1. Mass spectrometry program; *extract-msn* version 3, Bioworks 3.2 (Thermo Finnigan, USA).
2. MS/MS data analysis software (e.g., Mascot or Sequest). Mascot search algorithm (http://www.matrixscience.com/).
3. Mascot Daemon (http://www.matrixscience.com/daemon.html).
4. Java™ spectrum applet.
5. IPI human database (IPI.-HUMAN. current version; i.e., v.3.38, the number of entries were 70,757) from the European Bioinformatics Institute (EBI) (http://www.ebi.ac.uk/).

3. Methods

These methods assume the use of proper venipuncture technique for obtaining blood samples. For blood collection, standard protocols recommended by well-established organizations must be utilized (see Appendix A) (21). During phlebotomy, hemolysis can be caused by several factors including needle insertion for blood withdrawal (see Note 2). For routine venipuncture procedures, a 21-gauge needle is recommended to minimize hemolysis (see Note 1).

3.1. Blood Collection (See Notes 1–4)

1. It is important to obtain the required volume of blood using specific blood collection tubes. This is essential to ensure that the blood to anticoagulant ratio is not exceeded. Blood collection should be completed within 5 ± 2 min from the starting time.
2. After blood collection, gently mix the unit by inverting the tube eight to ten times.
3. Label the donor collection tube(s). If storage is required, do so immediately at −20°C.
4. Thawing of the plasma sample on the day of use should be performed at 37°C (not at RT or on ice) (see Notes 11 and 12). This is to prevent the formation of cryoprecipitate.
5. Protein concentration of the thawed plasma sample should be determined. For consistency, the bicinchoninic (BCA) protein assay, using bovine serum albumin (BSA) as a standard, should be used (22).

3.2. Centrifugal Ultrafiltration

1. Prepare centrifugal filter membranes according to the manufacturer's instructions by rinsing in 15 mL of HPLC-grade water at $2{,}000 \times g$ for 10 min (see Note 13). Set the centrifugal temperature to 20°C. Twist off the lock cap and remove the inner tube (filtrate collector). Make sure not to touch or bend the membrane. If the device is not to be used immediately, store it at 4°C with Milli-Q water covering the membrane surface.
2. Dilute 100 μL of thawed plasma with 900 μL of 10% (v/v) aqueous acetonitrile and allow to stand at RT for 2 min (see Note 9). Centrifuge each plasma sample (with a counterbalance) at $14{,}000 \times g$ for 2 min at RT to precipitate any insoluble material that may clog the filters.
3. Apply the supernatant to the prepared centrifugal filter(s) and place the samples in an M4 swing bucket rotor and centrifuge (with a counterbalance) at $4{,}000 \times g$ for 35 min at 20°C (see Note 6). A small aliquot (50 μL) of the sample is set aside to assess LMF recovery. This sample is stored at −80°C.

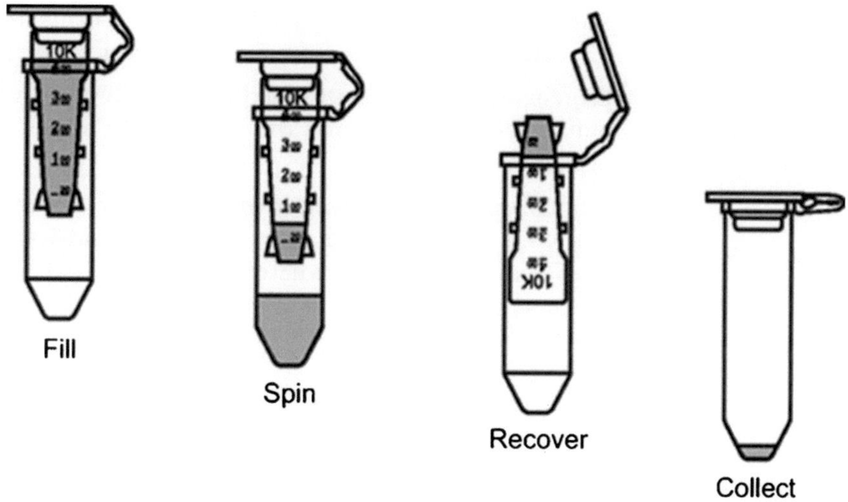

Fig. 1. Centrifugal filtration device. The assembly and operation of the centrifugal ultrafiltration device are shown, with retained volume (retentate, *upper*) and filtrated volume (filtrate, *lower*) indicated. Obtained with permission from the Millipore product catalog for protein purification and concentration (Reproduced with permission from http://www.millipore.com/catalogue/module/c82301).

4. The retentate (retained fraction, ~5% initial volume) should be removed and stored separately. The filtrate (flow-through fraction, ~90–95% initial volume) volume can be removed using a pipette or the filtrate is recovered by inverting the tube and centrifuging at 2,000 × *g* for 1 min (see Fig. 1).
5. The LMF recoveries of the filter membrane can be analyzed by BCA protein assay (22), comparing the initial plasma concentration to the concentration and volume of both the retained (retentate) and filtered (filtrate) samples. Typical recoveries for this experiment should be in the range of 94–97% (three experimental replicates). Retentate samples are stored at −80°C.
6. The plasma LMF filtrates are lyophilized to dryness by centrifugal lyophilization and resuspended in Laemmli nonreducing sample buffer.

3.3. SDS-PAGE Analyses

1. A plasma LMF protein sample (50 μg) is mixed with pre-warmed NuPAGE® LDS sample buffer (in a 2:1 ratio of sample: buffer).
2. The sample mixture is heated for 5 min on a heat block at 95°C and cooled (2 min) prior to sample loading.
3. Separation is performed using a precast 12-well SDS polyacrylamide gel (4–12% Bis–Tris precast gel).
4. 1 × MES SDS running buffer (500 mL) is prepared – approximately 200 mL in the upper (inner) buffer compartment and 300 mL in the lower (outer) buffer compartment.

5. Samples are loaded into defined gel lanes. Benchmark protein standards (5 μL) are used for molecular weight comparison.
6. Protein separation is performed at 150 V (constant voltage) until tracking dye reaches the bottom of the gel (approximately 75 min).
7. Immediately following electrophoresis, the gel should be washed with water and stained with colloidal Coomassie R-250, as described elsewhere (23) (Fig. 2) (see Note 14). Destain the background with water.

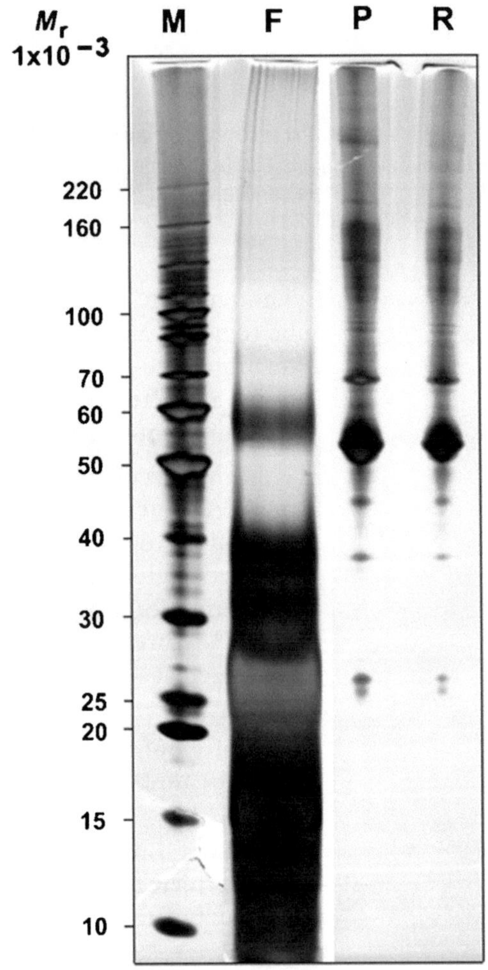

Fig. 2. 1-DE analysis of human plasma fractionated with centrifugal ultrafiltration. A volume of 100 μL of plasma was diluted 1:9 with 900 μL of 10% ACN, pH 8.5 v/v, as per Subheading 3.2. This sample was fractionated using Vivaspin-2 20K MWCO membrane filter at 4,000 × *g* until 95% of the input plasma had passed through the 20K filter. Aliquots of whole (*lane P*) or ultrafiltered plasma (filtrate, *lane F*, and retentate, *lane R*) were subjected to 1-DE and stained using silver staining. *Lane M*, benchmark molecular weight marker. Reproduced with permission from *Journal of Proteomics*.

3.4. In-Gel Digestion and Peptide Extraction

1. After staining, gel sections are excised (using either scalpel or gel-excision tool with slices ~1.0–1.5-mm thick) from a single lane.
2. The excised gel sections (23 sections in this study) are placed in a 96-well flat-bottom tissue culture plate (polypropylene, BD Biosciences) and digested with trypsin (0.05 μg).
3. The samples are first washed in 50 μL of 100 mM ammonium bicarbonate, followed by 5 μL of acetonitrile at 37°C for 20 min. This is repeated twice.
4. Each gel piece is subsequently treated with 50 μL of reduction buffer (10 mM DTT) at 37°C for 30 min and alkylated by incubation with 50 μL of alkylation buffer at 37°C for 30 min. After removal of excess buffer, the gel pieces are washed with 50 μL of 100 mM ammonium bicarbonate buffer at 37°C for 5 min, followed by 50 μL of acetonitrile at 37°C for 5 min. This washing step is repeated with 50 μL of 50 mM ammonium bicarbonate for 5 min. The wash is then removed as waste.
5. The gel pieces are dehydrated by the addition of 50 μL of acetonitrile and then dried for 10 min using centrifugal lyophilization.
6. Each gel section is rehydrated with 25 μL of the diluted trypsin stock solution (see Subheading 2.5). The plate should be sealed properly with adhesive plate seals or polypropylene plate covers.
7. Digestion is performed by incubating the plate at 37°C for 16 h.
8. To the gel pieces, add 60 μL of extraction buffer and incubate them at RT for 30 min. Carefully remove the extraction buffer containing generated tryptic peptides and place it into separate 100-μL glass autosampler vials. This is repeated twice with extraction/digestion buffer retained in the autosampler vial.
9. Extraction/tryptic digests are concentrated to 10 μL by centrifugal lyophilization in preparation for nano-LC–mass spectrometry (LC–MS).

3.5. Nano-LC Analysis

1. Peptide fractionation is achieved by capillary reversed-phase HPLC using the nano ACQUITY™ (C18) 150 × 1.0 mm I.D. RP-capillary column (nano ACQUITY™-C18, 1.8 μm) as detailed. The column is developed with a linear 60-min gradient from 0 to 100% B with a flow rate of 0.8 μL/min. The samples (~7 μL) are loaded onto the column via the autosampler.
2. The column temperature is maintained at 45°C and the eluent monitored for UV absorption at 215 and 280 nm.

3. The capillary HPLC is coupled online to the ESI-IT mass spectrometer for automated MS/MS analysis.
4. Tune and calibrate the LTQ Orbitrap according to the manufacturer's instructions (see Note 15).
5. Positive ion mode was used for data-dependent acquisition. Survey MS scans were acquired with the resolution set to a value of 30,000. Each scan was recalibrated in real time by co-injecting an internal standard from ambient air into the C-trap (24) (see Note 15). Up to five of the most intense ions per cycle were fragmented and analyzed in the linear trap. Target ions already selected for MS/MS were dynamically excluded for 180 s to optimize peptide coverage.

3.6. Data Processing and Analysis

1. The parameters used to generate the peak lists using *extract-msn* were as follows: minimum mass 700; maximum mass 5,000; grouping tolerance 0.01 Da; intermediate scans 200; minimum group count 1; 10 peaks minimum, and TIC of 100.
2. Peak lists for each LC–MS/MS run were merged into a single MGF file for Mascot searches using Mascot Daemon. Charge state of the selected ions was automatically determined from the survey scan.
3. Acquired MS/MS spectra were searched against the IPI human database (IPI.-HUMAN. current version; i.e., v.3.38, the number of entries were 70,757) from the EBI (http://www.ebi.ac.uk/).
4. Database search parameters were as follows: fixed modification, carboxymethylation of cysteine (+58 Da), variable modifications, NH_2-terminal acetylation (+42 Da), and methionine oxidation (+16 Da). Peptide mass tolerance was ±20 ppm, and #13C is defined as 1 with allowance for up to three missed tryptic cleavage sites.
5. Acceptance criteria based on ProteinScore, IonScore > Homology Score, and a <1% false discovery rate (FDR), in addition to manual verification of the spectra (Java™ spectrum applet), were determined in accordance with previously established guidelines for inclusion of true peptide identifications (25).
6. Proteins were submitted to BLAST (http://www.ncbi.nlm.nih.gov/BLAST/) with the criteria of >95% to remove redundancy.
7. Proteins were correlated with prediction of nonclassical protein secretion (SecretomeP) (http://www.cbs.dtu.dk/services/SecretomeP/) and also the Secreted Protein Database (http://spd.cbi.pku.edu.cn/spd_search.php).

8. Other resources to classify identified proteins based on several predictive algorithms included the SignalP (http://www.cbs.dtu.dk/services/SignalP/), TMHMM (http://www.cbs.dtu.dk/services/TMHMM/), Gene Ontology (GO) (http://www.geneontology.org/index.shtml?all/), UniProt (http://www.uniprot.org/) and Bioinformatic Harvester (http://harvester.fzk.de/harvester/) databases.

4. Notes

1. *Safety.* All blood and biological specimens and materials should be considered to be biohazards. Hence, it is important to use gloves, gowns, eye protection, other personal protective equipment, and controls to protect from blood splatter, blood leakage, and potential exposure to blood-borne pathogens. Use aseptic techniques at all times and sterile disposables (tubes, pipettes, etc.) throughout to prevent blood contamination. Risk factors for possible transfusion transmissible infections should be rigorously screened *prior* to blood collection. Handle the specimens as if they are capable of transmitting infection and dispose of with proper precautions in accordance with federal, state, and local regulations. Refer to your institutional regulations regarding the screening of blood for specific infectious disease markers (i.e., HIV, hepatitis B, hepatitis C, etc.). Discard all blood collection materials in biohazard containers approved for their disposal.
2. *Sample hemolysis.* The release of cellular material due to hemolysis into serum/plasma may introduce additional confounding factors. We recommend that if hemolysis (pink to red tinge in serum/plasma sample) is observed following centrifugation, this information should be recorded. It is recommended that hemolyzed samples should not be used for proteomic/peptidomic analyses.
3. *Monitoring pre/post-analytical variation.* In 2005, the HUPO PPP report detailed an extensive analysis of the variables that affect the stability of plasma (26). These included (a) the anticoagulant used in collection tube types (e.g., EDTA and ascorbate), (b) sample processing times, (c) temperatures at which blood specimens were processed and stored, (d) sample storage parameters, and (e) thaw/refreeze cycles, associated with obtaining human plasma and serum samples for proteomic analyses directed toward clinical research. It is of upmost importance that for diagnostic use, these variables are controlled and monitored at all times, from blood collection

as an anticoagulated or coagulated source, processing, handling, to storage (27–29). Recently, it has been shown that biomarker validation studies should use standardized collection conditions, and multiple control groups to detect and correct for potential biases associated with sample collection (30).

4. *Data points.* For blood handling, it is important to note also (a) the date and time of blood collection, (b) the number and volume of samples/aliquots prepared, (c) the date and time the samples are placed at –80°C, (d) the date and time of shipping, (e) any freeze–thaw cycles that occur, and (f) variations or deviations from the standard-operating protocol, and problems or issues that arise.

5. *Centrifugal ultrafiltration membrane devices.* A wide range of centrifugal filters are commercially available for concentrating and filtering protein solutions, removing small solutes, and/or buffer exchanging. These devices consist (mostly) of two chambers separated by a semipermeable membrane. These membranes can be composed of different chemistries and different orientations depending on their application (see Note 8). Under centrifugal force, solvent and solute molecules smaller than the NMWL readily pass through the membrane (filtrate) (see Note 7). Vertical or angular membrane configuration reduces concentration polarization (membrane fouling) and allows high flow rates for optimal solvent passage even with high proteinaceous solutions. The direction of the centrifugal force and flow rate of solute differ between the membrane devices used. Additional information can be obtained from http://www.millipore.com/ and http://www.sartorius.com/.

6. *Optimized centrifugal ultrafiltration.* Conditions for each plasma sample should be optimized. Conditions provided in this protocol are the combined effect of analyzing multiple filter membrane units, with conditions optimized with respect to plasma buffer and solvent compositions, centrifugal force, duration, and temperature (Fig. 2). Typically, plasma LMF should represent 95% of the initial supernatant applied to the filtration devices. The amount of protein recovered in the filtrate and retentate can be calculated as a percentage of the initial plasma protein concentration loaded.

7. *Appropriate membranes – selecting the NMWL.* Ultrafiltration membranes are not absolute in their pore size (NMWL) ratings. Separation occurs as a result of differences in the filtration rate of different components across the membrane in response to a given pressure. Unlike UF membranes, microporous membranes have a precisely controlled pore size that ensures quantitative retention of particles and biomolecules greater than the pore size of the membrane. In selecting the most effective membrane for filtration applications, a rule has

been developed to calculate the appropriate membrane pore size (NMWL) rapidly. It is a simple calculation based on the molecular weight of the desired protein to be concentrated or removed in the retentate unit (upper level of the membrane apparatus). The "rule of 1.5–2" requires a membrane cut-off approximately two times smaller than the desired protein's molecular weight. For example, to remove proteins of ~65,000 MW and greater, use a 30,000 NMWL-regenerated cellulose membrane. Typically, this results in >90–95% recovery of the filtrate, containing proteins/peptides <65,000 MW. Other factors to consider when determining an optimal membrane include flow rate, also known as flux, solute concentration, solute composition, and temperature.

8. *Centrifugal ultrafiltration membrane chemistries.* For a detailed overview, refer (31).

   Polyethersulfone – General purpose membrane, providing excellent performance with most solutions when retentate recovery is of primary importance. Polyethersulfone membranes exhibit no hydrophobic or hydrophilic interactions and are usually preferred for their low fouling characteristics, exceptional flux, and broad pH range.

   CTA – High hydrophilicity and very low nonspecific binding characterize this membrane. These membranes are preferred for sample cleaning and protein removal and when high recovery of the filtrate solution is of primary importance.

   Regenerated cellulose/Hydrosart – These membranes demonstrate the same properties as regenerated cellulose, but with the added benefit of enhanced performance characteristics and extremely low protein binding, making it the membrane of choice for applications such as concentration and desalting of immunoglobulin fractions.

9. *Disrupting protein–protein interactions.* A low concentration of organic solvent (typically, 5–10% acetonitrile) is added to buffers to disrupt high-M_r protein–protein interactions. For chemical compatibility of membranes, be careful to read each company's manual prior to operation (based on 2-h membrane contact time). Normally, small uncomplexed proteins and peptides (i.e., less than 30K) are rapidly cleared from the circulation through enzymatic degradation, uptake by the reticuloendothelial system, or by glomerular filtration, which discriminates on the basis of molecular size and charge (32). It is believed that the circulation half-life of the LMF is directly related to its binding affinity to large high-abundance carrier proteins (8, 11).

10. *Coomassie dye staining.* The Coomassie dyes (R-250 and G-250) bind to proteins through ionic interactions between

dye sulfonic acid groups and positive protein amine groups. Coomassie R-250, the more commonly used of the two dyes, can detect protein levels down to 0.1 μg. Additionally, Coomassie R-250 does not require methanol/acetic acid fixation and destaining.

11. *Cryoprecipitate formation.* A cryoprecipitate is often formed if the fresh-frozen plasma unit is slowly thawed at temperatures just above freezing (1–6°C), typically in a water bath or a refrigerator. The product is then centrifuged at low speed (typically 5,000 × *g*) to remove the majority of the precipitate. Formation of the cryoprecipitate can be avoided by thawing the plasma at 37°C.

12. *Plasma thaw process.* Thawing of plasma can be achieved in various ways, the most common of which uses a recirculating water bath. This carries a risk of bacterial contamination and must be maintained in a controlled sterile environment. Denaturation of plasma proteins can be avoided by using a dry heating apparatus.

13. *Prerinsing membranes.* Most ultrafiltration membrane devices contain trace amounts of glycerine/sodium azide. If this interferes with subsequent sample analysis, prerinse the device extensively with buffer or Milli-Q water through the concentrator. If interference still persists, rinse the membrane with 0.1 M NaOH, followed by repeated centrifugation with buffer or Milli-Q water.

14. *Silver staining protocol.* For protein visualization, we used the sensitive SilverSNAP® Stain Kit II gel stain, as described earlier (20). Briefly, the gels were washed (2×) for 5 min in deionized water, fixed (2×) [30% (v/v) aqueous ethanol containing 10% (v/v) acetic acid] for 15 min, and washed for 10 min first in 10% ethanol and then in deionized water for 10 min. The gels were incubated in SilverSNAP® Sensitizer solution for 1 min, stained with SilverSNAP® silver solution for 30 min, washed (2×) in deionized water for 1 min, developed in SilverSNAP® develop solution for approximately 3–5 min, and fixed in 5% (v/v) aqueous acetic acid. SilverSNAP® stained gels were imaged with a Personal Densitometer SI (Molecular Dynamics) with 100-μm pixel size.

15. *Mass accuracy and tuning.* For Orbitrap mass analyzers to maintain the ability to measure peptide masses accurately with less than 5 ppm error requires constant calibration every 2–3 days. The "Lock Mass" capability of the LTQ Orbitrap, which allows for real-time recalibration using polydimethylcyclosiloxane ions present in ambient air (24), is recommended.

References

1. Anderson, N.L. and N.G. Anderson (2002), The human plasma proteome: history, character, and diagnostic prospects. *Mol Cell Proteomics.* **1**, 845–67.
2. Omenn, G.S., D.J. States, M. Adamski, et al. (2005), Overview of the HUPO Plasma Proteome Project: results from the pilot phase with 35 collaborating laboratories and multiple analytical groups, generating a core dataset of 3020 proteins and a publicly-available database. *Proteomics.* **5**, 3226–45.
3. Putnam, R.W., (1975), The Plasma Proteins. New York: Academic Press.
4. Moritz, R.L., H. Ji, F. Schutz, et al. (2004), A proteome strategy for fractionating proteins and peptides using continuous free-flow electrophoresis coupled off-line to reversed-phase high-performance liquid chromatography. *Anal Chem.* **76**, 4811–24.
5. Bjorhall, K., T. Miliotis, and P. Davidsson (2004), Comparison of different depletion strategies for improved resolution in proteomic analysis of human serum samples. *Proteomics.* **5**, 307–17.
6. States, D.J., G.S. Omenn, T.W. Blackwell, et al. (2006), Challenges in deriving high-confidence protein identifications from data gathered by a HUPO plasma proteome collaborative study. *Nat Biotechnol.* **24**, 333–8.
7. Lowenthal, M.S., A.I. Mehta, K. Frogale, et al. (2005), Analysis of albumin-associated peptides and proteins from ovarian cancer patients. *Clin Chem.* **51**, 1933–45.
8. Zhou, M., D.A. Lucas, K.C. Chan, et al. (2004), An investigation into the human serum "interactome". *Electrophoresis.* **25**, 1289–98.
9. Lopez, M.F., A. Mikulskis, S. Kuzdzal, et al. (2005), High-resolution serum proteomic profiling of Alzheimer disease samples reveals disease-specific, carrier-protein-bound mass signatures. *Clin Chem.* **51**, 1946–54.
10. Albrethsen, J., R. Bogebo, S. Gammeltoft, et al. (2005), Upregulated expression of human neutrophil peptides 1, 2 and 3 (HNP 1-3) in colon cancer serum and tumours: a biomarker study. *BMC Cancer.* **5**, 8.
11. Liotta, L.A. and E.F. Petricoin (2006), Serum peptidome for cancer detection: spinning biologic trash into diagnostic gold. *J Clin Invest.* **116**, 26–30.
12. Drake, R.R., L. Cazares, and J.O. Semmes (2007), Mining the low molecular weight proteome of blood. *Proteomics Clin. Appl.* **1**, 758–68.
13. Tirumalai, R.S., K.C. Chan, D.A. Prieto, et al. (2003), Characterization of the low molecular weight human serum proteome. *Mol Cell Proteomics.* **2**, 1096–103.
14. Harper, R.G., S.R. Workman, S. Schuetzner, et al. (2004), Low-molecular-weight human serum proteome using ultrafiltration, isoelectric focusing, and mass spectrometry. *Electrophoresis.* **25**, 1299–1306.
15. Zheng, X., H. Baker, and W.S. Hancock (2006), Analysis of the low molecular weight serum peptidome using ultrafiltration and a hybrid ion trap-Fourier transform mass spectrometer. *J Chromatogr A.* **1120**, 173–84.
16. Hu, L., X. Li, X. Jiang, et al. (2007), Comprehensive peptidome analysis of mouse livers by size exclusion chromatography prefractionation and nanoLC-MS/MS identification. *J Proteome Res.* **6**, 801–8.
17. Jung, W.W., S. Phark, S. Oh, et al. (2009), Analysis of low molecular weight plasma proteins using ultrafiltration and large gel two-dimensional electrophoresis. *Proteomics.* **9**, 1827–40.
18. Di Girolamo, F., J. Alessandroni, P. Somma, et al. (2009), Pre-analytical operating procedures for serum Low Molecular Weight protein profiling. *J Proteomics.* **73**, 667–77.
19. Tammen, H., I. Schulte, R. Hess, et al. (2005), Peptidomic analysis of human blood specimens: comparison between plasma specimens and serum by differential peptide display. *Proteomics.* **5**, 3414–22.
20. Greening, D.W. and R.J. Simpson (2010), A centrifugal ultrafiltration strategy for isolating the low-molecular weight (</=25K) component of human plasma proteome. *J Proteomics* **73**, 637–48
21. *Internet pathology library for medical education.* Jan 2010; Available from: http://medlib.med.utah.edu/WebPath/TUTORIAL/PHLEB/PHLEB.html.
22. Smith, P.K., R.I. Krohn, G.T. Hermanson, et al. (1985), Measurement of protein using bicinchoninic acid. *Anal Biochem.* **150**, 76–85.
23. Neuhoff, V., N. Arold, D. Taube, et al. (1988), Improved staining of proteins in polyacrylamide gels including isoelectric focusing gels with clear background at nanogram sensitivity using Coomassie Brilliant Blue G-250 and R-250. *Electrophoresis.* **9**, 255–62.
24. Olsen, J.V., L.M. de Godoy, G. Li, et al. (2005), Parts per million mass accuracy on an

Orbitrap mass spectrometer via lock mass injection into a C-trap. *Mol Cell Proteomics.* **4**, 2010–21.

25. Kapp, E.A., F. Schutz, G.E. Reid, et al. (2003), Mining a tandem mass spectrometry database to determine the trends and global factors influencing peptide fragmentation. *Anal Chem.* **75**, 6251–64.
26. Rai, A.J., C.A. Gelfand, B.C. Haywood, et al. (2005), HUPO Plasma Proteome Project specimen collection and handling: Towards the standardization of parameters for plasma proteome samples. *Proteomics.* **5**, 3262–77.
27. Banks, R.E., A.J. Stanley, D.A. Cairns, et al. (2005), Influences of blood sample processing on low-molecular-weight proteome identified by surface-enhanced laser desorption/ionization mass spectrometry. *Clin Chem.* **51**, 1637–49.
28. Ferguson, R.E., D. Hochstrasser, and R. Banks (2007), Impact of preanalytical variables on the analysis of biological fluids in proteomic studies. *Proteomics Clin. Appl.* **1**, 739–46.
29. West-Norager, M., C.D. Kelstrup, C. Schou, et al. (2007), Unravelling in vitro variables of major importance for the outcome of mass spectrometry-based serum proteomics. *J Chromatogr B Analyt Technol Biomed Life Sci.* **847**, 30–7.
30. Thorpe, J.D., X. Duan, R. Forrest, et al. (2007), Effects of blood collection conditions on ovarian cancer serum markers. *PLoS ONE.* **2**, e1281.
31. Cheryan, M., (1998), Ultrafiltration and microfiltration handbook. 2nd ed. ed. Florida: CRC Press. 527.
32. Torchilin, V.P. and A.N. Lukyanov (2003), Peptide and protein drug delivery to and into tumors: challenges and solutions. *Drug Discov Today.* **8**, 259–66.

Chapter 7

High-Throughput Analysis of Glycoproteins from Plasma

Yan Li and Hui Zhang

Abstract

The glycoproteome is a major subproteome of plasma. To identify low-abundance plasma glycoproteins and specific glycoforms as disease biomarkers, high-throughput and reproducible methods for glycoproteomic analysis are needed. A glycoproteomic method, utilising solid-phase extraction of glycopeptides (SPEG) based on chemical immobilization of glycopeptides has been developed. The step-by-step protocol for this high-throughput SPEG method using multichannel liquid handling system is described. The protocol provides efficient enrichment for glycoproteins for proteomic identification using mass spectrometry.

Key words: Glycoprotein, Mass spectrometry, Solid-phase extraction of glycopeptides

1. Introduction

Glycosylation by oligosaccharides is one of most common posttranslational modifications of proteins expressed in an extracellular environment (1). It plays a key role in protein stability, folding, solubility, and activity, and affects both structural and functional properties of the protein (2–4). Numerous studies have demonstrated that glycosylation is involved in multiple cellular processes, such as cell signaling pathways, recognition, cell growth, and death (5–8). The aberrant glycosylation of proteins has been shown to correlate with the development of lung, liver, breast, prostate, ovarian, and other cancers (9–14). Thus, glycosylated proteins comprise an interesting subproteome, which needs to be explored by in-depth proteomic analyses to obtain a greater understanding of biological processes, typically in the areas of disease.

Richard J. Simpson and David W. Greening (eds.), *Serum/Plasma Proteomics: Methods and Protocols*,
Methods in Molecular Biology, vol. 728, DOI 10.1007/978-1-61779-068-3_7, © Springer Science+Business Media, LLC 2011

Plasma represents an important biological material, especially for monitoring and diagnosis of disease due to its easy accessibility. However, the wide dynamic range of plasma proteins spanning at least ten orders of magnitude becomes a major challenge in plasma protein identification and quantification (15). Low-abundance proteins (such as PSA, CA125, and Her2) are often undetectable in the presence of high-abundance proteins. Use of appropriate sample preparation methods to extract and concentrate glycoproteins in plasma is critical for identifying biomarkers.

Currently, a glycoproteomic method, utilising solid-phase extraction of N-linked glycopeptides (SPEG), has been established and successfully applied in tissue and plasma samples of prostate, breast, and ovarian cancer, and some other diseases (16–19). In this method, glycoproteins are captured by hydrazide resin. After washing away the nonglycopeptides, N-linked glycopeptides are released by peptide-*N*-glycosidase F (PNGase F) from the solid support by cleaving off the peptide portion from the immobilized glycan. The asparagine residues are converted to aspartic acid, causing 1 unit shift at the glycosylation site during the enzyme reaction. Therefore, peptide sequences and glycosylation sites can be identified in a single mass spectrometry analysis using a high-resolution mass spectrometer. To optimize this method, several studies have been published to extract N-linked glycopeptides from biological samples efficiently in a high-throughput manner (20–22).

2. Materials

2.1. Solutions

All chemicals are from Sigma–Aldrich except those specified in this section.

- Protein denaturing buffer: 8 M urea in 0.4 M NH_4HCO_3 with 0.1% (w/v) SDS, pH 8.0.

  Add 4.8 g of urea (Sigma–Aldrich, #U6504), 0.32 g of NH_4HCO_3 (Sigma–Aldrich, #A6141), and 50 μL of 20% SDS solution (Fisher Scientific #151-21-3) into a 15-mL tube, and then add dd (deionized) H_2O up to 10 mL. Prepare fresh for each use.
- 120 mM TCEP:

  Add 34.4 mg of TCEP (Fisher Scientific, #20490) into 1 mL of dd H_2O. Prepare fresh for each use.
- 160 mM Iodoacetamide:

  Add 30 mg of iodoacetamide (Sigma–Aldrich, #I1149) into 1 mL of dd H_2O. Prepare fresh.

- Sequencing grade trypsin (Promega, #V511X).
- Trypsin digestion buffer: 100 mM Tris–HCl, 1 mM $CaCl_2$, pH 8.0:

  Add 1.2 g of Tris (Sigma–Aldrich, #T1378), ~2 mL of 5 N HCl, and 0.011 g of $CaCl_2$ (Sigma–Aldrich, #C4901) in 100 mL of dd H_2O. Prepare fresh.
- 5 N Hydrochloric acid:

  Add 431 mL of concentrated HCl (37%) (Sigma–Aldrich, #258148) to 569 mL of dd H_2O, to make 1 L of 5 N HCl.
- 100 mM sodium periodate:

  Add 21 mg of $NaIO_4$ (Bio-Rad, #153-6055) in 1 mL of dd H_2O. Prepare fresh.
- PNGase F (New England Biolabs, #P0705L).
- Hydrazide resin (Bio-Rad, #156-0016).
- 1.5 M Sodium chloride:

  Add 8.76 g of NaCl (Sigma–Aldrich, #S3014) to 100 mL of dd H_2O.
- 0.1% (v/v) Aqueous trifluoroacetic acid (TFA) (Sigma–Aldrich, #T6508) in 80% acetonitrile (Sigma–Aldrich, #34967).
- 0.1% (v/v) Aqueous TFA in 50% acetonitrile.
- 0.1% (v/v) Aqueous TFA in deionized H_2O.
- 80% (v/v) Aqueous acetonitrile (Sigma–Aldrich, #34967).
- HPLC solution A: 0.1% (v/v) formic acid (Sigma–Aldrich, #33015) in dd H_2O.
- HPLC solution B: 0.1% (v/v) formic acid in acetonitrile.
- 0.4% (v/v) Acetic acid (Sigma–Aldrich, #338826) in dd H_2O.
- 96-Deepwell plates (Waters, #186001438).
- 96-Well microplate (Sigma–Aldrich, #CLS3363).
- 96-Well Plate Sep-Pak tC18 100 mg (Waters, #186002321).
- 96-Well Plate Sep-Pak tC18 μElution (Waters, #186002318).
- 96-Well Sirocco™ Protein Precipitation Plate, 1/pkg (Waters, #186003873).
- 1.5-mL Microcentrifuge tubes (Denville, #C-2170).
- pH paper (range 0–14) (EMD, #9590).
- 0.7-mL Glass vial inserts without caps (Waters, #186001437).

2.2. Equipment

SpeedVac with rotors suitable for microcentrifuge tubes and a 96-deepwell plate with glass inserts (Thermo Vavant, Model #SPD SpeedVac).

- Centrifuge (Kendro, Model #Sorvall Legend RT).
- Vortex apparatus (Fisher, Vortex Model #12-812).

- Incubator shaker (New Brunswick Scientific, Model #Innova4000).
- 37 and 60°C oven (Boekell, #133000).
- LTQ ion-trap mass spectrometer (Thermo Fisher).
- Hypersil GOLD HPLC columns, 3 μm particle size, 50×1 mm (Thermo Fisher, #25003-151030).
- TECAN Aquarius multichannel pipettor system with the autoloader (Mannedorf, Switzerland).
- Software for sequence assignment from MS/MS data: SEQUEST (Thermo Fisher).
- Software for statistical evaluation of peptide sequence assignment: PeptideProphet (23).

3. Methods

The SPEG method has been widely used in biomarker discovery studies (20, 24–30) with the following advantages: (a) most glycoproteins are secreted proteins, membrane proteins, and cell surface proteins and, therefore, are likely shed or secreted into blood. This explains the reason that the majority of currently approved tumor serum markers are known to be glycosylated (31). (b) Some of the high-abundance proteins, such as albumin which represents ~50% of the total protein content in serum (40–50 mg/mL), are not N-linked glycoproteins. The glycopeptide capture technique largely reduces the complexity of serum. (c) This method is able to concentrate glycopeptides about 100 times compared to normal blood concentrations. Overall, it significantly improves the analytical sensitivity compared to the analysis of whole serum samples, and the detection limit has been demonstrated to nanogram per milliliter scale (19).

Here, the protocol for glycoprotein capture using SPEG to enrich glycoproteins from plasma is provided. To improve the analytical throughput, a multichannel liquid handling system is used to process up to 96 samples simultaneously. The protocol using automatic liquid handling system not only increases the experimental throughput, but also improves the reproducibility for glycopeptide isolation. It holds great potential for glycoprotein extraction from a large set of clinical samples for biomarker discovery.

3.1. Digestion of Glycoproteins to Peptides

1. Centrifuge plasma samples at 10,000 ×*g* for 5 min at 4°C and transfer 20 μL of each sample (total protein amount about 1 mg, up to 96 samples per plate) to a 96-deepwell plate. Record sample information in the 96-deepwell plate.

2. From the TECAN buffer reservoirs transfer 90 μL of protein denaturing buffer (8 M urea/0.4 M NH_4 HCO_3 /0.1% SDS) and 10 μL of 120 mM TCEP to the 96-deepwell plate using the TECAN system, and incubate the sample plate at 60°C for 1 h.
3. Transfer 10 μL of 160 mM iodoacetamide to the 96-deepwell plate from TECAN reservoirs and incubate the plate at room temperature for 30 min in the dark with gentle shaking.
4. Add 230 μL of trypsin digestion buffer (100 mM Tris–HCl/1 mM $CaCl_2$) to the 96-deepwell plate to dilute the urea concentration to <2 M.
5. Add 20 μg of trypsin to the 96-deepwell plate from TECAN reservoirs and incubate the sample plate at 37°C for 4 h with gentle shaking (see Note 1).

3.2. Coupling Glycopeptides to Solid Support

1. Add 20 μL of 5 N HCl to the 96-deepwell plate from TECAN reservoirs to adjust sample pH <3.0.
2. Prepare C-18 SepPak 96-well plate by washing the plate with 1 mL of 50% acetonitrile/0.1% TFA twice and then with 1 mL of 0.1% TFA twice. Washing solution is spun down by centrifuging at 20×*g* for 5 min (see Note 2).
3. Transfer the peptide sample from the 96-deepwell plate to equilibrated C-18 SepPak 96-column plate. If necessary, sample could be reloaded once to improve the peptide adsorption.
4. Wash the C-18 SepPak 96-column plate with 1 mL of 0.1%TFA three times and remove the liquid by centrifuging at 20×*g* for 5 min.
5. Add 400 μL of 50% acetonitrile/0.1% TFA to the C-18 SepPak 96-column plate to elute peptide samples. A clean 96-deepwell plate is used to collect the eluted sample from the column.
6. Add 45 μL of 100 mM sodium periodate to each peptide sample in the plate and incubate the sample plate at 4°C for 1 h in the dark.
7. Add 3.6 mL of 0.1% TFA into the sample plate to dilute acetonitrile concentration to ~5% (see Note 3).
8. Prepare C-18 SepPak 96-column plate by adding 1 mL of 80% acetonitrile/0.1% TFA twice and 1 mL of 0.1% TFA twice using TECAN and remove the washing solution by centrifuging at 20×*g* for 5 min.
9. Transfer 4 mL of peptide sample from the 96-deepwell plates to the prepared C-18 SepPak 96-column plate. If necessary, sample could be reloaded once to improve peptide adsorption (see Note 4).

10. Wash the C-18 SepPak 96-column plate with 1 mL of 0.1% TFA three times and remove the washing solution by centrifuging at 20×*g* for 5 min.
11. Add 700 μL of 80% acetonitrile/0.1% TFA to the C-18 SepPak 96-column plate to elute peptide samples. A clean 96-deepwell plate is used to collect the eluted peptides.
12. Add 25 μL of hydrazide resin (50 μL of 50% slurry) to each well in a new 96-well filter plate and spin the filter plate at 20×*g* for 10 min to remove the liquid from the hydrazide resin. Wash the resin three times by adding 1 mL of H_2O, shaking the plate for 1 min, and centrifuge at 20×*g* for 10 min to remove water. A 96-vial seal is used to seal the plate.
13. Transfer the oxidized peptide sample to the 96-well filter plate loaded with hydrazide resin to conjugate the glycopeptides to hydrazide resin at room temperature for 4 h (see Note 5).

3.3. Releasing N-Linked Glycopeptide from Solid Support

1. Remove the 96-vial cap from the bottom of the filter plate after glycopeptide coupling, and remove unconjugated non-glycopeptides by centrifuging the filter plate at 20×*g* for 10 min.
2. Wash the resin three times each with 800 μL of 1.5 M NaCl, 80% acetonitrile, H_2O, and 0.1 M phosphate buffer (pH 7.5) and remove the washing solution by centrifuging at 20×*g* for 10 min after each wash.
3. Seal the bottom of the plate using a clean 96-vial cap and resuspend the resin in 25 μL of PNGase F digestion buffer 1× G7 (10× G7 stock solution was included in the PNGase F kit) and 2.5 mU of PNGase F. Incubate the plate at 37°C for 4 h with gentle shaking (see Note 6).
4. Remove the cap after PNGase F digestion and centrifuge the resin at 20×*g* for 10 min. Collect the solution with the released glycopeptides from the filter plate in a clean 96-well Microplate. Wash the resin with 100 μL of H_2O twice. Collect and combine the washing solution with the released peptides from the original PNGase F digestion.
5. Add 5 μL of 5 N HCl to each glycopeptide sample using TECAN to adjust the pH to <3.0.
6. Prepare a C-18-μ SepPak 96-column plate by washing each well with 0.5 mL of 80% acetonitrile/0.1% TFA twice and 0.5 mL of 0.1% TFA twice using TECAN. Remove the washing solution by centrifuging at 20×*g* for 5 min.
7. Transfer the peptide sample from the 96-well Microplate to the prepared C-18-μ SepPak 96-column plate. If necessary, sample could be reloaded once to improve peptide adsorption.

8. Wash the C-18-μ SepPak 96-column plate with 0.5 mL of 0.1% TFA three times and remove the washing solution by centrifuging the plate at 20×*g* for 5 min.
9. Add 700 μL of 80% acetonitrile/0.1% TFA to the C-18-μ SepPak 96-column plate to elute glycopeptides, and collect the peptides in a clean 96-deepwell plate with glass inserts until all solution pass through the column.
10. Dry the eluted glycopeptides in the 96-deepwell plate with glass inserts using a SpeedVac overnight.
11. Resuspend the peptides in 20 μL of 0.4% acetic acid for mass spectrometry analysis (see Note 7).

3.4. Identification of Glycopeptides by HPLC–MS

1. Protein identification by liquid chromatography tandem mass spectrometry (LCMS/MS) analysis of peptides was performed using an LTQ ion trap MS interfaced with a 2D-nanoLC system. For each HPLC-MS run, 1 μg of digested sample was loaded, and then fractionated on a reverse-phase HPLC C18 column with a 10-μm emitter using 4.5–45% acetonitrile/0.1% formic acid gradient from 0 to 95 min and then up to 90% at 105 min. The flow rate was 300 nL/min.
2. For each MS and MS/MS spectrum, peptide mass, monoisotopic peak area, cluster peak area, signal-to-noise ratio, chromatography elution time, fragment ion mass, and signal-to-noise ratio of each fragment ion peak are exported.
3. The MS/MS database searches are performed against the National Center for Biotechnology Information (NCBI) database using Thermo Scientific SEQUEST software.
4. The database search is set to the following parameters: (1) the mass tolerance of the precursor is set at 200 ppm; (2) the mass tolerance of the fragment ion is ±0.6 Da; (3) enzyme digestion is tryptic digestion; and (4) the presence of protein modifications including carboxymethylation of cysteines, oxidation of methionines, and an enzyme-catalyzed conversion of asparagine to aspartic acid at N-linked glycosylation site.
5. The dataset is further filtered to ensure that all the identified glycopeptides have a consensus N–*X*–S/T sequence (where *X* is any amino acid except proline), with the Mascot peptide ion score greater than 80%.

4. Notes

1. Trypsin digestion can be left covered at 37°C overnight. Protein digestion is a critical step to generate high yield of glycopeptides. The tryptic digestion of proteins is monitored

by SDS-PAGE to check the critical step of the procedure. The expected result for efficient tryptic digestion is that protein bands at high molecular weight essentially disappear upon trypsin digestion (most silver-stained bands >5 kDa present before digestion and disappear after the digestion).

2. To desalt the peptide samples using the C-18 SepPak 96-column plate, 1 mL of deionized water is added to each well in an empty C-18 SepPak 96-column plate to create a balance plate. It is critical to ensure that the balance plate has the same weight as the sample plate for centrifuging. An empty 96-deepwell plate could be placed under the C-18 SepPak 96-filter plate to collect the eluted solution.
3. The 96-deepwell plate can hold up to 2 mL of solution. Dividing the solution into two 96-deepwell plates may be needed if throughput is an issue.
4. The C-18 SepPak 96-column plate can hold up to 1 mL of solution. It may be necessary to add 1 mL of sample per time and repeat four times for sample loading. To speed up the processing procedure, the sample solution could be centrifuged at $20 \times g$ for 5 min after each sample reloading.
5. The hydrazide bead coupling can be left at room temperature overnight.
6. PNGase F digestion can be left at 37°C overnight.
7. Glycopeptide can be stored at −20°C.

References

1. Winterburn, P. J., and Phelps, C. F. (1972) The significance of glycosylated proteins. *Nature 236*, 147–51.
2. Mitra, N., Sinha, S., Ramya, T. N., and Surolia, A. (2006) N-linked oligosaccharides as outfitters for glycoprotein folding, form and function. *Trends Biochem Sci 31*, 156–63.
3. Imperiali, B., and O'Connor, S. E. (1999) Effect of N-linked glycosylation on glycopeptide and glycoprotein structure. *Curr Opin Chem Biol 3*, 643–9.
4. Wormald, M. R., and Dwek, R. A. (1999) Glycoproteins: glycan presentation and protein-fold stability. *Structure 7*, R155–60.
5. Vagin, O., Kraut, J. A., and Sachs, G. (2009) Role of N-glycosylation in trafficking of apical membrane proteins in epithelia. *Am J Physiol Renal Physiol 296*, F459–69.
6. Laderach, D. J., Compagno, D., Toscano, M. A. et al. (2009) Dissecting the signal transduction pathways triggered by galectin-glycan interactions in physiological and pathological settings. *IUBMB Life 62*, 1–13.
7. Apone, F., Tito, A., Carola, A. et al. (2009) A mixture of peptides and sugars derived from plant cell walls increases plant defense responses to stress and attenuates ageing-associated molecular changes in cultured skin cells. *J Biotechnol* **145**, 367–76.
8. Rudd, P. M., Elliott, T., Cresswell, P., Wilson, I. A., and Dwek, R. A. (2001) Glycosylation and the immune system. *Science 291*, 2370–6.
9. Peracaula, R., Barrabes, S., Sarrats, A., Rudd, P. M., and de Llorens, R. (2008) Altered glycosylation in tumours focused to cancer diagnosis. *Dis Markers 25*, 207–18.
10. Sinchaikul, S., Hongsachart, P., Sriyam, S. et al. (2008) Current proteomic analysis and post-translational modifications of biomarkers in human lung cancer materials. *Chang Gung Med J 31*, 417–30.
11. Mehta, A., and Block, T. M. (2008) Fucosylated glycoproteins as markers of liver disease. *Dis Markers 25*, 259–65.
12. Dwek, M. V., Ross, H. A., and Leathem, A. J. (2001) Proteome and glycosylation mapping

identifies post-translational modifications associated with aggressive breast cancer. *Proteomics* *1*, 756–62.

13. Saldova, R., Wormald, M. R., Dwek, R. A., and Rudd, P. M. (2008) Glycosylation changes on serum glycoproteins in ovarian cancer may contribute to disease pathogenesis. *Dis Markers* *25*, 219–32.
14. Goffard, A., and Dubuisson, J. (2003) Glycosylation of hepatitis C virus envelope proteins. *Biochimie* *85*, 295–301.
15. Anderson, N. L., and Anderson, N. G. (2002) The human plasma proteome: history, character, and diagnostic prospects. *Mol Cell Proteomics* *1*, 845–67.
16. Zhang, H., Li, X. J., Martin, D. B., and Aebersold, R. (2003) Identification and quantification of N-linked glycoproteins using hydrazide chemistry, stable isotope labeling and mass spectrometry. *Nat Biotechnol* *21*, 660–6.
17. Tian, Y., Zhou, Y., Elliott, S., Aebersold, R., and Zhang, H. (2007) Solid-phase extraction of N-linked glycopeptides. *Nat Protocols* *2*, 334–339.
18. Zhang, H., Liu, A. Y., Loriaux, P. et al. (2007) Mass spectrometric detection of tissue proteins in plasma. *Mol Cell Proteomics* *6*, 64–71.
19. Li, Y., Sokoll, L. J., Rush, J., Zou, N., Chan, D. W., and Zhang, H. (2009) Targeted detection of prostate cancer proteins in serum using heavy peptide standards and MALDI-TOF/TOF. *Proteomics-Clin. Appl.* *3*, 597–608.
20. Zhang, H., Yi, E. C., Li, X. J. et al. (2005) High throughput quantitative analysis of serum proteins using glycopeptide capture and liquid chromatography mass spectrometry. *Mol Cell Proteomics* *4*, 144–55.
21. Zhou, Y., Aebersold, R., and Zhang, H. (2007) Isolation of N-linked glycopeptides from plasma. *Anal Chem* *79*, 5826–37.
22. Zou, Z., Ibisate, M., Zhou, Y., Aebersold, R., Xia, Y., and Zhang, H. (2008) Synthesis and evaluation of superparamagnetic silica particles for extraction of glycopeptides in the microtiter plate format. *Anal Chem* *80*, 1228–34.
23. Keller, A., Nesvizhskii, A. I., Kolker, E., and Aebersold, R. (2002) Empirical statistical model to estimate the accuracy of peptide identifications made by MS/MS and database search. *Anal Chem* *74*, 5383–92.
24. Tian, Y., Gurley, K., Meany, D. L., Kemp, C. J., and Zhang, H. (2009) N-linked glycoproteomic analysis of formalin-fixed and paraffin-embedded tissues. *J Proteome Res* *8*, 1657–62.
25. Schiess, R., Mueller, L. N., Schmidt, A. et al. (2009) Analysis of cell surface proteome changes via label-free, quantitative mass spectrometry. *Mol Cell Proteomics* *8*, 624–38.
26. Carapito, C., Klemm, C., Aebersold, R., and Domon, B. (2009) Systematic LC-MS analysis of labile post-translational modifications in complex mixtures. *J Proteome Res* *8*, 2608–14.
27. Whiteaker, J. R., Zhang, H., Eng, J. K. et al. (2007) Head-to-head comparison of serum fractionation techniques. *J Proteome Res* *6*, 828–36.
28. Lewandrowski, U., Zahedi, R. P., Moebius, J., Walter, U., and Sickmann, A. (2007) Enhanced N-glycosylation site analysis of sialoglycopeptides by strong cation exchange prefractionation applied to platelet plasma membranes. *Mol Cell Proteomics* *6*, 1933–41.
29. Lewandrowski, U., Moebius, J., Walter, U., and Sickmann, A. (2006) Elucidation of N-glycosylation sites on human platelet proteins: a glycoproteomic approach. *Mol Cell Proteomics* *5*, 226–33.
30. Liu, T., Qian, W. J., Gritsenko, M. A. et al. (2005) Human plasma N-glycoproteome analysis by immunoaffinity subtraction, hydrazide chemistry, and mass spectrometry. *J Proteome Res* *4*, 2070–80.
31. Sokoll, L. J., and Chan, D. W. (2008) Biomarkers for Cancer Diagnostics, in *Abeloff's clinical oncology* (Abeloff, M. D., Armitage, J. O., Niederhuber, J. E., Kastan, M. B., and McKenna, W. G., Eds.), Elsevier Inc, Philadelphia, PA.

Part II

Blood Collection and Handling Strategies

Chapter 8

Minimizing Preanalytical Variation of Plasma Samples by Proper Blood Collection and Handling

Jizu Yi, David Craft, and Craig A. Gelfand

Abstract

Blood samples collected for proteome studies are subject to a variety of preanalytical instability, among which intrinsic proteolysis activities cause a broad spectrum of protein and peptide degradation. This chapter describes two MALDI MS-based methods for plasma peptidomic analyses; a direct MALDI-TOF MS and an LC MALDI-TOF MS. Using these methods, we compared peptides and their time-dependent changes in traditional serum, four plasma samples with different anticoagulants and additives: EDTA-based, citrate-based, or heparin-based, and EDTA-based with protease inhibitors. For minimizing plasma sample instability and preanalytical variation, we suggest using an optimized blood collection device, minimizing the dwell time during blood collection and handling, controlling centrifugation and handling at room temperature, and saving plasma samples for use at most one freeze/thaw cycle. We have optimized our protocol to achieve reproducibility in peptidomic analyses of plasma samples using MALDI-TOF MS by minimizing preanalytical and analytical variability.

Key words: Blood collection, Plasma preparation, Plasma peptidomics, Intrinsic proteolysis, MALDI-TOF MS, Peptide analysis, Preanalytical variability

1. Introduction

Human serum/plasma specimens are routinely used in clinical proteomics and diagnostics as a simple source of analytes that may be potential biomarkers for different disease states (1, 2). Blood has been the logical biospecimen and most-used biomarker discovery matrix to date due to its ability in sampling every tissue of the body, its noninvasive collection, and its ready availability in sufficient volume and quantity. The use of blood specimens is, however, subject to several challenges that particularly affect proteomic studies, such as the large dynamic range of plasma protein concentration (2), intrinsic enzymatic activity (3), and many

Richard J. Simpson and David W. Greening (eds.), *Serum/Plasma Proteomics: Methods and Protocols*,
Methods in Molecular Biology, vol. 728, DOI 10.1007/978-1-61779-068-3_8, © Springer Science+Business Media, LLC 2011

preanalytical variations and alternatives in how blood is collected and handled (4–6). These challenges limit overall reproducibility, sensitivity, and resolution in proteomics biomarker discovery efforts and are even more critical for translating biomarker discovery into clinical application. A prerequisite for plasma-based protein biomarker studies is to have a highly qualified specimen with an intact, in-vivo-like proteome, free or poor of platelets, no or low hemolysis, and minimal sensitivity to sample processing. Yet, how the plasma sample is collected and handled significantly affects the quality of blood specimen (3–6), which further complicates the already intense proteomics workflow.

There are many preanalytical variables and alternatives that impact virtually every clinical study, and the more studies are done, the more important these aspects are found to be with respect to proteomics and biomarker goals. Common variables during sampling and analysis include (1) choice of plasma versus serum samples, (2) addition of protease inhibitors or other additives, and (3) processing and handling of blood specimens. Only with an understanding of the challenges associated with developing a reproducible proteomics measurement system can one begin to understand the complexity involved in selecting, studying, and optimizing a serum/plasma sample. To this end, a detailed preanalytical strategy for sample handling is essential.

We began our studies in comparing plasma samples with focus on one particular source of variation – blood-borne protein variation and instability. We sought to observe whether proteins and peptides were being degraded or altered over time during the "preanalytical phase," i.e., from initial phlebotomy to sample analysis. Only after establishing reproducible analytical procedures, both for peptides and their degraded products, were we able to develop strategies to monitor preanalytical effects. Our results highlight the importance of the immediate inhibition of ex vivo plasma protease activities by including specific protease inhibitors in the primary evacuated blood collection tube (3, 7). More recently, it has been proposed by Villanueva et al. that tumor-derived proteases initiate cleavage of plasma proteins that can continue ex vivo, and this phenomenon combined with the physiological action of coagulation enzymes produces serum peptidome patterns potentially used for diagnosis of specific disease states (8, 9), and such studies and applications equally require careful sample collection and handling.

Peptides are a current focus as potential new biomarkers, specifically those discovered in circulation and associated with various cancers (10–12), metabolic diseases (13), and cardiovascular disease (14–16). In recent years, our group has developed optimized methods to facilitate reproducible analysis of plasma peptidome by MALDI-TOF MS, after careful evaluation of many preanalytical and analytical steps from the initial phlebotomy to final peptide detection. This chapter describes two such protocols for peptide analysis. The first is a straightforward method requiring minimal processing steps, and the second includes chromatography prior to the MS analysis to access more peptides. Although the details of these methods are specific for peptides, and our MALDI-TOF instrument, the care and level of detail included in these procedures can serve as a framework with which to understand preanalytical variability and further refine any measurement of peptides or proteins from blood samples.

2. Materials

2.1. Blood Collection and Plasma Preparation

1. BD Vacutainer® Push Button Blood Collection Set (# 367338, BD, Franklin Lakes, NJ).
2. BD™ P100* (*-for research use only) Blood Collection Kit (# 366455/366456): A P100 tube contains a lyophilized cocktail of protease inhibitors, which inhibit intrinsic proteolysis degradation of plasma proteins (see Note 1) (3), and a mechanical separator, which facilitates the removal of blood cells including platelets from plasma (unpublished data).
3. EDTA, heparin, and citrate plasma preparation tubes (BD Vacutainer® PPT), in which K_2EDTA (spray coded 1.76 mg/mL blood, # 367525), or lithium heparin (spray coded 90 USP units, # 366643), or sodium citrate buffer (0.105 M, # 369714) are included as anticoagulants respectively.
4. Serum preparation tube (BD Vacutainer® red top, # 366430): no additive (glass tube) is included.
5. A Sorvall RC3C + Centrifuge (Sorvall, USA); or a CellSep 6/720R centrifuge (SANYO, Japan).

2.2. Chemicals for Sample Processes

1. Trifluoroacetic acid (TFA ≥99.9%) from T.B. Baker (Phillipsburg, NJ) and acetonitrile (ACN >99.5%) from Sigma Chemical Co (Saint Louis, MO).
2. α-Cyano-4-hydroxycinnamic acid (CHCA or HCCA, Bruker Daltonics, Billerica, MA).
3. Ammonium phosphate (Thermo Fisher Scientific, Fairlawn, NJ).
4. Reagent-grade water (≥18.2 mΩ) using a Millipore MilliQ water purifier (Millipore, Billerica, MA).

2.3. Peptide Extraction

1. Size-based membrane separator: Microcone YM-3, YM-10 (Millipore, Billerica, MA).
2. Microcentrifuge 5417R (Eppendorf, Hamburg, Germany).
3. Generation solution (for solution preparation, see Note 2 in subsequent Chapter 10): 60% aqueous ACN, 5% aqueous TFA.

4. Equilibrium solution: 5% aqueous ACN, 0.1% aqueous TFA.
5. Washing buffer: 10% aqueous ACN, 0.1% aqueous TFA.
6. Elution buffer: 60% aqueous ACN, 0.1% aqueous TFA.
7. ZipTip® C18 Pipette Tips with 0.6 μL of C18 resin (Millipore, Billerica, MA).
8. MALDI sample target: a MTP 384 ground steel target (Bruker Daltonics, Billerica, MA).
9. Matrix solution: 5 mg/mL α-cyano-4-hydroxycinnamic acid (CHCA: Bruker Daltonics, Billerica, MA) dissolved in 50% aqueous ACN, 0.1% aqueous TFA solution. The matrix solution should be made fresh daily.

2.4. HPLC System for Peptide Separation

1. Sample column: a Zorbax 300SB-C18 5 μm, 5 mm × 0.3 mm trapping column (Agilent Technologies, Palo Alto, CA).
2. Nano-flow C18 separation column: a Chromolith CapRod RP-18 encapped column (Merck, Darmstadt, Germany) with dimensions of 150 mm × 0.1 mm.
3. HPLC system: a nano-flow HPLC1100 system (Agilent Technologies, Palo Alto, CA) was used to deliver accurate and stable flow of the mobile phase to the C-18 separation column. The system was operated and controlled with Chemstation software (Agilent Technologies, Palo Alto, CA).
4. Eluting solution A: 0.1% aqueous TFA.
5. Eluting solution B: 0.1% aqueous TFA in 99.9% ACN.
6. Sample target: A Disposable CHCA Prespotted AnchorChip target (#: PAC384, Bruker Daltonics, Bremen, Germany). Such a target can be loaded fully up to 384 sample spots or plasma fractions collected from nano-flow HPLC separation.
7. Auto sample spotter (fraction collector): the Proteineer FC (Bruker Daltonics, Billerica, MA), with installed Warp LC software (Bruker Daltonics), was connected with the output of the HPLC system. The spotter was set for automating sample collection process for MALDI-TOF MS analysis.

2.5. MALDI-TOF MS Analysis

1. Ultraflex II MALDI-TOF MS (Bruker Daltonics, Germany), the instrument is operated and controlled through Ultraflex control V2.4 (Bruker Daltonics, Germany).
2. Ground-Steel or polished steel MALDI targets (Bruker Daltonics, Germany).
3. Ultraflex Analysis V2.4 and ClinProTools 2.0 (Bruker Daltonics).

3. Methods

The blood-borne proteome has high interindividual variability, regardless of health or disease states. To minimize this natural biological variation from our experiments, blood was collected into different collection tubes from the same individual and from the same venipuncture event. Considering the complex effects caused by the sample processing and tube additives (17, 18), we analyzed freshly prepared samples to eliminate the effect of freeze–thaw variability.

Here, we describe two methods of peptide analyses, which we have used to investigate preanalytical variation. The first method contains a simplified procedure for peptide extraction followed by direct MS analysis. It can be used for side-by-side comparison of different samples (Fig. 1), and/or time-course study of peptide changes (3). In the second method, we developed a LC-MALDI TOF MS approach, creating a virtual 2D peptide display (Fig. 2), enabling a more comprehensive analysis of the peptidome (19). Both methods were designed for reproducibility, thus enabling investigation of plasma sample variation.

3.1. Blood Collection (see Note 2)

1. With the same venipuncture using a BD Vacutainer® Push Bottom Blood Collection Set with 21G butterfly needle and 12″ tubing, blood specimens are drawn directly into BD™ P100 tube, an EDTA plasma preparation tube, a citrated plasma tube, a heparin plasma tube, and a serum tube, or another alternative device (see Note 1) for specific biomarker studies (20).
2. For optimal dissolution and mixing of tube additives with the sample, slowly invert plasma (P100, EDTA, heparin) tubes ~8 times, or citrate tube 3–4 times, immediately after drawing blood. For serum collection without any additive (e.g., in a glass tube), no mixing is required.
3. For plasma collection, blood tubes are spun immediately (or as soon as possible), minimizing the dwell time in the whole-blood state (see Note 3), in a CellSep centrifuge for 15 min at 2,500 × g (rcf) at RT (see Note 4). For serum collection, the blood tubes are allowed to stand unperturbed at RT for 60 min to allow complete clotting and then spun at 2,500 × g for 15 min.
4. Remove the tube closures. Plasma or serum samples were removed from sample tubes into Eppendorf microcentrifuge tubes (1 mL/tube) for either direct analysis of the peptides or storage in a −80°C freezer until future and/or other usages (see Note 5).

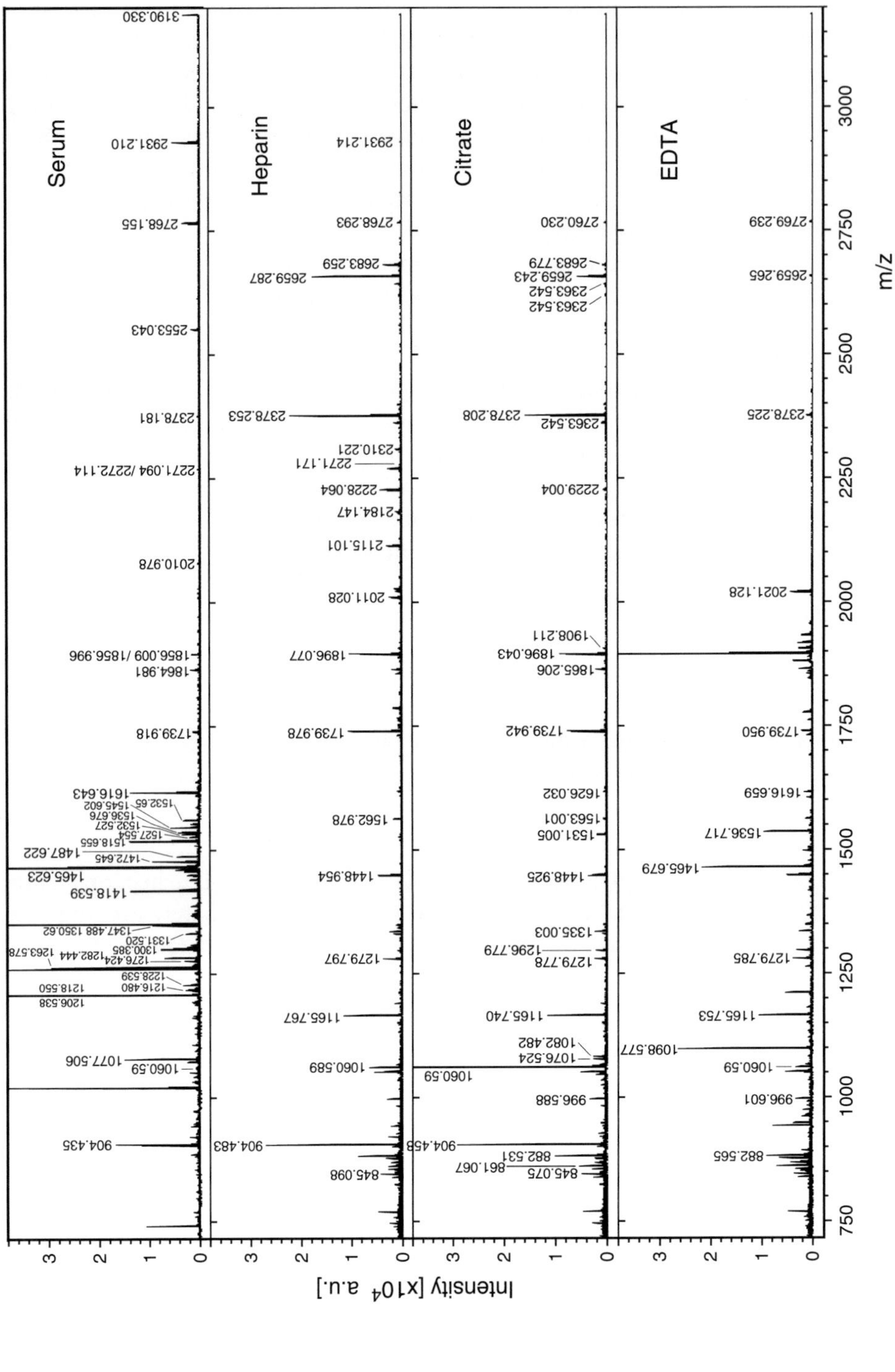

Fig. 1. Serum and plasma peptides. The samples collected from a single healthy subject were processed, and the peptides extracted were analyzed directly by MALDI-TOF MS. The differences observed here at time "0" (see Note 6) indicate that ex vivo peptide generation and digestion by intrinsic proteolysis cause the preanalytical variability of blood specimen (3).

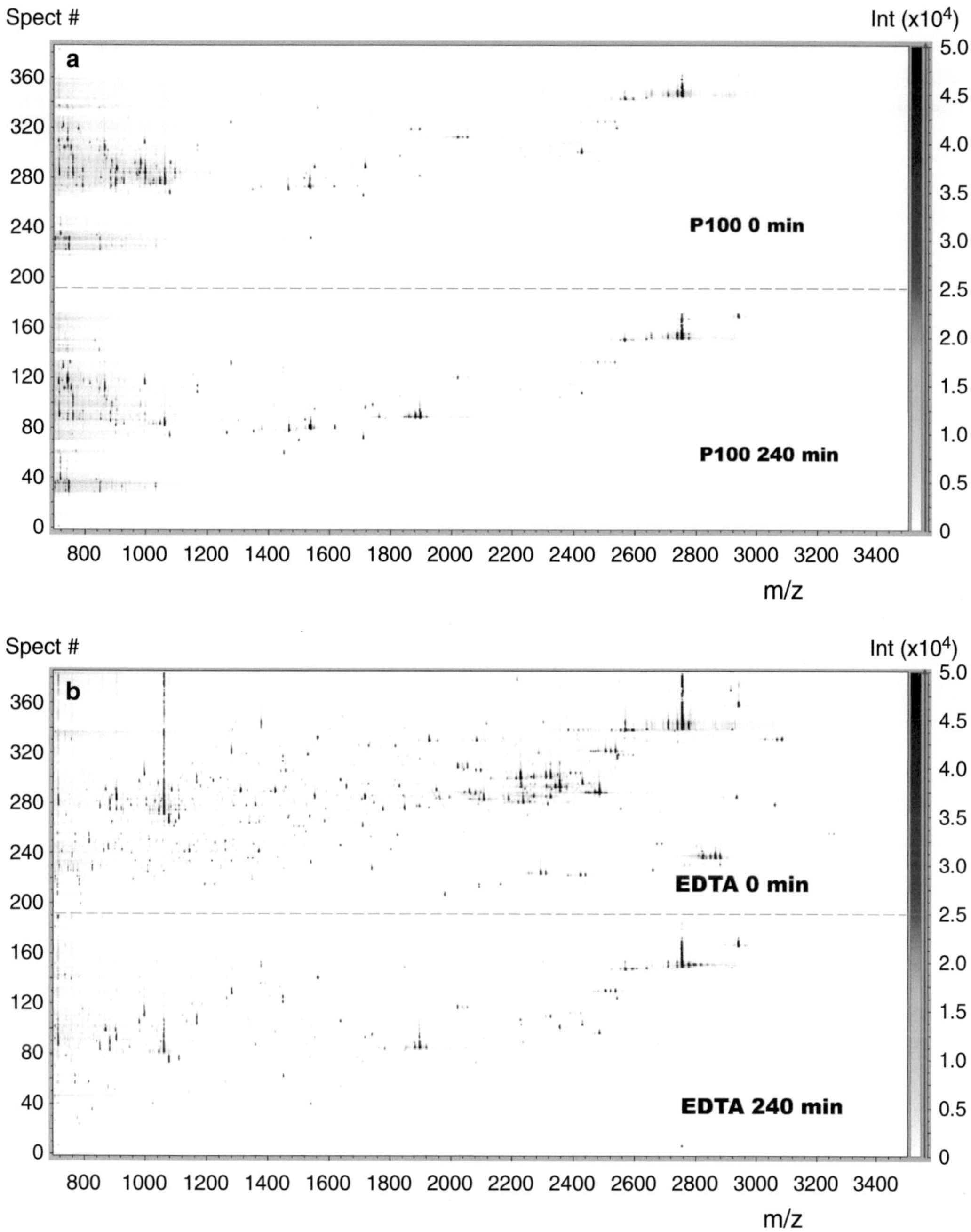

Fig. 2. Virtual two-dimensional view of plasma peptides. A blood specimen was drawn into P100 (**a**) or EDTA (**b**) tubes followed by plasma preparation, plasma incubation at RT for 0 and 4 h, peptide extraction, and LC-MALDI-TOF MS analysis. Less peptides at time "0" and less peptide variations between "0" and "240 min" were observed in P100 than in EDTA plasma (see Note 1) (19).

3.2. Sample Preparation for Direct MALDI-TOF MS Analysis

1. 45 μL of plasma or serum sample from above preparations were acidified by addition of 5 μL of 2% TFA solution to 0.2% final TFA concentration for quenching intrinsic proteolytic reactions. We define the moment of acidification as time "0" (see Note 6).
2. The quenched sample was loaded onto YM-3 filter and spun in the Eppendorf Centrifuge 5147R for 30 min at 15,000 × *g* at 10°C.
3. Peptides from the filtrates are extracted with Zip-Tip C18, using the following sequence of liquid passes:
   - 2 × 10 μL of generation solution (50% ACN, 5% TFA)
   - 3 × 10 μL of equilibrium solution (5% ACN, 0.1% TFA)
   - 15× with the filtrate sample by pipetting up and down to load peptides
   - 3 × 10 μL of washing buffer (10% ACN, 0.1% TFA)
   - 2 × 1.5 μL of elution buffer (60% ACN, 0.1% TFA).
4. The eluted peptide was mixed in a 1:1 ratio with CHCA matrix in an Eppendorf tube, and 1.5 μL of mixture was spotted on a ground steel MALDI target, air-dried for direct MALDI-TOF analysis.

3.3. Peptidome Sample Preparation for LC-MALDI-TOF MS Analysis

1. Three separate 50-μL aliquots of plasma were diluted with 450 μL of 20% ACN and 0.1% TFA.
2. The diluted solution was loaded onto three Ultracel YM-3 molecular weight cutoff (MWCO) centrifuge-tube filters (see Note 7) and spun under 12,600 × *g* for 90 min at 10°C, and three filtrates were collected.
3. The resulting peptide samples were dried by vacuum centrifugation at RT for 90 min to remove the acetonitrile.
4. Two ZipTip-C18 tips for each peptide filtrate were used sequentially (as described in Subheading 3.2) for cleanup and concentration of peptides.
5. The final peptide solutions from all six parallel processes per sample were pooled and centrifuged under vacuum to a final aqueous volume of 20 μL (see Note 7).

3.4. MALDI-TOF MS

MALDI spectra were collected on an Ultraflex II MALDI TOF/TOF MS instrument (see Note 8). For peptide analysis, the machine was set in the reflectron mode and positively charged ions were detected. The MS spectra of peptides between 750 and 4,000 *m/z* were collected in an automated mode. More detailed procedures for spectrum acquisition are described in Chapter 10 of this book, and peptide samples prepared from Subheading 3.2 were analyzed by following those procedures with sample results showed in Fig. 1.

3.5. LC-MALDI MS

1. The nano-flow HPLC system is primed and then washed with eluting solution A.
2. 5 μL of <3 kDa plasma peptidome sample prepared from the above sample process (Subheading 3.3) was loaded on reversed-phase column connected to the system. The flow rate was set at 2 μL/min, and separated peptides were monitored at 214 nm using an 80-nL flow cell during elution (19).
3. The gradient elution was performed with increasing solution B as follows:
   - Hold solution B at 2% for 7 min.
   - Linearly increase solution B from 2 to 30% B over 30 min.
   - Gradient increase of solution B from 30 to 90% over 20 min.
   - Hold 90% B for 10 min to clean the column.
4. Using the Proteineer FC controlled using Warp LC software, fractions from 15 s HPLC elution were directly spotted as one sample on a CHCA Prespotted AnchorChip target. This setting resulted in a 0.5-μL peptide fraction in each spot on the sample target.
5. The samples on target were air-dried for MALDI-TOF analysis.
6. Similar to the procedures as described in Subheading 3.4, a total of 2,000 laser shots are collected on each MALDI spot (Fig. 2a, b).

3.6. MALDI-TOF MS Analysis

1. All MALDI-TOF MS spectra can be processed by FlexAnalysis ver. 3.0 with moderate smoothing and baseline subtraction. Comparing peptide profiles among multiple plasma or serum samples by the alignment of their spectra using FlexAnalysis indicated that more peptides are found in serum than in plasma (Fig. 1). Furthermore, time-course MALDI-TOF MS analysis and MS/MS analysis indicated that intrinsic protease activities mediate these differences (3, 7, 19, 21), and specifically, thrombin may contribute these extra activities in serum (21).
2. The multiple spectra can be also analyzed using ClinProTools, which allows normalization, virtualization, and classification analysis of multiple MALDI-TOF mass spectra.
3. The virtual 2D plots of the LC-MALDI-MS data can be created using Warp LC SurveyViewer (Fig. 2a, b).

4. Notes

1. *Selection of sample collection device.* There are many choices of blood collection devices, including traditional serum tubes and plasma tubes with either EDTA or citrate or heparin

added as an anticoagulant (3, 4, 6). One important source of preanalytical variation, the intrinsic protease activity in blood, has been identified and characterized in our previous work (3). While intrinsic sample proteolysis causes ex vivo damage of sample proteins and peptides, each blood collection tube results in a unique peptide spectrum (Fig. 1), since each anticoagulant provides inhibitory effects directly or indirectly on different enzymes, specifically clotting cascade enzymes (3, 6). Accordingly, standardization of sample type makes sense for plasma proteomics, and we, as well as Human Proteome Organization-Plasma Proteome Project (HUPO-PPP) initiative, have suggested that EDTA plasma has the least intrinsic protease activity of the standard blood sample (3, 22), despite the fact that measurable proteolytic damage of proteins and peptides is observed even in EDTA samples. To further minimize proteolytic degradation of plasma proteins, we developed a new device, BD P100* (* = for research use only) a blood collection tube that includes a cocktail of protease inhibitors, in addition to EDTA. Our research demonstrates that P100 plasma enhances stability of proteins and peptides compared to EDTA plasma (Fig. 2), and minimizes the inter-individual variability (3, 7, 19). It is challenging to preserve the entire plasma proteome with protease inhibitors for an extended period of time, given the complexity of the blood proteome. However, we have demonstrated general stabilization with our broadly active-inhibitor formulation within reasonable sample-processing time frames. It is more feasible to optimize a protease inhibitor cocktail to specifically inhibit the classes of enzymes that may cleave specific proteins/peptides of interest. In this regard, we have achieved protection of GLP-1 in BD™ P700*, and a family of metabolic peptide biomarkers including GLP-1, GIP, glucagon, and ghrelin in BD™ P800* (* = for research use only, not for diagnostic use) (20).

2. *Drawing blood.* In addition to considerations of specific requirements for the designed test, such as disease states, and donor's gender and age, there are also general considerations when drawing blood, even from a healthy individual:

   - All samples must be collected under the informed consent of the subjects, and under any IRB and/or other requirements.
   - Since several metabolic products of food can increase or change due to postabsorptive hormonal effect and other analytes may be disturbed by turbidity due to chylomicronemia present in postprandial blood samples, a fasting state (of approximately 12 h) of the subjects is suggested before the blood drawing (23).

- Whether from a healthy or diseased subject, blood should always be handled under "usual precautions," more specifically with the presumption that any sample could carry pathogens or infectious disease. Anyone drawing or handling blood samples should be fully trained and equipped with proper personal protective equipment (gloves, lab coat, goggles, etc.).
- When multiple tube types are used to collect blood samples from the same subject, such as in this study, the tubes are filled in a random order for any one subject to avoid any sequence-specific bias.

3. *Dwell time during sample processing*. Ideally, sample processing should occur immediately after blood draw to eliminate time-dependent ex vivo damage. However, under clinical settings, "immediately" is not always possible, so a more feasible goal is to minimize the dwell time between any two processing steps. These dwell periods include but are not limited to the following:
   - The time a blood-filled tube stands after phlebotomy and/or before centrifugation (whole blood time).
   - The time of the separated plasma or serum and the time it may stand in the centrifuge, after centrifugation.
   - The time for sample transportation from the draw/process site to the research lab.
   - Aliquot time: moving plasma/serum samples from primary collection tubes into storage tubes.
   - Time for transferring sample into the freezer (e.g., –80°C), and freeze-down time, or time to get the sample onto a test platform if used fresh.
   - Shipping time, if collection and proteomics analyses are in different places.
   - Time during plasma sample thawing and transportation for experimental analysis.
4. *Blood centrifugation*:
   - Before and during centrifugation, cold temperatures (e.g., 4°C) should be avoided to limit platelet activation in whole blood (24, 25). We maintain our primary centrifuges at ~25°C or at RT.
   - The speed or *g*-force of centrifugation is controlled to maximize removal of cellular content for optimal plasma quality, or to minimize variability of residual cell contamination in plasma samples.
5. *Freezing–thawing process and plasma storage*. The process of freezing and thawing may cause red cell lysis (26), mass

spectroscopic changes of plasma proteins (17), and/or a range of poorly defined phase transitions that can destabilize proteins (18, 27). We suggest using freshly prepared plasma whenever possible, and if this is not feasible, limit the sample to only one freeze–thaw process (4). To facilitate this, aliquot the sample into single-use volumes and freeze the multiple individual aliquots at –80°C.

6. *Time "0"*. Time "0" is the time point when plasma sample is ready for analysis, e.g., immediately after primary centrifugation, with as little time delay as possible from phlebotomy to this point. At this point, the sample has been ex vivo for a minimum of 15 min (since the centrifugation takes at least that long) or as much as 90 min for clotting in serum after drawing.
7. *Three parallel sample processes.* Recombination of these three parallel-processed samples helps to average out any operator errors or variations, but then requires the drying step to concentrate the peptides and also to reduce acetonitrile in the sample for LC-MALDI MS analysis.
8. *Ultraflex II.* Ultraflex II (Bruker Daltonics) is a MALDI-TOF/TOF MS. We mainly used its MALDI-TOF MS function for exploring peptide profiles of serum or plasma samples. For particular peptides of interest, e.g., those identified through our experiments (Fig. 1) (3), we used its TOF/TOF MS function for peptide sequence identification. Our MS/MS analysis further led us to understand how plasma proteins and peptides were digested (3, 7).

Acknowledgment

The authors appreciate the stimulation and encouragement from James Down and Robert Pierce, the critical discussions from Patrick O'Mullan, David Warunek, Bruce Haywood, and Maria Saluta, and support from Lisa Gevirtz and Lena Khumush.

References

1. Omenn, G. S. (2004) Advancement of biomarker discovery and validation through the HUPO plasma proteome project, *Disease Markers* *20*, 131–134.
2. Anderson, N. L., Anderson, N. G. (2002) The human plasma proteome-History, character, and diagnostics preospects, *Mol. Cell. Proteomics* *1*, 845–867.
3. Yi, J., Kim, C., and Gelfand, C. A. (2007) Inhibition of intrinsic proteolytic activities moderates preanalytical variability and instability of human plasma, *J Proteome Res* *6*, 1768–1781.
4. Rai, A. J., Gelfand, C. A., Haywood, B. C. et al. (2005) HUPO Plasma Proteome Project specimen collection and handling: towards the standardization of parameters for plasma proteome samples, *Proteomics* *5*, 3262–3277.
5. West-Nielsen, M., Høgdall, E. V., Marchiori, E., et al. (2005) Sample Handling for Mass

Spectrometric Proteomic Investigations of Human Sera, *Analy Chem 77*, 5114–5123.

6. Bowen, R. A. , Hotin, G., Csako, G., et al. (2010) Impact of blood collection devices on clinical chemistry assays., *Clin Biochem*, *43*, 4–25.
7. Yi, J., Liu, Z., Craft, D., et al. (2008) Intrinsic peptidase activity causes a sequential multi-step reaction (SMSR) in digestion of human plasma peptides, *J Proteome Res 7*, 5112–5118.
8. Findeisen, P., Peccerella, T., Post, S,, et al. (2008) Spiking of serum specimens with exogenous reporter peptides for mass spectrometry based protease profiling as diagnostic tool, *Rapid Commun Mass Spectrom 22*, 1223–1230.
9. Villanueva, J, Nazarian, A., Lawlor, K., Yi, S. S., Robbins, R. J., Tempst, P. (2008) A sequence-specific exopeptidase activity test (SSEAT) for "functional" biomarker discovery, *Mol Cell Proteomics 7*, 509–518.
10. Petricoin, E. F., Belluco, C., Araujo, R. P., and Liotta, L. A. (2006) The blood peptidome: a higher dimension of information content for cancer biomarker discovery, *Nature Reviews. Cancer 6*, 961–967.
11. Villanueva, J., Shaffer, D. R., Philip, J. et al. (2006) Differential exoprotease activities confer tumor-specific serum peptidome patterns, *Journal of Clinical Investigation 116*, 271–284.
12. Bensalah, K, Lotan, Y., Karam, J. A., Shariat, S. F. (2008) New circulating biomarkers for prostate cancer, *Prostate Cancer Prostatic Dis 11*, 112–121.
13. Bierhaus, A., and Nawroth, P. P. (2005) Posttranslational modification of lipoproteins â€“ A fatal attraction in metabolic disease?: Commentary on: Hone et al., Alzheimer's disease amyloid-beta peptide modulates apolipoprotein E isoform specific receptor binding, *Journal of Alzheimer's Disease 7*, 315–317.
14. Elbatarny, H. S., Netherton, S. J., Ovens, J. D., Ferguson, A. V., and Maurice, D. H. (2007) Adiponectin, ghrelin, and leptin differentially influence human platelet and human vascular endothelial cell functions: implication in obesity-associated cardiovascular diseases, *Eur J Pharmacol 558*, 7–13.
15. Lorenzo, O., Martín-Ventura, J. L., Blanco-Colio, L. M., et al. (2009) The proteomic approach in the development of prognostic biomarkers in atherothrombosis. *Recent Pat Cardiovasc Drug Discov 4*, 25-31.
16. Fu, Q, Van Eyk, J. E. (2006) Proteomics and heart disease: identifying biomarkers of clinical utility. *Expert Rev Proteomics 3*, 237–250.
17. Mitchell, B. L., Yasui, Y., Li, C. I., et al. (2005) Impact of Freeze-thaw Cycles and Storage Time on Plasma Samples Used in Mass Spectrometry Based Biomarker Discovery Projects, *Cancer Informatics 1*, 98–104.
18. Hsieh, S.-Y., Chen, R.-K., Pan, Y.-H., and Lee, H.-L. (2006) Systematical evaluation of the effects of sample collection procedures on low-molecular-weight serum/plasma proteome profiling, *Proteomics 6*, 3189–3198.
19. Craft, D., Yi, J., and Gelfand, C. A. (2009) Time-Dependent and Sample-to-Sample Variations in Human Plasma Peptidome are Both Minimized Through Use of Protease Inhibitors, *Analytical Letters 42*, 1398–1406.
20. Yi, J, Liu, Z. X., Gelfand C. A, Craft, D. (2011) Investigation of Peptide Biomarker Stability in Plasma Sample using Time-course MS Analysis, Ed. Simpson Richard J, Greening David, *Serum/Plasma Proteomics: Method and Protocols. Humana Press.*
21. O'Mullan, P., Craft, D., Yi, J., and Gelfand, C. A. (2009) Thrombin induces broad spectrum proteolysis in human serum samples, *Clin Chem Lab Med. 47*, 685–697.
22. Tammen, H., Schulte, I., Hess, R. et al. (2005) Peptidomic analysis of human blood specimens: comparison between plasma specimens and serum by differential peptide display, *Proteomics 5*, 3414–3422.
23. Guder W G, N. S., Wilsser H, Zawta B. (1996) *Samples: From the Patient to the Laboratory GIT VERLAG.*
24. Slichter, S. J. (1978) Preservation of platelet viability and function during storage of concentrates., *Prog Clin Biol Res 28*, 83–101.
25. Ayukawa, O., Nakamura, K., Kariyazono, H. et al. (2009) Enhanced platelet responsiveness due to chilling and its relation to CD40 ligand level and platelet-leukocyte aggregate formation, *Blood Coagul Fibrinolysis 20*, 176–185.
26. Gulliksson, H., van der Meer, P. F. (2009) Storage of whole blood overnight in different blood bags preceding preparation of blood components: in vitro effects on red blood cells., *Blood Transfus 7*, 210–235.
27. Rossignol, P., Cambillau, M. l., Bissery, A. et al. (2008) Influence of blood sampling procedure on plasma concentrations of matrix metalloproteinases and their tissue inhibitors. *Clinical & Experimental Pharmacology & Physiology 35*, 464–469.

Chapter 9

Collection and Handling of Blood Specimens for Peptidomics

Harald Tammen and Rüdiger Hess

Abstract

Pre-analytical variables can alter the analysis of blood-derived samples. In particular, sample collection and specimen preparation can alter the validity of results obtained by modern multiplex assays (e.g., LC–MS). Low-molecular weight proteins (peptides) as products of proteolytic cleavage events exhibit a close connection to protease activity and function. Altered proteolytic activity during sample collection, preparation, handling, and storage is mirrored by alterations in abundance of specific peptides. Awareness of clinical practices in medical laboratories allows for the identification of specific variables that may affect the results of a peptidomic study. Knowledge of pre-analytical variables is a prerequisite to understand and control their impact.

Key words: Blood, Plasma, Serum, Peptidomics, HUPO, Specimen, Pre-analytical variables

1. Introduction

Peptidomics (1–5) is defined as the systematic, comprehensive, qualitative, and quantitative multiplex (e.g., mass spectrometry) analysis of endogenous peptides in a biological sample at a defined time point and location (6–11). The term peptide is typically used for polypeptides in the range up to approximately 20 kDa molecular weight. The origin of the term "peptide" (derived from the Greek terms "peptos," meaning digestible, and "poly-," referring to its composition of two or more amino acids) reflects that peptides are usually products of proteolytic processing of larger precursor molecules (e.g., cleavage of activation peptides from zymogens). Proteases (12) can initiate, modulate, and terminate many important biological functions by specific substrate cleavage.

Richard J. Simpson and David W. Greening (eds.), *Serum/Plasma Proteomics: Methods and Protocols*, Methods in Molecular Biology, vol. 728, DOI 10.1007/978-1-61779-068-3_9,

Owing to its ease of access and its central role in biological functions, much emphasis has been placed on analyzing blood specimens for biomarker discovery. Blood, regarded as a complex liquid tissue comprising cells and extracellular fluid, provides the major link between cells and tissues of an organism. One critically important, and often underestimated, factor modulating the likelihood of success in discovering relevant peptides in blood specimens is the pre-analytical procedures used in handling the samples. Pre-analytical procedures comprise the processes done prior to the actual analysis of the sample and include steps needed to obtain the primary sample (e.g., blood) and the analytical specimen (e.g., plasma, serum, and cells). Legal or ethical issues (e.g., importance of informed consents) or potential risks of phlebotomy (e.g., bleeding) are not covered in this article.

Pre-analytical procedures can significantly alter the results derived from the analysis of blood-derived samples (13), especially by introducing unwanted systematic biases, which could lead to increased false-positive results. The Plasma Proteome Project (PPP) of the Human Proteome Organization (HUPO) (14) is an international effort for establishing guidance in the form of standard operating procedures for choice of specimen to collect and details of specimen collection, processing, handling, aliquoting, freezing, and thawing. Standardization of blood sample collection is a prerequisite to avoid systemic biases (13, 15). The choice of suitable specimen is also crucial to minimize the occurrence of ex vivo processes (e.g., proteolysis or platelet activation) during specimen collection and preparation (16). Although serum is one of the most frequently analyzed blood specimens, its utility bears several caveats. The generation of serum is associated, for example, with the activation of coagulation cascade and the complement system (17). These processes influence the composition of the samples since they result in cell lysis (e.g., thrombocytes and erythrocytes) and protease activation. The activation of proteases leads to the generation of ex vivo generated peptides. An example of peptides as messengers of ex vivo protease activity is the pronounced differences in the peptidome between plasma and serum samples (14, 15, 18). In serum, numerous peptides derived from the coagulation cascade (e.g., activation peptides) or other proteolytic events (19) are present which are not detectable in plasma, since ion-dependent proteases are inhibited by the addition of chelate-forming agents. On the contrary, a fast degradation of analytes (e.g., hormones) may occur (20). As a consequence of cell lysis, the concentration of components in the extracellular fluid such as aspartate aminotransferase, serotonin, neuron-specific enolase, and lactate dehydrogenase is increased in serum (21). It has also been reported that serum undergoes significant time-dependent changes, which, among other things, affect the reproducibility of sample collection and analysis (22). A reason to use serum as a specimen is based on the notion that the proteome or peptidome of serum may reflect biological events (23). Post-sampling proteolytic cleavage products might reflect cancer type-specific differences (24). However, it has been reported that more protein expression changes occur in serum than in plasma (22). Thus, it can be expected that the reproducibility of such ex vivo proteolytic events is comparatively low.

To counteract the activation of the coagulation cascade (25), citrate, EDTA, or heparin are included during blood collection and/or processing. Citrate and EDTA inhibit coagulation and other enzymatic processes by inhibiting ion-dependent enzymes via chelate formation with ions, while heparin inhibits coagulation through the activation of antithrombin III. The main concern associated with heparinized plasma for proteomic studies is that heparin is a polydisperse charged molecule that binds many proteins nonspecifically (26, 27). Furthermore, heparin may also influence separation procedures and mass spectrometric detection of peptides and small proteins due to its similar molecular weight (18).

In contrast to many of the challenges of using serum as a sample, plasma offers several advantages. Plasma acquisition is less time consuming than the serum acquisition. Separation of cells and the liquid phase, which includes the depletion of thrombocytes below 10/nL to obtain platelet-poor plasma, can be performed subsequent to sample collection since no clotting time (30–60 min) is required. In comparison, the amount of plasma generated from blood is approximately 10–20% higher (28). Additionally, the protein content of plasma is also higher than that of serum, because of the presence of clotting factors and associated components (28). Furthermore, proteins may be bound to the clot, resulting in a diminished protein concentration during the formation of the clot.

On the peptidomic level, more peptides are observed in serum compared to that in plasma (14, 15) due to the level of protease activation in each specimen. Avoiding ex vivo degradation of proteins and peptides is beneficial, since the complexity of the resulting sample is markedly reduced, and for the analysis of in vivo protease cleavage sites, it is mandatory to preserve the integrity of the peptidome.

Conversely, it is known that the complement system is also activated in vitro in EDTA and citrate plasma (29). Furthermore, spontaneous clot formation might occur during storage or freeze–thaw cycles (28).

1.1. Collection of Blood Samples (see Note 1)

It has been reported that the most frequent faults in the preanalytical phase are the result of erroneous procedures for sample collection (e.g., blood drawing from an infusive line, resulting in sample dilution) (30). The design of blood collection devices may aid in correct sampling: evacuated containers sustain the draw of the accurate quantity of blood to ensure the correct concentration of additives or the correct dilution of the blood, such as in the case of citrated plasma. The speed of the blood draw is also controlled and restricts the mechanical stress. The favored site of collection is the median cubital vein, which is generally easily found and accessed. As such, it will be most comfortable for the patient and should not evoke additional stress. Preparation of the collection site includes proper cleaning of the skin with an alcohol (2-propanol). The alcohol must be allowed to evaporate, since commingling of remaining alcohol with the blood sample may result in hemolysis, raise the levels of distinct analytes, and cause interferences. The position of the patient (standing, lying, or seated) can affect the hematocrit (31) and hence change the concentration of analytes. The most common position is lying. The tourniquet should be applied 3–4 in. above the site of venipuncture and should be released as soon as blood begins flowing into the collection device. The duration of venous occlusion (>1 min) can affect the sample composition: prolonged occlusion may result in hemoconcentration and subsequently result in an increase in miscellaneous analytes (e.g., total protein levels). Blood should be collected from fasting patients in the morning between 7 and 9 am, because ingestion or circadian rhythms can alter the concentration of analytes considerably (e.g., total protein, hemoglobin, and myoglobin) (32).

1.2. Processing of Blood Samples (see Note 2)

A quick separation of cells from the plasma is favorable, since cellular constituents may liberate substances that alter the composition of the sample. Generally, it is recommended that blood be centrifuged at 1,300–2,000 × *g* for 10 min within 30 min of collection of the sample. The temperature should generally be maintained at 15–24°C (33), unless recommended otherwise for distinct analytes such as gastrin or A-type natriuretic peptide. Processing at 4°C appears to be attractive, because enzymatic degradation processes are reduced at low temperatures (34). However, platelets become activated at low temperatures (35) and release intracellular proteins and enzymes, which affect the sample composition. Thus, processing at low temperatures should be performed only after thrombocytes have been removed. Since one centrifugation step may not be sufficient enough for the depletion of platelets below 10 cells/nL, a second centrifugation step (2,500 × *g* for 15 min at room temperature, 20°C) or filtration step may be required to obtain platelet-poor plasma. This procedure is applicable only to plasma since platelets in serum are already activated.

1.3. Protease Inhibitors

Protease inhibitors would be attractive, but commonly used protease cocktails may introduce difficulties due to interference with mass spectrometry and formation of covalent bonds with proteins, resulting in shifting the isoform pattern (36). Protease inhibitors have been considered and investigated as additives in proteome research to prevent or slow down proteolytic processes and thereby provide a means of more sensitive detection of markers in blood (37). For example, for analysis of peptides susceptible (38) to degradation by dipeptidyl peptidases (DPP (39)), it might be useful to add DPP inhibitors to the sample collection tubes (40).

Even though protein integrity has been shown to be maintained by the addition of 15 commercially available protease inhibitors, the usefulness of protease inhibitors in overall protein stabilization of blood samples remains to be investigated in more detail (13). The presence of certain protease inhibitors in whole blood is toxic to live cells. Stressed, apoptotic, or necrotic cells release substances, and it may be argued that this impacts the composition of serum or plasma until the cellular and soluble fractions of blood are separated. However, careful selection of an appropriate protease inhibitor may solve this problem.

2. Materials

1. 18–20-Gauge needles and the appropriate adapter (e.g., Sarstedt, Nümbrecht, Germany) or a vacutainer system (BD bioscience, Franklin Lakes, USA).
2. Alcohol (2-propanol) in spray flask.
3. Swabs.
4. Examination gloves.
5. Tourniquet or sphygmomanometer.
6. Citrate- or EDTA-containing blood collection tubes (e.g., Sarstedt, Nümbrecht, Germany).
7. Centrifuge with a swing bucket rotor (e.g., Sigma 4K15, Sigma Laborzentrifugen, Osterode, Harz).
8. 2-mL Cryovials.

3. Methods

3.1. Venipuncture

Venipuncture of a cubital vein is performed using an 18- or 20-gauge needle (diameter: 1.27–0.9 mm, e.g., butterfly system; maximum tubing length: 6 cm). If a tourniquet is applied, it should not remain in place for longer than 1 min (due to the risk of falsifying results due to hemoconcentration) (see Note 1).

3.1.1. Initial Blood Sample

Prior to blood collection for peptidomic analysis, blood is aspirated into an initial container (e.g., 2.7-mL S-Monovette, Sarstedt, Nümbrecht, Germany). This is done to flush all surfaces and remove initial traces of contact-induced coagulation. This sample is not useful for analysis.

3.1.2. Blood Sampling

Afterward, blood is drawn into a standard EDTA- or citrate-containing syringe (e.g., 9 mL EDTA-Monovette, Sarstedt, Nümbrecht, Germany). Depending on the ease of blood flow, several samples can be collected.

3.2. First Centrifugation

After venipuncture, plasma is obtained by centrifugation for 10 min at 2,000 × *g* at *room temperature*. Start centrifugation within 30 min after blood collection. The resulting plasma sample is now separated from red and white blood cells in an efficient and gentle way (see Note 4). Nevertheless, a significant number of platelets (~25%) are still present in the sample. The platelets will lyse as soon as the sample is refrigerated, thereby releasing intracellular substances. Removal of platelets below 10/nL requires an additional preparation step.

3.2.1. Platelet Removal by Centrifugation

The plasma sample is transferred into a second vial for another centrifugation for 15 min at 2,500 × *g* at *room temperature*. After this centrifugation, the supernatant is transferred in aliquots of 1.5 mL into cryovials.

3.3. Storage

The samples are transferred to a –80°C freezer within 30 min. Storage is at –80°C. Transport of samples is done on dry ice (see Note 3).

4. Notes

1. Considerations during blood withdrawl.
    - The patient was not fasting (i.e., had eaten prior to sampling with 12 hrs).
    - Blood was drawn from an infusive line. This may lead to dilution of analytes and may induce coagulation.
    - Blood was drawn in other position (e.g., supine or upright).
    - To avoid hemolysis, free flow with mild aspiration has to be assured.
    - To avoid hemoconcentration, the tourniquet has to be released at least partially as soon as blood flows into the container. If more time is required, the tourniquet has to

be released so that circulation resumes and normal skin color returns to the extremity. Usage of a sphygmomanometer may help to apply the optimal pressure by controlling the peripheral pulse (A. brachialis) in parallel.

- The consumables used were different to those recommended.
- The expiry date of consumables was already reached.
- The tubes were not properly filled.
- The tubes were agitated vigorously (shaken instead of gentle movement to dissolve the anticoagulant).
- The blood sample tubes were not consistently kept at room temperature.
- The sample tubes were placed on ice or in a refrigerator.

2. Considerations during laboratory handling and processing
   - Centrifugation was delayed more than 30 min after blood withdrawal.
   - The temperature of a cooling centrifuge was adjusted below room temperature.
   - The centrifugation speed was incorrect (e.g., rounds per minute were set instead of *g*-force).
   - The centrifugation time was incorrect.
   - Removal of blood plasma by pipetting was done without proper caution. Consequently, the buffy coat or the red blood cells were churned up.
   - The second centrifugation of recovered plasma samples was delayed after the end of the first centrifugation.
3. Considerations during sample storage
   - Storage of the samples was delayed (maintained at room temperature).
   - The storage temperatures were above –80°C.
   - The labeling of sample containers is unreadable or confusing.
   - Attachment of the labels to the sample containers was not sufficient during storage or handling, resulting in loss of labels.
4. General recommendations
   - A proper first centrifugation should produce a visible white blood cell layer (buffy coat) between red blood cells and plasma. If not, centrifugation speed or time may be was incorrect.
   - Icteric plasma or plasma exhibiting signs of hemolysis should be discarded. One should check with a specialist if this may be due to that particular disease.

References

1. Schrader, M. and Schulz-Knappe, P. (2001) Peptidomics technologies for human body fluids. *Trends Biotechnol.* **19**, 55–60
2. Baggerman, G., Verleyen, P., Clynen, E., et al. (2004) Peptidomics. *J Chromatogr.B Analyt. Technol.Biomed.Life Sci.* **803**, 3–16
3. Soloviev, M. and Finch, P. (2005) Peptidomics, current status. *J Chromatogr.B Analyt.Technol. Biomed.Life Sci.* **815**, 11–24
4. Tammen, H., Zucht, H. D., and Budde, P. (2007) Oncopeptidomics - A commentary on opportunities and limitations. *Cancer Lett.* **249**, 80–86
5. Schrader, M. and Schulz-Knappe, P. (2001) Peptidomics technologies for human body fluids. *Trends Biotechnol.* **19**, S55–S60
6. Schulz-Knappe, P., Zucht, H. D., Heine, G., et al. (2001) Peptidomics: the comprehensive analysis of peptides in complex biological mixtures. *Comb.Chem High Throughput Screen.* **4**, 207–217
7. Pierson, J., Norris, J. L., Aerni, H. R., et al. (2004) Molecular profiling of experimental Parkinson's disease: direct analysis of peptides and proteins on brain tissue sections by MALDI mass spectrometry. *J Proteome.Res.* **3**, 289–295
8. Pasinetti, G. M., Ungar, L. H., Lange, D. J., et al. (2006) Identification of potential CSF biomarkers in ALS. *Neurology* **66**, 1218–1222
9. Sasaki, K., Sato, K., Akiyama, Y., et al. (2002) Peptidomics-based approach reveals the secretion of the 29-residue COOH-terminal fragment of the putative tumor suppressor protein DMBT1 from pancreatic adenocarcinoma cell lines. *Cancer Res.* **62**, 4894–4898
10. Zhang, X., Wei, D., Yap, Y., et al. (2007) Mass spectrometry-based "omics" technologies in cancer diagnostics. *Mass Spectrom Rev.* **26**, 403–431
11. Lopez, M. F., Mikulskis, A., Kuzdzal, S., et al. (2007) A novel, high-throughput workflow for discovery and identification of serum carrier protein-bound Peptide biomarker candidates in ovarian cancer samples. *Clin Chem.* **53**, 1067–1074
12. Overall, C. M. and Blobel, C. P. (2007) In search of partners: linking extracellular proteases to substrates. *Nat.Rev Mol Cell Biol.* **8**, 245–257
13. Rai, A. J. and Vitzthum, F. (2006) Effects of preanalytical variables on peptide and protein measurements in human serum and plasma: implications for clinical proteomics. *Expert Rev Proteomics* **3**, 409–426
14. Omenn, G. S., States, D. J., Adamski, M., et al. (2005) Overview of the HUPO Plasma Proteome Project: results from the pilot phase with 35 collaborating laboratories and multiple analytical groups, generating a core dataset of 3020 proteins and a publicly-available database. *Proteomics.* **5**, 3226–3245
15. Rai, A. J., Gelfand, C. A., Haywood, B. C., et al. (2005) HUPO Plasma Proteome Project specimen collection and handling: towards the standardization of parameters for plasma proteome samples. *Proteomics* **5**, 3262–3277
16. Hsieh, S. Y., Chen, R. K., Pan, Y. H., and Lee, H. L. (2006) Systematical evaluation of the effects of sample collection procedures on low-molecular-weight serum/plasma proteome profiling. *Proteomics* **6**, 3189–3198
17. Mollnes, T. E., Garred, P., and Bergseth, G. (1988) Effect of time, temperature and anticoagulants on in vitro complement activation: consequences for collection and preservation of samples to be examined for complement activation. *Clin.Exp.Immunol.* **73**, 484–488
18. Tammen, H., Schulte, I., Hess, R., et al. (2005) Peptidomic Analysis of Human Blood Specimens: Comparison between Plasma Specimens and Serum by Differential Peptide Display. *Proteomics* **5**
19. Koomen, J. M., Li, D., Xiao, L. C., et al. (2005) Direct tandem mass spectrometry reveals limitations in protein profiling experiments for plasma biomarker discovery. *J Proteome Res.* **4**, 972–981
20. Evans, M. J., Livesey, J. H., Ellis, M. J., and Yandle, T. G. (2001) Effect of anticoagulants and storage temperatures on stability of plasma and serum hormones. *Clin Biochem.* **34**, 107–112
21. Guder W.G., N. S. W. H. Z. B. (2003) Samples: From the Patient to the Laboratory. Wiley-VCH 3rd Edition
22. Tammen, H., Schulte, I., Hess, R., et al. (2005) Prerequisites for peptidomic analysis of blood samples: I. Evaluation of blood specimen qualities and determination of technical performance characteristics. *Comb Chem High Throughput Screen* **8**, 725–733
23. Villanueva, J., Shaffer, D. R., Philip, J., et al. (2006) Differential exoprotease activities confer tumor-specific serum peptidome patterns. *J Clin Invest.* **116**, 271–284
24. Liotta, L. A. and Petricoin, E. F. (2006) Serum peptidome for cancer detection: spinning biologic trash into diagnostic gold. *J Clin Invest.* **116**, 26–30

25. Banfi, G., Salvagno, G. L., and Lippi, G. (2007) The role of ethylenediamine tetraacetic acid (EDTA) as in vitro anticoagulant for diagnostic purposes. *Clin.Chem.Lab Med.* **45**, 565–576
26. Landi, M. T. and Caporaso, N. (1997) Sample collection, processing and storage. *IARC Sci Publ.* 223–236
27. Holland, N. T., Smith, M. T., Eskenazi, B., and Bastaki, M. (2003) Biological sample collection and processing for molecular epidemiological studies. *Mutat.Res.* **543**, 217–234
28. Lundblad R.L. (2004) Considerations for the Use of Blood Plasma and Serum for Proteomic Analysis. *The Internet Journal of Genomics and Proteomics*
29. Pfeifer, P. H., Brems, J. J., Brunson, M., and Hugli, T. E. (2000) Plasma C3a and C4a levels in liver transplant recipients: a longitudinal study. *Immunopharmacology* **46**, 163–174
30. Plebani, M. and Carraro, P. (1997) Mistakes in a stat laboratory: types and frequency. *Clin Chem.* **43**, 1348–1351
31. Burtis, C. A. A. E. R. (2001) Fundamentals of Clinical Chemistry W.B. Saunders Company.
32. Winkel, P., Statland, B. E., and Bokelund, H. (1975) The effects of time of venipuncture on variation of serum constituents. Consideration of within-day and day-to-day changes in a group of healthy young men. *Am.J.Clin. Pathol.* **64**, 433–447
33. Favaloro, E. J., Soltani, S., and McDonald, J. (2004) Potential laboratory misdiagnosis of hemophilia and von Willebrand disorder owing to cold activation of blood samples for testing. *Am J Clin Pathol.* **122**, 686–692
34. Heins, M., Heil, W., and Withold, W. (1995) Storage of serum or whole blood samples? Effects of time and temperature on 22 serum analytes. *Eur.J.Clin.Chem.Clin.Biochem.* **33**, 231–238
35. Mustard, J. F., Kinlough-Rathbone, R. L., and Packham, M. A. (1989) Isolation of human platelets from plasma by centrifugation and washing. *Methods Enzymol.* **169**, 3–11
36. Schuchard, M. D., Mehigh, R. J., Cockrill, S. L., et al. (2005) Artifactual isoform profile modification following treatment of human plasma or serum with protease inhibitor, monitored by 2-dimensional electrophoresis and mass spectrometry. *Biotechniques* **39**, 239–247
37. Jeffrey D. Hulmes, Deidra Bethea, Keith Ho, Shu-Pang Huang, Deborah L. Ricci, Gregory J. Opiteck, and Stanley A. Hefta (2004) *An Investigation of Plasma Collection, Stabilization, and Storage Procedures for Proteomic Analysis of Clinical Samples* Humana Press.
38. Balkan, B., Kwasnik, L., Miserendino, R., et al. (1999) Inhibition of dipeptidyl peptidase IV with NVP-DPP728 increases plasma GLP-1 (7-36 amide) concentrations and improves oral glucose tolerance in obese Zucker rats. *Diabetologia* **42**, 1324–1331
39. Lambeir, A. M., Durinx, C., Scharpe, S., and De, M., I. (2003) Dipeptidyl-peptidase IV from bench to bedside: an update on structural properties, functions, and clinical aspects of the enzyme DPP IV. *Crit Rev.Clin.Lab Sci.* **40**, 209–294
40. Theodorakis, M. J., Carlson, O., Muller, D. C., and Egan, J. M. (2004) Elevated plasma glucose-dependent insulinotropic polypeptide associates with hyperinsulinemia in impaired glucose tolerance. *Diabetes Care* **27**, 1692–1698

Chapter 10

Investigation of Peptide Biomarker Stability in Plasma Samples Using Time-Course MS Analysis

Jizu Yi, Zhaoxia Liu, Craig A. Gelfand, and David Craft

Abstract

Peptide biomarkers in plasma or serum are subject to proteolytic degradation caused by intrinsic peptidase activities, resulting in a potential barrier in translating a discovered biomarker into clinical application. This chapter describes a method using time-course MALDI-TOF MS analysis to investigate the stability of a plasma peptide biomarker under a variety of preanalytical situations. A synthesized peptide with the same primary sequence as a potential endogenous biomarker is spiked into a blood sample, and the sample is incubated over time at r.t. (25 ± 1°C) or other preanalytical situations. At a specific period of incubation time, the sample is quenched with the addition of acid with or without an internal control peptide. The spiked peptides in the sample are extracted with one of three procedures for highly soluble, moderately soluble, or essentially insoluble peptides. The peptide samples are then analyzed using MALDI-TOF MS. The abundance changes of the peptide biomarker are monitored by time-course changes of the mass spectra. These changes over-time are measured and fitted to a first-order degradation reaction so that stability of the peptide biomarker (half-life) can be calculated. Kinetics analysis of both parent and shorter (daughter) peptides are also possible by fitting to a sequential multiple-step reaction (SMSR) model. This optimized method facilitates evaluation of biomarker stability, and helps to define sample handling and analytical processing steps that contribute to instability of measured peptide biomarker(s).

Key words: Peptide biomarker, Stability, Degradation, Proteolysis, Plasma, Time-course MS, MALDI-TOF, FPA, GLP-1, GIP

1. Introduction

Human blood specimens are extensively utilized traditionally as a sample resource containing a variety of disease biomarkers, and blood-derived plasma and serum samples are readily available for proteomic and/or clinical analyses. Research surrounding the human plasma proteome has most recently led to the growing identification and characterizations of up to total 3,206 plasma proteins (1–4), with many more proteins presumed to be present.

Richard J. Simpson and David W. Greening (eds.), *Serum/Plasma Proteomics: Methods and Protocols*, Methods in Molecular Biology, vol. 728, DOI 10.1007/978-1-61779-068-3_10, © Springer Science+Business Media, LLC 2011

This work has further stimulated (5) the successful discovery of protein and peptide biomarkers associated with various diseases (6), including Alzheimer's disease (7), cancer (8–11), diabetes (12), and cardiac and heart diseases (13–15), among many other published works. However, there has been notably less success in translation of the discoveries into clinical utility (16, 17). This poor conversion rate of discovered peptide biomarkers into the clinical setting could be largely attributed to a lack of biomarker stability and/or to the preanalytical variables related to sample collection (17, 18). Minimizing these preanalytical variations should be the focus for proteomic applications translating biomarker discovery toward clinical application (18).

We have previously described that intrinsic protease and peptidase activity causes instability of peptide biomarkers (17, 20). To overcome this barrier, we introduced and optimized a cocktail of protease inhibitors for plasma preparation, which provides immediate inhibition of intrinsic proteolysis, and broad protection of the "proteome." Yet, stabilization of *all* plasma proteins or peptides for an extended period of time is virtually impossible, given the complexity of the proteome and the enzymatic content of blood (17). However, our research in the past few years demonstrated that stabilization of specific peptide biomarkers (e.g., metabolic peptides: GLP-1, GIP, glucagon, and ghrelin) was feasible using specific inhibitory reagents. To this end, we developed MS-based assays for each of these peptides, allowing us to investigate the instability and variability caused as a consequence of sample handling and analytical processing. This chapter provides protocols of these assays using time-course MALDI-TOF MS analysis. The described method can be used to evaluate biomarker stability (half-life), also to optimize protocols from phlebotomy through MS analysis for improving assay reproducibility, and thus to reach better and more useful peptide biomarkers for clinical applications.

2. Materials

2.1. Blood Collection and Plasma Preparation

1. Blood collection kit, plasma, and serum tubes were from BD (Franklin Lakes, NJ) and previously listed in chapter 8 of this book (19).
2. BD™ P800* blood collection tube (#366420; 2.0 mL, or #366421; 8.5 mL). The new blood collection device is an EDTA plasma preparation tube with a cocktail of enzymatic inhibitors for stabilization of metabolic peptides, including GLP-1, GIP, glucagon, and ghrelin (http://www.bd.com/proteomics).

* For Research Use Only, Not For Clinic Diagnostics. BD P800 tube is to stabilize metabolic bio-active peptides such as GLP-1, GIP, Glucagon and Ghrelin.

3. A Sorvall RC3C+ Centrifuge (Sorvall, USA); or a CellSep 6/720R centrifuge (SANYO, Japan).

2.2. Peptides

Peptides used in this study were purchased from Sigma–Aldrich. Isotopically labeled peptides (see Note 1), e.g., AQUA GLP-1, were custom-synthesized by Sigma–Aldrich (Woodlands, TX). The synthesized peptides were purified by HPLC to reach >95%, and their purities were confirmed by amino acid analysis and ESI MS analysis.

1. AQUA Fibrinogen Peptide A (*FPA*): ADSGEGD$\underline{\text{F}}$LA EGGGVR FPA (hereafter the underlined residue carries the stable isotope label) was dissolved into dH_2O in 400 fM as a stock solution.
2. *GLP-1* Analogs including

   GLP-1(7-37): HAEGTFTSDVSSYLEGQAAKEFIAW LVKGRG-OH

   GLP-1(7-36A): HAEGTFTSDVSSYLEGQAAKEFIA WLVKGR-NH_2

   AQUA GLP-1: HAEGTFTSDVSSYLEGQAA$\underline{\text{K}}$EFIAW LVKGR-OH.

   Each of these GLP-1 analogs was dissolved into 50% acetonitrile (ACN) solution in a 4-μM stock for time-course experiment. Both GLP-1(7-36A) and GLP-1(7-37) were used as peptide biomarkers while AQUA GLP-1 was used as a control with underlined lysine residue in stable isotope labeling.
3. *GIP*: YAEGTFISDYSIAMDKIHQQDFVNWLLAQK-GKKNDWKHNITQ GIP was dissolved into 50% aqueous ACN solution in a 40-μM stock.
4. *PYY*: YPIKPEAPGEDASPEELNRYYASLRHYLQLVTRQ-RYNH_2 PYY, used as control, was dissolved into 50% aqueous ACN solution in a 20-μM stock.

2.3. Materials for Peptide Extraction

1. Centrica Concentrator (Labconco, Kansas City, MO) was used with a Centrica Cold Trap (Labconco) and a vacuum pump, Welch 8907 (Welch, Niles, IL). The system was used to vacuum dry suspended peptide solutions under controlled temperature, centrifuge speed, and time.
2. Centrifuge: one Micro centrifuge 5417R with a fixed angle rotor FA45-24-11 (Eppendorf, Hamburg, Germany).
3. Pipettes: a set of calibrated pipettes (VWR, West Chester, PA): 0.5–10, 10–100, 20–200, and 100–1,000 μL.
4. Solution for peptide stock solution (see Note 2): 50% (v/v) aqueous ACN (≥99.9%, Sigma–Aldrich, Saint Louis, MO).
5. Size-based membrane filter (separator): Microcon YM-3, YM-10 (Millipore, Billerica, MA), or other size filters which

allow the tested peptide to pass through based on relative molecular weight.

6. Zip-Tip C18 (Millipore, Billerica, MA).
7. Solutions for Zip-Tip purification:
   S1. Generation solution: 60% (v/v) aqueous ACN, containing 5% (v/v) aqueous trifluoroacetic acid (TFA≥99.9%, Baker, Phillipsburg, NJ).
   S2. Equilibrium solution: 5% aqueous ACN, containing 0.1% TFA.
   S3. Washing buffer: 10% aqueous ACN, containing 0.1% TFA.
   S4. Elution buffer: 60% aqueous ACN, containing 0.1% TFA.
8. Solutions for quenching samples after incubations:
   Q1. 50% aqueous ACN, containing 2% TFA.
   Q2. 25% aqueous ACN, containing 0.2% TFA.
   Q3. 50% aqueous ACN, containing 30% (v/v) acetone, containing 0.2% (v/v) acetic acid (>99.7%, Sigma–Aldrich, Saint Louis, MO).
9. Solution for peptide resuspension: 15% aqueous ACN, containing 0.1% (v/v) acetic acid.
10. Matrix solution for MALDI-TOF MS: 5 mg/mL CHCA (Bruker-Daltonics, Billerica, MA) in 50% ACN, containing 0.1% (v/v) TFA (the matrix solution should be made fresh/daily before use).

2.4. MALDI-TOF MS Analysis

MS instrument: Ultraflex II MALDI TOF/TOF MS (Bruker-Daltonics, Billerica, MA) was used in this study. A similar instrument or another type of mass spectrometer, either MALDI or ESI, could be adapted to this method.

2.5. Data Analysis

1. Ultraflex Analysis V2.4.
2. MS Office Excel.
3. Metlab 2.3.

3. Methods

Intrinsic plasma exopeptidases may truncate a peptide from the amino and/or carboxyl termini. Our kinetics studies indicated that this reaction fits a sequential multiple-step reaction (SMSR) model and each step reaction is observed to be a first-order degradation (Figs. 1 and 2). This data indicates that stability (half-life) of these peptides is independent on its initial concentration (20). These results rationalize the assumption that stability measured

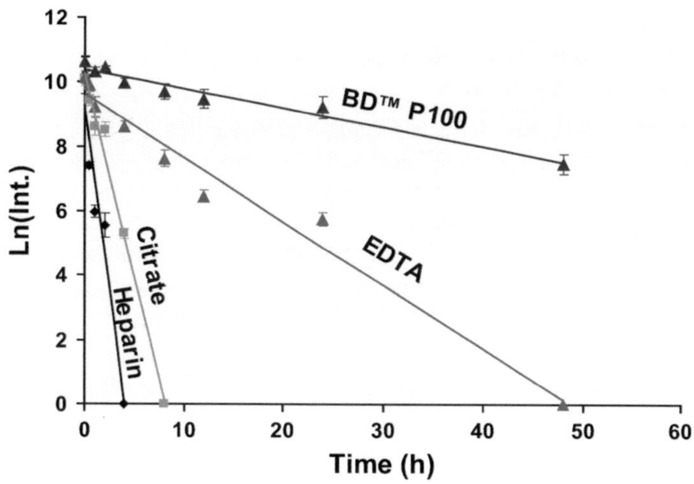

Fig. 1. Stability of FPA in plasma samples. AQUA FPA was spiked into each of the inhibited plasma (P100) and three traditionally used plasma samples for time-course incubations. At specified time, the samples were quenched, filtrated by YM-3, followed by peptide extraction with Zip-Tip C18, and MALDI-TOF MS analysis. The linear plots indicate that intrinsic peptidase activities cause first-order degradation of the peptide, and that the stabilities of FPA in plasma samples are in an order of P100 > EDTA > citrate > heparin (*20*).

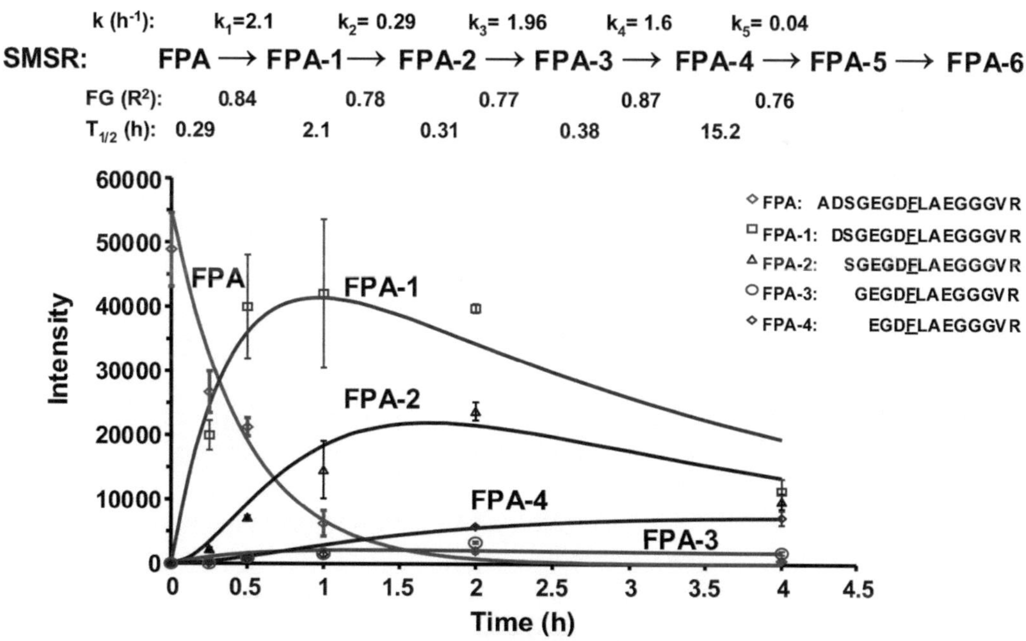

Fig. 2. Modeling SMSR of FPA in heparin plasma sample. FPA was spiked into heparin plasma, followed by time-course MS analysis. FPA and its fragments were detected simultaneously within the timeframe of a single experiment. Their intensities were simulated according to SMSR model (*20*), and their stabilities were determined.

using a synthetic peptide spiked into a plasma sample at higher concentrations than endogenous levels, to make detection easier, is a valid representation of stability in vitro.

For MS-based proteomics, protein preparation and separation from a complex sample, such as serum or plasma is critical for down-stream analysis. A variety of methods such as multidimensional chromatography (21), bead-based extraction (22–24), and centrifugal ultra filtration (4, 25) have been reported for low molecular weight plasma proteins or peptide sample preparation. For peptide biomarker analysis using MALDI-TOF MS, removing or minimizing matrix effects from a complex sample such as plasma is recommended for obtaining high-quality spectra (26). The sample preparation processes may differ due to the different solubility and/or physical characteristics of each individual peptide. We provide herein a method with three typical sample preparation protocols for three types of peptide biomarkers: (a) soluble and able to pass through a permeable filter, (b) less-soluble and unable to pass the permeable filter, and (c) insoluble peptides. Each of these protocols is followed by peptide extraction by reversed-phase media (Zip-Tip C18) and MS analysis. This optimized method facilitates a relatively simple, fast, cost-effective, and highly reproducible analysis of time-dependent changes in peptide levels in human blood (Fig. 3).

3.1. Time-Course Incubation of Peptides in Samples

1. Blood collection and plasma sample preparation can be performed according to the protocols described in chapter 8 of this book (19).
2. All peptide stock solutions were made for time-course experiments according to their solubility, which was determined through a pretest (see Note 3). Soluble peptides, e.g., FPA, are dissolved into dH_2O; less-soluble or insoluble peptides, e.g., GLP-1 or GIP, are dissolved into 50% ACN solution.
3. An appropriate aliquot of frozen sample (at –80°C) was thawed in a water bath (23 ± 2°C) for 5 min. The fast warming provided by the water bath may reduce protein denaturation and/or precipitation during thawing (27).
4. The thawed plasma sample (1.5-mL eppendorf tube) was spun in a temperature-controlled centrifuge (e.g., an eppendorf centrifuge 5417R) at 25,000 × *g* (RFC) and 15°C for 10 min, and clear supernatant was used. This centrifugation removes possible air bubbles and precipitants generated during the sample storage and/or thawing process.
5. Into ~1.0 mL of either a serum or plasma sample (e.g., EDTA, P100, or P800 plasma), a peptide stock solution containing the peptide biomarker of interest was added at a (v/v) ratio of 1/10 (peptide/plasma) to an appropriate final concentration

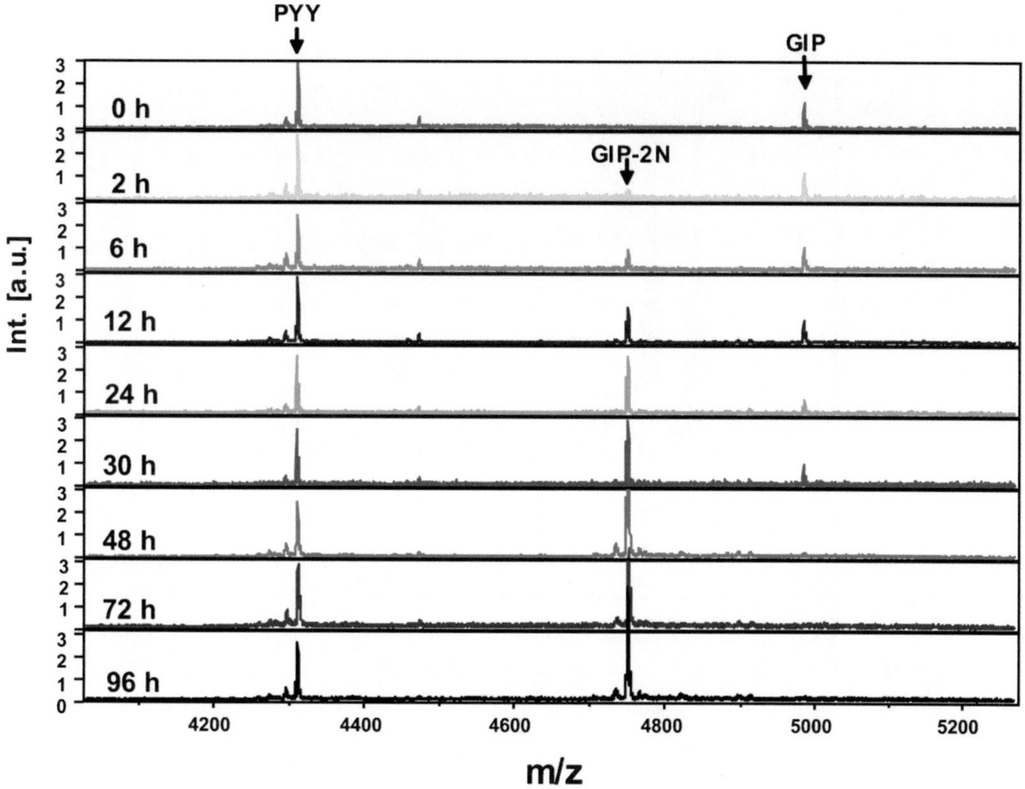

Fig. 3. Time-course stability of GIP in EDTA plasma. GIP was incubated in EDTA sample at r.t. for time-course experiment. At specific time, PYY was added as a control peptide, immediately followed by quenching and protein precipitation. Clean solution was vacuum dried, and peptides were extracted for MS analysis. Over time, both GIP and its fragment (GIP-2N) with two N-terminal residues removed presumably by dipeptidyl peptidase-4 (DPP-4) were detected with the half-life of GIP < 24 h in EDTA plasma.

(e.g., 200–400 fM for FPA, 4 μM for GLP-1). The peptide was mixed by pipetting the sample up and down (5×). In the case of a new or unknown peptide biomarker, its initial plasma concentration at spiking was determined by one pretest (see Note 4).

Simultaneously testing two peptide biomarkers (e.g., GLP-1 (7-37) and GLP-1 (7-36A), see Fig. 4a, b) is feasible if both peptides have similar solubility properties and share the same procedures for peptide extraction as described (Subheadings 3.1, step 8 and 3.2, step 1, see Note 5).

6. The spiked samples were divided into 0.6-mL eppendorf tubes, for a series of time-course experiments, typically: 0, 1, 2, 4, 8, 12, 24, 48, 72, and 96 h, with four replicates (see Note 6) for each time point and 10–45 μL sample for each replicate. Samples were incubated at defined temperatures (e.g., 4°C or r.t.) and/or under the conditions for biomarker assay development.

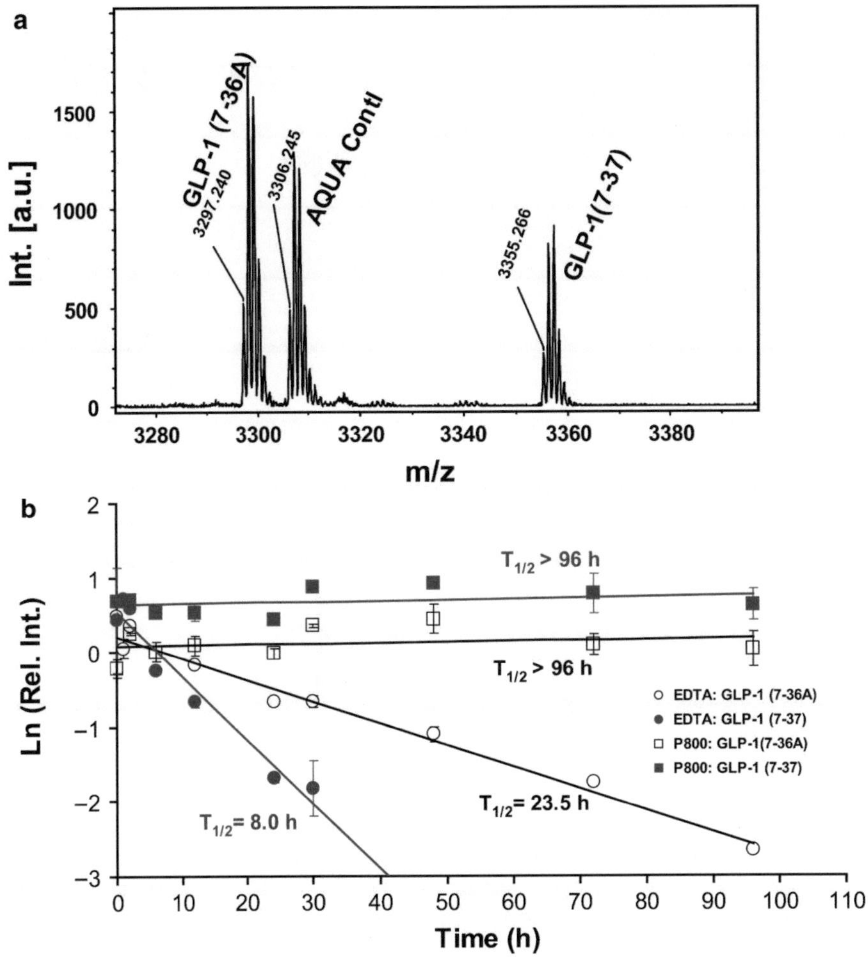

Fig. 4. (**a**) A typical spectrum of three GLP-1 peptides. Two active forms: GLP-1 (7-36A) and GLP-1(7-37), were spiked into an EDTA plasma sample. At specific time points, the sample was spiked with the quenching solution containing AQUA GLP-1 as the control peptide, followed by peptide extraction, and MALDI-TOF MS analysis. Three peptides are fully resolved in a single spectrum. (**b**) Stability comparison in EDTA and inhibited plasma. The peak heights of two active peptides relative to the control were fitted to the first-order degradation and their stabilities in $T_{1/2}$ were then determined. The results indicate that two active GLP-1 forms being instable in EDTA are stabilized in inhibited (P800) plasma.

7. Into the incubation tubes at the designed time points of incubation, add the control peptide for relative quantitation (see Note 7).
8. Immediately following step 7, add a quenching solution (as described in steps 8a–c) to stop further degradation reactions. The immediacy of quenching is essential (see Note 7). For the time "0" experiment (see Note 8), the incubation tubes were preloaded with quenching solution before adding the plasma sample (step 6).
   (a) For soluble peptides, into 45 μL peptide sample (assuming the peptide, e.g., FPA, can pass through a size-based

membrane filter (see Note 9)), add 5 μL of 50% ACN and 2% TFA solution (Q1) to reach a final concentration of 5% ACN and 0.2% TFA (continue on Subheading 3.2, step 1a).

(b) For less soluble peptides, into 7 μL peptide sample (assuming the peptide, e.g., GLP-1, cannot pass through a size-based membrane filter (see Note 9)), add 43 μL solution with 24% ACN and 0.2% TFA to reach ~20% ACN as a final concentration in the plasma (continue on Subheading 3.2, step 1b).

(c) For insoluble peptides, into a 10-μL peptide sample (e.g., GIP), add ~40 μL of 50% ACN, 30% acetone, and 0.2% acetic acid (Q3) to reach a final concentration of 40% ACN and 24% acetone, 0.16% acetic acid. This step resulted in significant precipitation of plasma proteins (continue on Subheading 3.2, step 1c).

The quenched samples were then supplied for further peptides extraction.

3.2. Sample Preparation for Biomarker MS Analysis

Due to the difference in peptide solubility, three different quenching solutions have been used. Accordingly, three different extraction procedures are recommended.

1a. For a soluble peptide capable of passing through a molecular weight size cut-off filter (see Note 9), load the quenched sample (~45–60 μL) on the top of the MWCO filter (e.g., YM-3, or YM-10, Millipore), and spin in a microcentrifuge at 14,500 × *g* and at 10°C for 35 min. The filtrate is then used for Zip-Tip C18 peptide extraction.

1b. For a less soluble peptide which cannot pass through a membrane-based filter, the quenched sample is directly provided for Zip-Tip extraction.

1c. For an insoluble peptide:

- Spin the quenched samples at r.t., 25,000 × *g* for 15 min to remove the precipitates. Transfer 30–60 μL of clear supernatant to another eppendorf tube.
- The solution is vacuum dried using Centrica Concentrator, at 60°C for 30 min.
- Into each tube containing dry peptides, add 35 μL resuspending solution (15% ACN, 0.1% TFA), followed by moderate mixing by pipetting up and down.
- Spin the tubes at 25,000 × *g* and r.t. for 15 min to remove precipitate, and take 30 μL of the clear supernatant for desalting by Zip-Tip C18.

2. The three peptide samples generated from the above procedures share the same desalting and/or purification steps, as follows:

   - Pre-activate Zip-Tip (with 0.6 μL C18 resin) by 3× with 10 μL generation solution (S1).
   - Rebalance the Zip-Tip by 3 × 10 μL of washing solution (S2).
   - Pipette the Zip-Tip in the sample solution for ~15×.
   - Wash the sample in the Zip-Tip by 3 × 10 μL of washing solution (S3).
   - Elute the bound peptides from the Zip-Tip with 1.5 μL elution solution (S4) into a 0.6-mL eppendorf tube.

3. Into each eluted peptide solution, add 1.5 μL of 5 mg/mL CHCA as the MALDI matrix (see Note 10) (in 1:1 v/v ratio), and mix them by pipetting up and down two to three times. Load 1.5 μL of the mixed sample onto the sample target for MALDI-TOF MS (see Note 11). A mixture of peptide standards with known molecular weight is also loaded onto separate spots close to the sample spots for external calibration. The loaded drops are allowed to air-dry at r.t. (for ~20 min), and then analyzed by MALDI-TOF MS.

3.3. MALDI-TOF MS

1. A flexcontrol method for MALDI-TOF mass spectrometer (Ultraflex II, Bruker-Daltonics) was established and tuned for peptide analysis approximately using 25.0, 22.0, 9.6, 20.3, and 13.9 kV for ion source 1, ion source 2, lens, reflector, and reflector 2, respectively.
2. Load the target with dry samples into the mass spectrometer, followed by auto-teaching for positioning properly (adjusted manually if necessary).
3. For quality spectra, laser power and detection gain (reflector) were first tested and selected manually for a qualified spectrum in terms of balance between sensitivity and peak resolution.
4. Based on above selected parameters, calibrate the machine and the target using the loaded peptide standards with molecular masses ranging from 750 to 5,000 Da, and then acquire sample spectra by the autocontrol method.
5. To minimize the effect of heterogeneous crystals of samples, during the spectrum acquisition we controlled three variables:

   - Setting a qualification filter for each spectrum: the spectrum from 100 laser shots was evaluated in terms of its peak resolution, signal/noise (s/n) ratio. When a spectra with a peak resolution less than 1,000 and/or an s/n < 5, the spectrum was filtered out, and only a qualified spectrum was added into the accumulated spectrum.

- Each spectrum contained a large number of laser shots (e.g., total 3,000): the final spectrum was accumulated from 30 qualified spectra, each of which was from 100 laser shots.
- Controlling laser targeted sites: during acquisition each site was allowed for only 100 laser shots by moving the sample target.

3.4. Data Analysis

1. Sample spectra as a group of files were open using analysis software, e.g., flexAnalysis for Ultraflex II (Bruker-Daltonics). The spectra were smoothed moderately, followed by base-line subtraction.
2. The peak lists from flexAnalysis were exported into Microsoft Excel or to a text file.
3. In the case of FPA or when a control for quantitative analysis was not necessary (see Note 7), the peak height was directly plotted in Ln (intensity) versus time (Fig. 1). In the case of GLP-1 or GIP or when a control was used, the peak height of the targeted peptides was analyzed as a ratio to that of the control. The ratios were plotted against the incubation time in a semi-log plot (Fig. 4b). In both cases, these plots were linear, and from the slopes, the stabilities (half-life) of the peptide biomarkers were determined.
4. The fragments from digestion of FPA were detected, and analyzed using our developed SMSR model as shown in Fig. 2 (20).

4. Notes

1. *Peptide as biomarkers*: Ideally the peptide biomarker of interest may be synthesized to create an isotopically labeled analog. This isotopically labeled peptide spiked into a plasma sample can be easily distinguished from its endogenous counterpart. However, in cases when an endogenous peptide is too low to be detectable by mass spectrometry, a nonlabeled native peptide either highly purified or from synthesis, can be spiked in to plasma representing the biomarker at the beginning of the time-course experiment.
2. *Solutions and stock solution*: The solutions (10–100 mL) used in this study were made by (v/v)% mixing ACN with dH_2O first, followed by adding the rest components using a pipette (1). To make 20 mL of 50% ACN solution containing 0.1% TFA, firstly, pipette 10 mL of dH_2O into a VWR sample

borosilicate glass vial (22–30 mL, with a cap), secondly, add 10 mL of pure ACN into the water, and then add 0.1 mL of 20% TFA solution into the vial. This solution can be stored on the bench, tightly capped, for up to 6–8 weeks for usage. (2) To make 5 mg/mL CHCA solution with 50% ACN and 0.1% TFA, pipette 2 mL of 50% ACN and 0.1% TFA solution, prepared as described above, into 10 mg of CHCA, and mix by pipetting the solution up and down until CHCA is fully dissolved. The matrix solution should be made by freshly dissolving powder CHCA with the solution before usage.

3. *Peptide solubility test*: Peptide powder is dissolved into dH_2O up to 1 mg/mL (or 50–100 μM) as stock, followed by serial dilutions of the stock, and centrifugation at 25,000×*g* for 10 min. 1 μL of the supernatant is mixed with the same volume of matrix solution, and analyzed by MALDI-TOF MS. The appearance or disappearance of the tested peptide peak will show whether this peptide is soluble or not and at what concentration in dH_2O, if it is soluble. If the above test shows that the peptide is insoluble in pure water, or if no peptide peak is observed in any concentration, into the above stock add 50–100% ACN to different concentrations from 5 to 50% ACN, followed by mixing, centrifugation, and MALDI-TOF MS test of the clear solution. Results will show at which ACN concentration the peptide is dissolved. To date, all of our tested peptides (fewer than 100 μM) are soluble in 50% ACN solution.
4. *Initial peptide concentration for spiking*: The optimal peptide spiking concentrations can be tested before a time-course incubation experiment. Peptide stock solutions in concentrations from 200 fM to 20 μM are spiked into a plasma sample, immediately followed by quenching, peptide extraction, and MS analysis. The peptide concentrations are targeted to provide total ion counts of the peak from 2,000 to 5,000 (a.u.) at time "0" (see Note 8), which we find to be optimal for time-course measurement.
5. *Simultaneous test of two peptides*: Two peptide biomarkers can be tested simultaneously in the same sample, with this test providing a direct comparison of their stability relative to each other (Fig. 4a, b). The initial two peptide concentrations spiked in plasma samples for time-course incubations are not necessarily close to each other. These concentrations need to be determined by a pretest (see Note 4) and adjusted until their peak-height ratio is close to 1:1, to ensure minimal ion suppression of one to the other in MS.
6. *Replicates*: Four or more replicates were performed, in order to achieve data as quantitative as possible. In this case, if a

"hot-spot" occurs in MALDI-TOF MS, the outlying intensity from one replicate can be removed from the average prior to data analysis.

7. *Control peptide*: An ideal control peptide is a stable isotopically labeled peptide synthesized with the identical sequence of the tested biomarker (Fig. 4a). When the native (either intrinsic or spiked) peptide levels can be detected, the labeled control peptide is spiked to achieve an initial peak height of approximately 50% of the native peptide at "time 0." This initial ratio of two peptides can reduce ion suppression each other, and favoring the peptide measurements along which it is also considered that the native peptide will decrease in time-course experiments while the control peptide will remain constant (Fig. 3). When it is not possible to synthesize a stable isotopically labeled peptide, a different peptide with a molecular mass close to the tested biomarker can be used as the control (Fig. 3). To avoid degradation of the control peptide by intrinsic plasma proteolysis, it is recommended to add the control peptide either right after incubation followed immediately by quenching the sample, or right after quenching. If the control is soluble in quenching solution, both can be added together by dissolving the control peptide into the quenching solution (i.e., combining Subheading 3.1, steps 7–8). In the case where a linear relationship has been established between peak height of a peptide and its plasma concentration (e.g., FPA (20)), the peptide intensities can be directly analyzed for stability, possibly avoiding using a control peptide (Fig. 1).
8. *Time "0"*: Time "0" herein represents the beginning of a peptide–plasma incubation right after spiking the peptide biomarker, with immediate acid quenching of the sample. For other time points, the acid quenching defines the "stop" time for the incubation.
9. *A pretest to see if a peptide passes a size-based membrane filter*: A proper membrane filter is chosen according to the molecular weight of the peptide to be tested. The peptide is first dissolved into 15% ACN solutions in four concentrations (e.g., 0.1, 1, 10, 100 pM). 40–50 μL of each solution is loaded onto a sized-based membrane filter (e.g., YM-3 from Millipore), followed by centrifugation at 15,000×g for 10 min. 1.0 μL aliquot from each filtrate and nonfiltrate solution above the membrane (or solution before adding to the YM-3) is mixed with 1.0 μL of matrix and analyzed by MALDI-TOF MS. The peptide peak heights of two samples can be compared to see the efficiency of the peptide passing through the membrane. Also, comparison of spectra of four

different concentration samples provides an estimate of the sensitivity of peptide detection using MS. If the initial peptide is from a matrix sample complicated with salts or other compounds, a purification step (e.g., using Zip-Tip C18, Subheading 3.2, step 2) is required prior to MALDI-TOF MS analysis.

10. *Matrix for MALDI-based MS*: CHCA was selected for all peptides MALDI-TOF MS analysis. We observed that a saturated CHCA solution (e.g., 10 mg/mL in 50% ACN and 0.1% TFA) could easily precipitate upon mixing with plasma peptides and form heterogeneous crystals, and/or matrix "hot-spots." We also observed that using 5.0 mg/mL CHCA solution generated much better homogeneous crystals for MALDI-based MS analysis.

11. *MS instrument*: MALDI-TOF MS was chosen for this study with its advantage of spectrum simplicity (due to mostly a singly charged ion), cost- and time-efficiency. However, MALDI-based MS gives ~2 orders of magnitude in quantitative analysis. For a better quantification, other instruments, such as ESI-based MS, can be easily adapted in present method and procedures.

Acknowledgments

We are grateful for the encouragement and support of our proteomics program from Robert Pierce. We are also thankful for continuous supports and helps from Frank Augello, Dr. Kesava Nagar-Anthal, Dr. Paul Holmes, Dr. David Warunek, Lena Khumush, and Lisa Gevirtz.

References

1. Anderson NL, Anderson NG (2002) The Human proteome-history, character, and diagnostic prospects. *Mol. Cell Proteomics* **1**, 845–866.
2. Omen GS, States DJ, Adamski M et al (2005) Overview of the HUPO Plasma Proteome Project: results from the pilot phase with 35 collaborating laboratories and multiple analytical groups, generating a core dataset of 3020 proteins and a publicly available database. *Proteomics* **5**, 3226–3245.
3. States DJ, Omenn GS, Blackwell TW et al (2006) Challenges in deriving high-confidence protein identifications from data gathered by a HUPO plasma proteome collaborative study. *Nat Biotechnol.* **24**, 333–338.
4. Greeing DW, and Simpson RJ (2009) A centrifugal ultrafiltration strategy for isolating the low-molecular weight (≤25 k) component of human plasma proteome. *J. Proteomics*, **73**, 637–648.
5. Aebersold R, Anderson L, Caprioli R et al (2005) Perspective: a program to improve protein biomarker discovery for cancer. *J. Proteome Res.* **4**, 1104–1109.
6. Hortin GL (2006) The MALDI-TOF mass spectrometric view of the plasma proteome and peptidome. *Clin. Chem.* **52**, 1223–1237.
7. Ewers M, Mielke MM, Hampel H (2010) Blood-based biomarkers of microvascular pathology in Alzheimer's disease. *Exp. Gerontol*, **45**, 75–79.

8. Villanueva J, Martorella AJ, Lawlor K et al (2006) Serum peptidome patters that distinguish metastatic thyroid carcinoma from cancer-free controls are unbiased by gender and age. *Mol. Cell Proteomics* **5**, 1840–1852.
9. Mor G, Visintin I, Lai Y et al (2007) Serum protein markers for early detection of ovarian cancer. *PNAS* **102**, 7677–7682.
10. Lin Y, Goedegebuure PS, Tan MCB et al (2006) Proteins associated with disease and clinical course in pancreas cancer: a proteomic analysis of plasma in surgical patients. *J. Proteomce Res.* **5**, 2169–2176.
11. Bensalah K, Lotan Y, Karam JA et al (2008) New circulating biomarkers for prostate cancer. *Prostate Cancer and Prostatic Diseases* **11**, 112–120.
12. Metz TO, Qian W-J, Jacobs JM et al (2008) Application of proteomics in the discovery of candidate protein biomarkers in a diabetes autoantibody standardization program sample subset. *J. Proteome Res.* 7, 698–707.
13. Lorenzo O, Martín-Ventura JL, Blanco-Colio LM et al (2009) The proteomic approach in the development of prognostic biomarkers in atherothrombosis. *Recent Pat. Cardiovasc Drug Discov.* **4**, 25–30.
14. Fu Q, Van Eyk JE (2006) Proteomics and heart disease: identifying biomarkers of clinical utility. *Expert Rev. Proteomics* **3**, 237–49.
15. Rosalki SB, Roberts R, Katus HA et al (2004) Cardiac biomarkers for detection of myocardial infarction: perspectives from past to present. *Clin. Chem.* **50**, 2205–2213.
16. Diamandis EP (2006) Peptidomics for cancer diagnostics: present and future. *J. Proteome Res.* **5**, 2079–2082.
17. Yi J, Kim C, Gelfand CA (2007) Inhibition of intrinsic proteolytic activities moderates preanalytical variability and instability of human plasma. *J. Proteome Res.* **6**, 1768–1781.
18. Rai AJ, Gelfand CA, Haywood BC, et al (2005) HUPO plasma proteome project specimen collection and handling: towards the standardization of parameters for plasma proteome samples. *Proteomics* **5**, 3262–3277.
19. Yi J, Craft D, Gelfand CA (2011) Minimizing Pre-analytical and Analytical Variability in Plasma Proteomics using MALDI-TOF MS. Ed. Simpson RJ, Greening DW, *Serum/Plasma Proteomics, Methods Mol Biol* ***728***, Humana Press.
20. Yi J, Liu Z, Craft D et al (2008) Intrinsic peptidase activities causes a Sequential Multi-Step Reaction (SMSR) in digestion of human plasma peptides. *J. Proteome Res.* 7, 5112–5118.
21. Fujii K, Nakano T, Kawmura T et al (2004) Multidimensional protein profiling technology and its application to human plasma proteome. *J. Proteome Res.* **3**, 712–718.
22. Baumann S, Ceglarek U, Fiedler GM et al (2005) Standardized approach to proteome profiling of human serum based on magnetic bead separation and matrix-assisted laser desorption/ionization time-of-flight mass spectrometry. *Clin. Chem.* **51**, 973–980.
23. Villanueva J, Shaffer DR, Phiip I et al (2006) Differential exoprotease activities confer tumor-specific serum peptidome patters. *J. Clin. Invest.* **116**, 271–284.
24. Penno MAS, Ernst M, Hoffmann P (2009) Optimal preparation methods for automated matrix-assisted laser desorption/ionization time-of-flight mass spectrometry profiling of low molecular weight proteins and peptides. *Rapid Comm. Mass Spec.* **23**, 2656–2662.
25. Tirumalai RS, Chan KC, Prieto DA et al (2003) Characterization of the low Molecular Weight Human Serum Proteome. *Mol. Cell Proteomics* **2**, 1096–1103.
26. Annesley TM (2003) Ion suppression in mass spectrometry. *Clin. Chem.* **49**, 1041–1044.
27. Pikal-Cleland KA, Rodriguez-Hornedo N, Amidon GL et al (2000) Protein denaturation during freezing and thawing in phosphate buffer systems: monomeric and tetrameric beta-galactosidase. *Arch. Biochem. & Biophy.* **384**, 398–406.

Part III

Methods for Quantitative Assaying of Blood-Based Biomarkers

Chapter 11

Biomarker Validation in Blood Specimens by Selected Reaction Monitoring Mass Spectrometry of *N*-Glycosites

Reto Ossola, Ralph Schiess, Paola Picotti, Oliver Rinner, Lukas Reiter, and Ruedi Aebersold

Abstract

Targeted mass spectrometry using selected reaction monitoring (SRM) has emerged as the method of choice for the validation in blood serum, plasma, or other clinically relevant specimens of biomarker candidates arising from comparative proteomics or other discovery strategies. Here, we describe a method in which *N*-glycosites are selectively enriched from biological specimens by solid phase capture and PNGase F release, and then analyzed by SRM. Focusing the highly sensitive targeted mass spectrometry method on a subproteome enriched for secreted and shed proteins reproducibly identifies and quantifies such proteins in serum and plasma at the low nanogram per milliliter (ng/mL) concentration range. This protocol is intended to give an introduction to SRM-based targeted mass spectrometry with a special focus on the validation of biomarker candidates.

Key words: Glycoproteins, N-linked glycosylation, Biomarker validation, Mass spectrometry, Selected reaction monitoring, Selected proteotypic peptides

1. Introduction

Despite dramatic advances in cancer research and clinical research in general, the stage of detection of the disease remains a major predictor and determinant for treatment outcome, thus explaining the need for biomarkers for early cancer detection. Reliable biomarkers are also important for drug development, monitoring the efficacy of treatments, detecting the re-occurrence of the disease after treatment, and for identifying disease subtypes, commonly referred to as disease stratification.

The field of proteomics with its objective to comprehensively identify and quantify proteomes immediately raised high expectations for plasma biomarker discovery. However, to date biomarker

Richard J. Simpson and David W. Greening (eds.), *Serum/Plasma Proteomics: Methods and Protocols*, Methods in Molecular Biology, vol. 728, DOI 10.1007/978-1-61779-068-3_11, © Springer Science+Business Media, LLC 2011

discovery studies in which the proteome of disease affected and healthy control (serum or plasma) samples were compared, have achieved only modest success (1). Conventional mass spectrometry (MS)-based techniques only sample a relatively small fraction of the plasma proteome that mostly consist of the relatively abundant proteins. Both the large dynamic concentration range of up to 12 orders of magnitude for plasma proteins as well as the presence of very high abundance proteins such as serum albumin (35–50 mg/mL) and immunoglobulins (5–18 mg/mL), which mask the lower abundance plasma proteins, present major challenges for comprehensive plasma proteome analysis, especially for proteins below the microgram per milliliter concentration (2, 3). Furthermore, comparative plasma proteome analyses of disease affected and control samples are faced with the discrimination between disease relevant and coincidental proteome pattern differences. The latter can occur due to sample processing, instrument variability, or poor experimental design (4).

To overcome some of these limitations we developed a novel two-stage biomarker discovery and validation strategy. It consists firstly of the identification of candidate proteins by comparative tissue analysis via the selectively enriched *N*-glyco subproteome and, secondly, of the detection and quantification of biomarker candidates in plasma via selected reaction monitoring (SRM) (3). The detection of differentially abundant tissue glycoproteins is considerably less challenging than the detection of differentially abundant glycoproteins in plasma, due to the much more favorable protein composition in tissue, the closer proximity of the analyzed sample to the disease site, and the lower dynamic range (six orders of magnitude) of the tissue compared to the plasma proteome. We have developed a solid-phase extraction method for the isolation of *N*-glycosites (SPEG, deglycosylated peptides that are N-glycosylated in the intact protein) from plasma (5), tissues (6) and, more recently, also from the surface of intact cells (7). The extension of the *N*-glycosite isolation method to tissue and cell surface proteins is significant for the biomarker strategy because it facilitates the detection of disease-specific changes in these important subproteomes and to subsequently detect the same peptides in plasma by SRM. The focus on the *N*-glycoproteome not only reduces the complexity of the analytes, but also targets an information-rich subproteome which is relevant for remote sensing of the state of disease affected tissue in plasma (3). Using this approach we are able to consistently and reproducibly measure proteins in the low ng/mL concentration range (8).

SRM, a MS technique for the targeted quantification of selected proteotypic peptides (PTPs) in a complex sample matrix, is also known as multiple reaction monitoring (MRM) (9). PTPs serve as surrogates for the candidate protein and they have known fragmentation properties, a unique amino acid sequence for the protein, and are easily detectable in a mass spectrometer (9–11). The successful implementation of SRM requires multiple preparatory steps comparable to the generation of immunoassays for a set of proteins, except that SRM assays are generated and deployed in a triple quadrupole (QQQ) mass spectrometer and that no affinity reagents (i.e., antibodies) are required. Specifically, for each targeted protein the PTPs that provide optimal detectability and selectivity for the targeted proteins have to be identified and for each one of these peptides the optimal precursor ion to fragment ion transitions need to be determined. These parameters, along with the precursor ion mass, its predominant charge state, relative ion intensities, and the chromatographic retention time, constitute a definitive assay for the detection and, if a suitable reference peptide is available, the accurate absolute quantification of the respective peptide in a complex sample. Importantly, these assays have to be developed only once and can then be deployed indefinitely. We and others have achieved significant progress in the conceptualization and implementation of this technology on a proteome-wide scale. It has reached a state at which sets of targeted proteins can be confidently detected and quantified in complex samples, like whole cell lysates and human serum/plasma at unprecedented speed (reasonably up to thousand transitions, covering 100 proteins in a single analysis), sensitivity, and reproducibility (12–15).

There are several ways to assemble a menu of SRM assays for the quantitative analysis of biomarker candidates:

(a) The assays are already available and can be retrieved from publicly available resources. Our group is undertaking a substantial effort to generate and make publicly accessible SRM assays for thousands of human proteins. These assays are accessible through a data portal called MRMAtlas (16) (http://www.mrmatlas.org/). These downloadable SRM assays were generated using synthetic peptides as reference standards (17). MRMAtlas can be queried by protein description, by specific peptide sequences, or even by members of a certain pathway. From this data repository, one can assemble SRM assays for the proteins/peptides of interest and use them directly for the respective measurements or as a starting point for the optimization or refinement of SRM assays on the particular instrument used for the biomarker measurements.

(b) SRM assays are generated from LC–MS/MS data, spectral libraries, or predicted peptide sequences. In cases where no validated SRM assay is publicly available, the optimal transitions need to be defined from LC–MS/MS fragment ion spectra or by in silico selection of suitable peptides and transitions. Preferably, the peptide of interest has been already sequenced by a tandem mass spectrometer using CID. Large databases of quality controlled CID-spectra for peptides are publicly available, like the PeptideAtlas project (18, 19) or spectral libraries (20, 21). Alternatively, peptides identified by any LC–MS/MS experiment can be used. The most intense b- or y-fragment ions are selected from these data and used directly for the assay generation process. If no tandem MS data are available, transitions for peptides can be generated in silico. As a first step, PTPs are selected from protein sequences. Several tools are available that use the amino acid composition in order to predict the suitability of a peptide for LC–MS analysis (10, 11, 22). In a second step, the most intense fragment ions are selected. Such generated SRM assays need to be validated, by the measurement of endogenous peptides with a full fragment spectra on a QQQ instrument. If the endogenous peptides are present at low concentrations and in a complex sample matrix their use for SRM assay development is a challenging task and the use of synthetic peptide libraries is recommended (see below).

(c) Libraries of synthetic peptides are beneficial for the generation and optimization of SRM assays because they provide abundance equalized samples of known composition. Our group developed a workflow for multiplexed and fast SRM assay generation using unpurified synthetic peptide libraries (17). Mixtures of several hundred synthetic peptides are analyzed first by LC–MS/MS to generate CID spectra for each peptide. Applying directed mass spectrometry at this stage maximizes peptide coverage while minimizing the number of required LC–MS/MS runs (23). The LC–MS/MS data are searched using a common search engine against a sequence database and the peptide hits are validated using the Trans-Proteomic Pipeline (TPP) (24). Transitions for SRM analysis are finally extracted directly from CID-spectra, or from consensus spectra libraries (20, 21).

The latter approach allows fast developments of experimentally verified SRM assays, which can be used directly to analyze and quantify whole proteomes (15, 25) or to profile biomarker candidates in complex sample matrices. Here, we describe a detailed step-by-step procedure for the isolation of *N*-glycosites from human plasma/serum samples on the one hand, and for the development, validation, and verification of SRM assays for candidate proteins on the other hand (Fig. 1). More detailed information about SRM than covered by this chapter can be found in several publications (9, 26, 27).

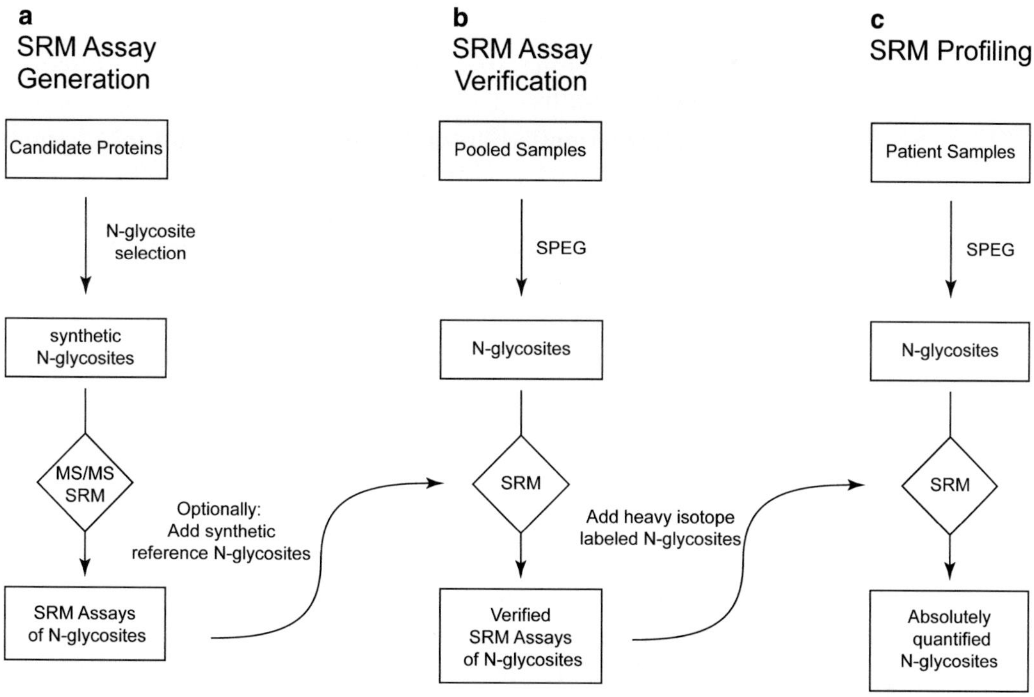

Fig. 1. The workflow is subdivided into three parts. SRM assays of biomarker candidate proteins are generated using synthetic peptides. This process (**a**) is done only once and assays can be used for all sample types though validated for matrix effects/noise. Secondly (**b**), the SRM assays are verified in pooled biological/clinical samples of interest. Assays for which no endogenous peptides were found can be removed from the used SRM assay library. Finally (**c**), the candidates are profiled by LC-SRM in individual samples using the verified SRM assays library.

2. Materials

All chemicals are from Sigma, if not specified otherwise.

2.1. Isolation of N-Glycosites from Human Serum or Plasma Glycoproteins

1. Denaturation buffer: 8 M urea, 0.1% SDS (w/v) 0.2 M Tris–HCl, pH 8.5. Has to be prepared upon usage.
2. Dithiothreitol (DTT) stock solution: Dissolve weighed dithiothreitol in HPLC-grade water to a final concentration of 1 M. Can be stored frozen for several months.
3. Iodoacetamide stock solution: Dissolve weighed iodoacetamide in HPLC-grade water to a final concentration of 0.5 M. Keep in dark; has to be prepared upon usage.
4. *N*-Acetyl-cysteine stock solution: Dissolve weighed *N*-acetyl-cysteine in HPLC-grade water to a final concentration of 0.5 M. Has to be prepared upon usage.
5. Sequencing grade modified trypsin, frozen (Promega).
6. SepPak™ C18 cartridge, 500 mg (Waters).

7. Vacuum concentrator.
8. Tris–HCl buffer: 0.2 M Tris–HCl, pH 8.5. Can be stored for several months.
9. Oxidation buffer: 20 mM sodium acetate, 100 mM sodium chloride, pH 4.5. Can be stored for several months.
10. Sodium meta-periodate stock solution: Dissolve weighed sodium meta-periodate (Thermo Fisher Scientific) with HPLC-grade water to a final concentration of 50 mM. Keep in dark, has to be prepared upon usage.
11. Coupling buffer: 100 mM sodium acetate, 1 M sodium chloride, pH 5.0. Can be stored for several months.
12. Affi-Prep Hz Hydrazide resin (Bio-Rad).
13. Sodium chloride solution: 5 M sodium chloride. Can be stored for several months.
14. Methanol, HPLC-grade.
15. HPLC-grade water or equivalent.
16. Phosphate buffer: 100 mM disodium-hydrogen phosphate, 25 mM EDTA, pH 7.1. Can be stored for several months.
17. *N*-Glycosidase F (PNGase F) of Flavobacterium meningosepticum, recombinant from *E. coli*, lyophilizate (Roche Applied Science). Dissolve in HPLC-grade water to a 1-U/μL concentration.
18. MicroSpin C18 columns, Vydac C18 (The Nest group Inc., Southborough, USA).
19. MS sample buffer: 5% (v/v) aqueous acetonitrile, 0.1% (v/v) trifluoroacetic acid. Can be stored for several months.

2.2. SRM Assay Generation and Targeted MS Analysis

1. SpikeTides™, nonpurified synthetic peptides (JPT Peptide Technologies GmbH, Berlin, Germany).
2. Fused silica capillaries, 10 cm×75 μm (BGB Analytik AG, Boeckten, Switzerland).
3. C18 AQ 5 μm resin (Michrom BioResources, Auburn, CA).
4. Applied Biosystem 4000 QTrap instrument (AB Sciex, Toronto, Canada).
5. Tempo nano LC system (Applied Biosystems Inc., Foster City, CA).
6. Solvent A: 3% (v/v) acetonitrile, 0.1% (v/v) formic acid in water, all HPLC-grade qualities. Can be stored for several months.
7. Solvent B: 3% (v/v) water, 0.1% (v/v) formic acid in acetonitrile, all HPLC-grade qualities. Can be stored for several months.

3. Methods

3.1. Selective Isolation of N-Glycosites

1. Dilute 50 μL serum or plasma with 300 μL denaturation buffer containing 8 M urea, 0.1% (w/v) SDS, 0.2 M Tris–HCl, pH 8.5.
2. Reduce the disulfide bonds with final 5 mM DTT for 1 h at 37°C. Add 1.8 μL of the prepared 1 M DTT stock solution to the sample.
3. Alkylate the free thiol-groups with final 25 mM iodoacetamide at room temperature and protected from light for 30 min. Add 18.5 μL of the freshly prepared 0.5 M iodoacetamide stock solution to the sample.
4. Let the excess of iodoacetamide react with final 30 mM *N*-acetyl-cysteine at room temperature for 15 min. Add 23.7 μL of the prepared 0.5 M *N*-acetyl-cysteine stock solution to the sample.
5. Dilute the sample with 0.2 M Tris–HCl buffer pH 8.5 to get a final urea concentration of 2 M.
6. Add trypsin in a protein:protease ratio of 100:1 to the sample, and incubate at 37°C overnight (16 h).
7. Clean up the digest using a reverse-phase SepPak™ 500 mg C18 (Waters) cartridge according to the supplier's protocol (see Note 1).
8. Concentrate the sample on a vacuum concentrator to a volume of 10 μL (see Note 2).
9. Dilute the sample with oxidation buffer containing 20 mM sodium acetate, 100 mM sodium chloride, pH 4.5 to final 100 μL (see Note 3).
10. Oxidize the sugar moiety of glycosylated peptides by incubating the sample with final 8 mM sodium meta-periodate. Add 19 μL of the freshly prepared 50 mM sodium meta-periodate solution and incubate the sample at 6°C and protect it from light for 60 min.
11. Remove the excess sodium meta-periodate by reverse-phase SepPak™ 500 mg C18 (Waters) according to the manufacturer's protocol, and finally concentrate the sample again on a vacuum concentrator to a volume of 10 μL (see Notes 1 and 2).
12. Dilute the oxidized sample with coupling buffer containing 100 mM sodium acetate, 1 M sodium chloride, pH 5.0 to a final volume of 100 μL (see Note 3).
13. Equilibrate 50% (w/v) hydrazide slurry in coupling buffer (see Notes 4–6).

14. Combine 500 µL of the equilibrated hydrazide slurry with the oxidized sample and shake the slurry at room temperature overnight (16 h).
15. Remove the supernatant (nonglycosylated peptides) and wash the resin thoroughly with 400 µL of 5 M sodium chloride (ten times), water (five times), 80% (v/v) acetonitrile in water (ten times), methanol (ten times), water (five times) and finally with phosphate buffer (100 mM disodium-hydrogen phosphate, 25 mM EDTA, pH 7.1) (ten times) (see Notes 5 and 7).
16. Resuspend the resin in 400 µL phosphate buffer containing 100 mM disodium-hydrogen phosphate, 25 mM EDTA, pH 7.1, and add 5 µL of a 1-U/µL solution of PNGase F double-distilled water (see Note 8).
17. Incubate the sample at 37°C overnight (16 h).
18. Collect the supernatant and wash the resin twice with 400 µL 80% (v/v) acetonitrile in water (see Note 5).
19. Combine the washes with the collected supernatant and concentrate the sample using a vacuum concentrator to final 10 µL volume (see Note 2).
20. Perform a final sample clean up using a MicroSpin C18 column (The Nest Group Inc., Southboro, USA) according to the manufacturer's protocol (see Note 1).
21. After evaporating the organic from the eluate, dilute the sample to 50 µL with sample buffer (5% acetonitrile, 0.1% TFA) and inject 1 µL per LC–MS analysis.

3.2. Generation of the Peptide Library for SRM Analysis

1. Select PTPs between 7 and 22 amino acids of length (see Note 9).
2. Select peptides previously observed in shotgun proteomics experiments, either on the basis of own acquired data, or based on the information present in public proteomic data repositories, such as PeptideAtlas (see Note 10).
3. For proteins not present in PeptideAtlas, select a set of at least five peptides with good MS properties by bioinformatic prediction, using the publicly available tools (10, 22).
4. Order synthetic peptides of the final set of *N*-glycosites (17). Pay attention to order the deamidated sequence, in which the asparagines carrying the carbohydrate are converted to aspartic acids.
5. Resuspend the peptides in 20% (v/v) aqueous acetonitrile, 1% (v/v) formic acid, vortex for 20 min, and sonicate for 15 min (see Note 11).
6. Pool up to 100 peptides, evaporate the acetonitrile on a vacuum centrifuge, and dilute them in 0.1% (v/v) formic acid to get acceptable ion counts on the mass spectrometer (see Note 2).

7. Optionally, spike a set of synthetic peptides with elution times spanning the whole solvent gradient into each sample to facilitate the correlation of relative retention times between LC–MS/MS runs (see Notes 11 and 12).

3.3. Generation of SRM Assays

1. Calculate one transition for the doubly and triply charged ion for every peptide, corresponding to the first fragment ion of the y-series with m/z greater than $[m/z\ \text{precursor} + 20\ m/z]$.
2. Analyze peptide samples on a hybrid triple quadrupole/ion trap mass spectrometer 4000 or 5500 QTrap, equipped with a nanoelectrospray ion source. Chromatographic separation of peptides is performed on a Tempo nanoLC system (Applied Biosystems) coupled to a 15-cm fused silica emitter, 75 μm diameter, packed with Magic C18 AQ 5 μm resin (see Note 13).
3. Load the peptides on the column from a cooled (4°C) Tempo autosampler and separate them with a linear gradient of acetonitrile/water, containing 0.1% (v/v) formic acid, at a flow rate of 300 nL/min. Use a gradient from 5 to 30% (v/v) acetonitrile in 45 min.
4. Calculate the collision energies according to the formulas: $CE = 0.044 \times m/z + 5.5$ and $CE = 0.051 \times m/z + 0.55$ (CE, collision energy, m/z, mass-to-charge ratio of the precursor ion) for doubly and triply charge precursor ions, respectively (9).
5. Operate the mass spectrometer in selected reaction-monitoring mode, triggering acquisition of a full MS2 spectrum upon detection of an SRM trace (threshold 500–1,000 ion counts).
6. Perform SRM acquisition with Q1 and Q3 operated at unit resolution (0.7 m/z half maximum peak width) with 200 or 300 transitions (dwell time 10 or 7 ms/transition, respectively) per run.
7. Acquire MS2 spectra in enhanced product ion (EPI) mode for the highest SRM transitions, using dynamic fill time, a low Q1 resolution, a scan speed of 4,000 amu/s, m/z range of 300–1,400, and two scans summed.
8. Generate a decoy database from the used peptides and append it to the target database, consisting of the concatenated peptide sequences. Search the MS2 data with a search engine against this database. Set the precursor mass tolerance to 2.0 Da and fragment mass tolerance to 0.8 Da.
9. Validate the search results using a false-discovery rate of less than 1% ($FDR < 1\%$) according to Elias and Gygi (28) (see Note 14).
10. Create a spectral library from the spectrum-peptide matches using SpectraST (20, 21). Use these consensus fragment ion spectra as a reference to derive the optimal coordinates of

each SRM assay (e.g., best responding fragments, fragment relative intensities, peptide elution time) (see Notes 15 and 16).

11. For each consensus-spectrum, select the most intense five to ten fragment ions as optimal SRM transitions. Exclude fragments due to neutral loss from the precursor. Discard fragments with m/z values close to the precursor ion m/z ($|m/z_{Q1} - m/z_{Q3}| \leq 5\, m/z$), as such transitions result in high noise levels.
12. Derive collision energies associated to each transition from the formulas given above (step 4). Extract additional features, such as fragment relative intensities and peptide elution for each peptide from the corresponding MS/MS data.

3.4. Validation and Optimization of SRM Assays

1. Analyze the peptide pools again preferably in time-resolved SRM mode using the same conditions as described above with a 30-min instead of a 45-min gradient (see Note 17).
2. Validate the acquired transitions for each transition group by considering co-elution of transitions, their relative intensities, and shapes of the peaks as qualifiers (see Note 18) (9, 29).
3. If possible, determine the exact elution time of the peptides in relation to the spiked reference peptides (see Note 19).
4. Optimize the LC method so that the gradient length results in peak widths of about 0.3 min.
5. Optionally, split the full SRM assay library to several acquisition methods. Consider peak widths, number of co-eluting transitions, and the sample complexity (see Note 20).
6. Optionally, optimize MS instrument parameters according to the work of Sherwood et al. (30) (see Notes 21–23). Put the main focus on the collision energies. Use the calculated collision energies as a starting point, and ramp the volt number by −8, −4, +4, and +8 V (17).

3.5. Verification of SRM Assays in Serum or Plasma Samples

1. Prepare pools of patient samples, which represent the sample cohort as closely as possible.
2. If available, spike heavy labeled synthetic peptides to the pooled samples (see Note 24) and subsequently analyze them in order to uniquely identify the endogenous peptides at low signal-to-noise level. Use starting concentrations of 10–100 fmol synthetic peptide per microliter in cases where the expected peptide abundance and the ionization efficiency are unknown (see Notes 11 and 25).
3. Optionally, spike the reference peptides used for retention time normalization into the sample, and correct the elution times of SRM assays based on the elution times of these peptides (see Notes 11 and 12).

4. Analyze the samples by time-resolved SRM as described.
5. Confirm the endogenous peptides by the following criteria: co-elution of all transitions of the peptide, relative intensity ratio of the transitions compared to the expected ratios, and the variation of the measured from the expected retention time (29).
6. Remove SRM assays for endogenous peptides that have not been detected in the sample from the library (see Note 26).

3.6. Quantitative Analysis of Biomarker Candidates in Serum or Plasma

7. Spike heavy-isotope labeled and quantified peptides into the individual samples (see Notes 11, 24, 25, and 27). Adjust the spiking amount on the basis of the measured endogenous peptide abundances in the pooled samples.
8. Optionally, spike the reference peptides used for retention time normalization to the sample, and correct the elution times of the SRM assays based on the elution times of these peptides (see Notes 11 and 12).
9. Analyze each sample, process and analyze the resulting data files as described.
10. Calculate the absolute endogenous protein concentrations according to the ratio of the known spiked amount of heavy-isotope labeled peptides (see Note 28).

4. Notes

1. For the various C18 clean-up procedures, we use not more than 5% acetonitrile in the aqueous buffer for column equilibration, sample loading, and column washing. Furthermore, we use a peptide elution buffer of 40% acetonitrile, 0.1% formic acid to elude the peptides and to minimize the elution of hydrophobic lipids.
2. If the sample was dried completely, use sonication and some organic solvent (up to 20% (v/v) acetonitrile) to assist resolubilization. After resolubilization, evaporate organic and concentrate sample using a vacuum concentrator, or dilute the sample with 0.1% (v/v) formic acid in water.
3. If necessary, adjust the pH in the sample using either a 1-M sodium acetate stock solution, or 10% (v/v) acetic acid in water.
4. Use filter tubes (e.g., Mobicol™ from MoBiTec) or 96-well filter plates (e.g., Sirocco filter plate from Waters) to reduce the loss of resin during the whole procedure and to increase sample throughput.

5. For washing and equilibrating the resin, always add the new solvent to the resin, vortex the slurry thoroughly, and remove the supernatant.
6. Try the alternative Affi-Gel Hydrazide immunoaffinity resin (Bio-Rad) in cases where the Affi-Prep hydrazide resin causes blockage of filter tubes. However, we recommend using the Affi-Prep product since the polymeric resin is less sticky and more stable.
7. In order to improve the isolation specificity for *N*-glycosites, we recommend extensive vortexing for every washing step and to use prewarmed washing solutions and solvents (37°C). Additionally, an overnight (16 h) pre-incubation of the resin in phosphate buffer with three additional washes using the same conditions also improves the specificity according to our experience.
8. Alternatively, purified and glycerol-free PNGase F from New England Biolabs or a combination of thereof can be used in order to improve the yield and capture specificity.
9. Select either at least three experimentally observed PTPs or at least five predicted PTPs per protein for the initial set of peptides.
10. Do not consider peptide identifications that are derived from experiments involving labeling reagents other than cysteine alkylation reagents. Such modifications have the potential to alter the nature of the peptide significantly and thus alter the fragmentation patterns.
11. Attention has to be paid to the following points in order to minimize the loss of synthetic peptides during pipetting. It is beneficial to keep the peptide in a solvent containing some degree of organic, e.g., 20% (v/v) acetonitrile. Furthermore, try to keep peptides as concentrated as possible before adding them to the sample. Using glass vials and minimizing the pipetting steps also helps to reduce peptide binding to surfaces. Also avoid multiple thaw and freeze cycles.
12. If no preferred peptides for retention time normalization are available, we recommend using a set of nine synthetic peptides (AAVYHHFISDGVR, HIQNIDIQHLAGK, TEVSSNHVLIYLDK, GGQEHFAHLLILR, TEHPFTVEEFVLPK, TTNIQGINLLFSSR, NQGNTWLTAFVLK, LVAYYTLIGASGQR, ITPNLAEFAFSLYR) (17).
13. As an alternative to the workflow described in this protocol using the AB Sciex QTrap, peptides can be sequenced on any tandem mass spectrometer, either in automated selection of precursor ions for fragmentation, or using inclusions lists (23), as long as suitable measures are taken to confirm the specificity in the QQQ instrument used for SRM quantification analysis.

14. Search results are validated like in conventional shotgun data analysis. Decoy hits are enumerated as function of the score (e.g., ion score in Mascot or Sequest score). The proportion of hits to decoy peptides*2 is the false discovery rate. For small databases care must be taken that the scoring is not biased toward the target sequences. Sequest, for instance, uses a delta-Cn score which can become artificially high if only few candidate hits are present. In this case, it is better to append the target peptide sequences to a larger database, for instance, a bacterial proteome.
15. Several software packages are available to create transition lists from spectral libraries, examples are Skyline (31), which can also be used to analyze the MRM data and MaRiMba (32) for transition list generation
16. As an alternative, well-matching individual fragment spectra can be used to generate SRM assays.
17. Any other QQQ mass spectrometer can be used for the SRM experiments.
18. Recently, concerns regarding redundancy in m/z-pairs for individual PTPs came up (33). Considering not only m/z-pairs as qualifiers, but also other physico-chemical properties like co-elution of transitions especially in combination with heavy labeled reference peptides will likely overcome this issue.
19. Having accurately determined elution times allow the use of time-resolved SRM, in which many more transitions can be analyzed in a single run, compared to nontime-resolved SRM. In nontime-resolved SRM, the mass spectrometer cycles through the whole list of transition. If the list of transitions is long, either very long cycle times or very low dwell times (decreasing the sensitivity) are required. In time-resolved SRM, also called scheduled SRM or dynamic SRM, an elution range is defined for each transition, and the mass spectrometer acquires data for these particular peptides only during this specific time window. Therefore, either a higher number of transitions or more favorable dwell times can be used in time-resolved SRM.
20. Too many co-eluting peptides will either reduce the dwell time for the transitions (if the cycle time is kept constant) or will reduce the number of measure points on the peak (if the dwell times are kept constant). However, a gradient that is too long will result in peak broadening and increased analysis time. Assuming peak widths of 0.3 min and a constant cycle time of 2 s, 200 co-eluting transitions will result in a dwell time of about 10 ms for each transition and about eight to ten measure points on every transition peak.

21. There are two different processes in a mass spectrometer which need to be optimized. After the ionization by ESI the ions need to be pushed into the mass spectrometer, a process that is accomplished by applying a voltage between the ESI-Emitter and the entrance of the mass spectrometer. A too high voltage leads to increased in-source fragmentation, which results in a loss of precursor ion intensity. In contrast, if the voltage is too low a loss of precursor ions due to low transfer efficiency is observed. The generation of fragment ions is performed by colliding the precursor ion with gas molecules, like nitrogen or helium, in the second quadrupole. The precursor ions get accelerated right before the collision by another voltage, called the collision energy (CE). Having this voltage set too high leads to increased fragmentation, whereas insufficient fragmentation results from too low voltage.
22. To get the best optimization, one can infuse synthetic peptides directly into the mass spectrometer one-by-one and ramp each instrument parameter of interest. Optimizing hundreds of peptides in this way adds up in time considerably.
23. Alternatively, vendor-specific workflows (e.g., Themo Fisher – Pinpoint™, AB Sciex – MRM Pilot™, Agilent – MassHunter Optimizer™, or others) or open-source packages, like Skyline (31), can be used to optimize MRM assays.
24. A heavy-isotope labeled peptide has the same chemical property as the endogenous peptide, but differs in the mass and is therefore distinguishable in a mass spectrometer. Spiking such heavy-isotope labeled peptides in a sample allows the quantification through the comparison of the peak areas/intensities of the endogenous and heavy-isotope labeled peptides. Absolute quantification is possible, if the heavy-isotope labeled peptides are in very pure states and the quantities are certified (e.g., by amino acid analysis).
25. Always check the heavy labeled peptides for isotopical purity. Generally, adapt the spiking amount of the individual reference peptides to the amount of endogenous target peptides to avoid false-positive signals.
26. Peptides with unfavorable elution profiles, such as bad resolution, or interfering noise in the heavy or light transitions are excluded from further data analysis.
27. Special care has to be taken at which step the heavy-isotope peptides are inserted to the sample processing workflow. In general, the heavy-isotope labeled peptides are spiked to the sample as early as possible. However, since we focus in this workflow, on *N*-glycosites, spiking the heavy-isotope labeled peptides before the sample preparation process is not possible. We therefore add the peptides earliest before the last clean up by reverse-phase columns.

28. Use the ratio of peak heights from heavy and light in order to correct for spray efficiency and ionization differences between runs using the following formula:

$$\frac{\mathrm{Lx}(\mu\mathrm{L})}{\mathrm{Hx}(\mu\mathrm{L})} \times r \times [H(\mathrm{fmol}/\mu\mathrm{L})] \times [\mathrm{Pr}(\mathrm{kDa})] = [L(\mathrm{ng}/\mathrm{mL})]$$

where Lx is the ion count light transition, Hx is the ion count heavy transition, r is the yield, the average recovery of the *N*-glycosite peptides was determined to be 50% (r=0.5) despite the involved sample preparation protocol as demonstrated by spiking known amounts of a specific protein into the plasma prior to its processing (8), [H] is the concentration of AQUA peptide in fmol/μL, [Pr] is the molecular weight (MW) of protein in kDa, and [L] is the concentration of endogenous protein in fmol/μL.

References

1. Ahn, S.M., and Simpson, R.J. (2007) Body fluid proteomics: Prospects for biomarker discovery, *Proteomics Clin. Appl.* **1**, 1004–1015.
2. Anderson, N. L., and Anderson, N. G. (2002) The human plasma proteome: history, character, and diagnostic prospects, *Mol Cell Proteomics* **1**, 845–867.
3. Schiess, R., Wollscheid, B., and Aebersold, R. (2009) Targeted proteomic strategy for clinical biomarker discovery, *Mol Oncol* **3**, 33–44.
4. White, C. N., Chan, D. W., and Zhang, Z. (2004) Bioinformatics strategies for proteomic profiling, *Clin Biochem* **37**, 636–641.
5. Zhang, H., Li, X. J., Martin, D. B., and Aebersold, R. (2003) Identification and quantification of N-linked glycoproteins using hydrazide chemistry, stable isotope labeling and mass spectrometry, *Nat Biotechnol* **21**, 660–666.
6. Zhang, H., Liu, A. Y., Loriaux, P. et al. (2007) Mass spectrometric detection of tissue proteins in plasma, *Mol Cell Proteomics* **6**, 64–71.
7. Wollscheid, B., Bausch-Fluck, D., Henderson, C. et al. (2009) Mass-spectrometric identification and relative quantification of N-linked cell surface glycoproteins, *Nat Biotechnol* **27**, 378–386.
8. Stahl-Zeng, J., Lange, V., Ossola, R. et al. (2007) High sensitivity detection of plasma proteins by multiple reaction monitoring of N-glycosites, *Mol Cell Proteomics* **6**, 1809–1817.
9. Lange, V., Picotti, P., Domon, B., and Aebersold, R. (2008) Selected reaction monitoring for quantitative proteomics: a tutorial, *Mol Syst Biol* **4**, 222.
10. Mallick, P., Schirle, M., Chen, S. S., et al. (2007) Computational prediction of proteotypic peptides for quantitative proteomics, *Nat Biotechnol* **25**, 125–131.
11. Kuster, B., Schirle, M., Mallick, P., and Aebersold, R. (2005) Scoring proteomes with proteotypic peptide probes, *Nat Rev Mol Cell Biol* **6**, 577–583.
12. Addona, T. A., Abbatiello, S. E., Schilling, B. et al. (2009) Multi-site assessment of the precision and reproducibility of multiple reaction monitoring-based measurements of proteins in plasma, *Nat Biotechnol* **27**, 633–641.
13. Anderson, L., and Hunter, C. L. (2006) Quantitative mass spectrometric multiple reaction monitoring assays for major plasma proteins, *Mol Cell Proteomics* **5**, 573–588.
14. Kuzyk, M. A., Smith, D., Yang, J. et al. (2009) Multiple reaction monitoring-based, multiplexed, absolute quantitation of 45 proteins in human plasma, *Mol Cell Proteomics* **8**, 1860–1877.
15. Picotti, P., Bodenmiller, B., Mueller, L. N. et al. (2009) Full dynamic range proteome analysis of S. cerevisiae by targeted proteomics, *Cell* **138**, 795–806.
16. Picotti, P., Lam, H., Campbell, D., Deutsch, E. W. et al. (2008) A database of mass spectrometric assays for the yeast proteome, *Nat Methods* **5**, 913–914.

17. Picotti, P., Rinner, O., Stallmach, R. et al. (2010) High-throughput generation of selected reaction-monitoring assays for proteins and proteomes, *Nat Methods* 7, 43–46
18. Desiere, F., Deutsch, E. W., Nesvizhskii, A. I. et al. (2005) Integration with the human genome of peptide sequences obtained by high-throughput mass spectrometry, *Genome Biol* **6**, R9.
19. Deutsch, E. W., Lam, H., and Aebersold, R. (2008) PeptideAtlas: a resource for target selection for emerging targeted proteomics workflows, *EMBO Rep* **9**, 429–434.
20. Lam, H., Deutsch, E. W., Eddes, J. S. et al. (2007) Development and validation of a spectral library searching method for peptide identification from MS/MS, *Proteomics* 7, 655–667.
21. Lam, H., Deutsch, E. W., Eddes, J. S. et al. (2008) Building consensus spectral libraries for peptide identification in proteomics, *Nat Methods* **5**, 873–875.
22. Fusaro, V. A., Mani, D. R., Mesirov, J. P., and Carr, S. A. (2009) Prediction of high-responding peptides for targeted protein assays by mass spectrometry, *Nat Biotechnol* **27**, 190–198.
23. Schmidt, A., Claassen, M., and Aebersold, R. (2009) Directed mass spectrometry: towards hypothesis-driven proteomics, *Curr Opin Chem Biol* **13**, 510–517.
24. Keller, A., Eng, J., Zhang, N., Li, X. J., and Aebersold, R. (2005) A uniform proteomics MS/MS analysis platform utilizing open XML file formats, *Mol Syst Biol* **1**, 2005 0017.
25. Malmstrom, J., Beck, M., Schmidt, A. et al. (2009) Proteome-wide cellular protein concentrations of the human pathogen Leptospira interrogans, *Nature* **460**, 762–765.
26. Kitteringham, N. R., Jenkins, R. E., Lane, C. S. et al. (2009) Multiple reaction monitoring for quantitative biomarker analysis in proteomics and metabolomics, *J Chromatogr B Analyt Technol Biomed Life Sci* **877**, 1229–1239.
27. Pan, S., Aebersold, R., Chen, R. et al. (2009) Mass spectrometry based targeted protein quantification: methods and applications, *J Proteome Res* **8**, 787–797.
28. Elias, J. E., and Gygi, S. P. (2007) Target-decoy search strategy for increased confidence in large-scale protein identifications by mass spectrometry, *Nat Methods* **4**, 207–214.
29. Reiter, L., Rinner, O., Picotti, P., et al. (2011) mProphet: A general and flexible data model and algorithm for automated SRM data processing and statistical error estimation, *Nat Methods*, submitted.
30. Sherwood, C. A., Eastham, A., Lee, L. W. et al. (2009) Rapid optimization of MRM-MS instrument parameters by subtle alteration of precursor and product m/z targets, *J Proteome Res* **8**, 3746–3751.
31. MacLean, B., Tomazela, D. M., Shulman, N. et al. (2010) Skyline: an open source document editor for creating and analyzing targeted proteomics experiments, *Bioinformatics* **26**, 966–968.
32. Sherwood, C. A., Eastham, A., Lee, L. W. et al. (2009) MaRiMba: a software application for spectral library-based MRM transition list assembly, *J Proteome Res* **8**, 4396–4405.
33. Sherman, J., McKay, M. J., Ashman, K., and Molloy, M. P. (2009) How specific is my SRM?: The issue of precursor and product ion redundancy, *Proteomics* **9**, 1120-1123.

Chapter 12

A Fluorescent Microsphere-Based Method for Assay of Multiple Analytes in Plasma

Oliver K. Bernhard, Rommel A. Mathias, Thomas W. Barnes, and Richard J. Simpson

Abstract

Measurement of multiple analytes can provide increased sensitivity and specificity for the detection and management of disease. The *e*nzyme-*l*inked *i*mmuno*s*orbent *a*ssay (ELISA) is currently the "gold standard" for protein quantification; however, individual assays for each analyte must be performed, placing demand on sample volume. On the contrary, multiplex assays using microsphere-based technologies allow for multiple analytes to be simultaneously assayed within a single sample. Here, we present a protocol for the preparation and development of a multiple-analyte assay in human plasma using the BioPlex 200 platform (Bio-Rad), which incorporates xMAP technology (Luminex).

Key words: Fluorescent beads, Multiplex assay, Biomarker validation, Flow cytometry, Plasma

1. Introduction

The ability to quantify proteins in a complex mixture is a technical challenge in clinical chemistry, especially when faced with the problem of assaying multiple analytes. While the *e*nzyme-*l*inked *i*mmuno*s*orbent *a*ssay (ELISA) is the "gold standard" for routine analyte quantitation in clinical samples (1), multiplex microsphere-based assays are gaining support, due largely to their ability to quantitate several analytes in a single biospecimen sample simultaneously (for a review, see ref. 2).

The ELISA, first developed during the 1970s (1), typically employs a double-antibody sandwich method, in which an immobilised antibody is used to "capture" the analyte of interest. A "detection" antibody, which is coupled to an enzyme, then binds the analyte, and the enzyme catalyses the conversion of an added substrate to a coloured product. The colour intensity is a

Richard J. Simpson and David W. Greening (eds.), *Serum/Plasma Proteomics: Methods and Protocols*, Methods in Molecular Biology, vol. 728, DOI 10.1007/978-1-61779-068-3_12, © Springer Science+Business Media, LLC 2011

proportional measure of protein concentration. A key advantage of ELISAs is the reproducibility of results (1); however, since they were never designed for multiple-analyte measurement, their usefulness is limited in instances where the quantity of biospecimens is scarce, e.g. plasma samples for biospecimen research.

1.1. Multiplex Microsphere-Based Assays

The past 30 years have seen the development of solid-phase assays that utilise capture antibodies immobilised on microspheres. The first of these, described by Horan and Wheeless in 1977, was used to determine analytes in serum and other fluids (3). Technology has since progressed, allowing microspheres to be used for multiplex applications. Microsphere-based assays are estimated to have greater sensitivity (100 fg/mL) than conventional ELISAs (50 pg/mL) (4); given that the microsphere–antibody complex is suspended in liquid, this facilitates liquid-phase reaction kinetics (5). Currently available systems include the cytometric bead assay (BD Biosciences), coupled particle light scattering (Diasorin), and xMAP technology (Luminex) (for a review, see ref. 2). Of these systems, Luminex technology can detect the most number of analytes simultaneously. It has been applied recently to a number of clinical applications, including cytokine and hormonal quantification (6, 7).

1.2. Luminex xMAP Technology

Luminex xMAP technology utilises 5.6-μm diameter polystyrene microspheres, each "set" of which is internally "dyed" with a different ratio of red and fluorescent fluorophores, to create a unique spectral signature. The first version of this technology, released in 1997, had 64 microsphere sets and was theoretically capable of detecting 64 different analytes from a single sample (8, 9). More recently, this expanded to 100 different microsphere sets. Due to the recent incorporation of a third internal dye, up to 500 analytes may be simultaneously assayed.

Various surface chemistries may be employed to link a capture reagent of interest to the microsphere surface. For example, antibodies are immobilised on microspheres via an amide bond, which forms from the condensation reaction of a carboxylate group attached to the microsphere and an amine group present on lysine or arginine residues within the antibody.

The assay is performed as a double-antibody sandwich format. The capture antibody is immobilised on the bead, and the detection antibody is biotinylated, enabling a streptavidin–phycoerythrin (SA–PE) conjugate to bind (Fig. 1). Fluorescent dyes such as phycoerythrin have been demonstrated to have increased sensitivity when compared to the coloured product formed during the ELISA (10).

The Luminex system is essentially a modified flow cytometer that uses two lasers focused at the same point in the detection chamber. Suspended microspheres are channelled in a single file

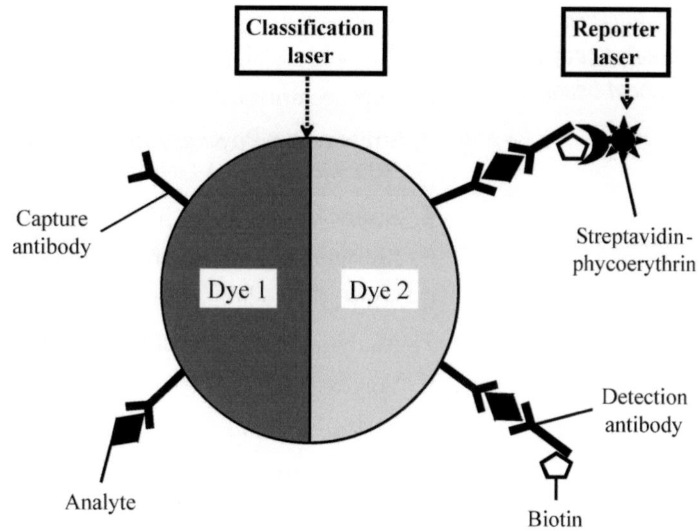

Fig. 1. Fluorescent microsphere-based assay. Each bead contains two dyes at defined concentrations to give them a unique spectral signature detectable by the classification laser. The capture antibody is conjugated to the bead, and the plasma sample added for analyte binding. The biotinylated detection antibody then binds the analyte at a secondary site, and addition of streptavidin–phycoerythrin enables detection by the reporter laser. Analyte concentration is calculated from the mean fluorescence intensity.

into the detection chamber. The classification laser is used to identify the microsphere class present, while the reporter laser concurrently excites the phycoerythrin conjugate. Pre-installed software then deconvolutes the signals detected by both lasers to generate a measurement of mean fluorescence intensity (MFI) for each individual microsphere.

This protocol describes the measurement of the concentration of an analyte in human plasma using Luminex technology and in the BioPlex 200 system (Bio-Rad). Absolute analyte concentration can be determined by comparing sample MFI against that of a calibration curve made from standards.

2. Materials

Assay buffer (required throughout the protocol) containing 1× PBS, pH 7.4, 0.5% (w/v) BSA, and 0.1% (w/v) sodium azide. Dissolve 10 g BSA (Sigma) and 2 g sodium azide in 2.0 L PBS. Filter through a 0.22-μm filter and keep sterile. This can be made up in bulk and stored at 4°C.

2.1. Generation of Calibration Standard

1. Purified protein standard (see Note 1).
2. Sterile PBS.
3. 1.5-mL microcentrifuge tubes.

2.2. Conjugation of Capture Antibody to Fluorescent beads

1. Fluorescent beads, Cat. #171-606xx (Bio-Rad), Cat. #LC10xxx-01 (Luminex) (x, bead number) (see Note 2).
2. Capture antibody (see Note 3).
3. Streptavidin-Phycoerythrin (SA-PE), Cat. #554061 (BD Biosciences) (see Note 4).
4. Amine coupling kit, Cat. #171-406001 (Bio-Rad).
5. Refrigerated microcentrifuge (set at either RT or 4°C).
6. Sonicating water bath (at RT).
7. Analytical balance with 0.1-mg precision.
8. Vortex.
9. Aluminium foil.
10. 1.5-mL microcentrifuge tubes.
11. Micro Bio-Spin 6 Column, Tris chromatography columns, Cat. #732-6221 (Bio-Rad).
12. 1-Ethyl-3-(3-dimethylaminopropyl) carbodiimide, Cat. #22980 (Pierce).
13. *N*-hydroxysulfosuccinimide, Cat. #24510 (Pierce).
14. Light microscope and haemacytometer or electronic cell counter (Invitrogen "Countess™ automated cell counter").
15. 96-well ELISA Microplate, Flat-bottom Cat. #353279 (BD Falcon).

2.3. Plasma Sample Preparation

1. Human plasma (see Note 5) or other samples for analysis (see Note 6).
2. 1.5-mL microcentrifuge tubes.
3. Refrigerated microcentrifuge.

2.4. Assay Protocol

1. 96-well Filter Plate, Cat. #40-008 (Millipore), Cat. #5029 (Pall) (see Note 7).
2. Aurum™ Vacuum Manifold, Cat. #732-6470 (Bio-Rad).
3. Biotinylated detection antibody (see Note 8).
4. SA-PE conjugate, as per Section 2.2, Step 3.
5. Fixation solution: Add 1.0 mL of 37% paraformaldehyde stock solution to 36 mL PBS.
6. Bio-Plex calibration, Cat. #171-203060, and validation, Cat. #171-203001, kits (Bio-Rad).
7. Vortex.
8. Aluminium foil.
9. Multichannel pipette.
10. Pipette tips (see Note 9).
11. 1.5-mL microcentrifuge tubes.
12. Microplate shaker.

2.5. Data Acquisition and Analysis

1. Bio-Plex 200 system (Bio-Rad) with Bio-Plex Manager 5.0 software.
2. Bio-Plex Manager user guide.

3. Methods

3.1. Generation of Calibration Standard

This should be performed prior to running the assay. Prepare a 100 μg/mL master stock solution of the analyte in sterile PBS. Store at −80°C in 100 μL aliquots. Thaw one aliquot and dilute 100 μL master stock with 9.9 mL assay buffer to create a 1 μg/mL working stock solution. Freeze in 100 μL aliquots (see Note 10).

3.2. Conjugation of Capture Antibody to Fluorescent Beads

Coupling of the capture antibody to beads must be completed prior to running the assay. Conjugate the antibody to the selected functional beads by following the manufacturer's protocol (see Notes 11 and 12).

3.3. Plasma Sample Preparation

1. Thaw samples on ice. Check that all the samples are transparent/clear. If the samples are cloudy or contain a precipitate, spin for 10 min at maximal speed in a microcentrifuge, and transfer the supernatant into a fresh tube without disturbing the pellet. Discard the pellet.
2. Dilute the plasma samples fourfold in assay buffer by adding 20 μL plasma to 60 μL assay buffer.

3.4. Assay Protocol

1. Perform a one in three dilution series of the calibration standard. Thaw one vial of the working stock (1 μg/mL). Mark this vial #1. Add 100 μL assay buffer to seven additional vials marked #2–#8. Make a serial dilution by adding 50 μL of one stock to the second vial and vortex, with 50 μL of the #2 solution removed and added to #3. Continue diluting each standard until vial #8.
2. Determine the number of wells required. Calibration standards should be measured in duplicate, there should be six blank samples, and plasma samples are measured in triplicate. Figure 2 shows a typical layout of a plate in which six samples are to be measured. A total of 40 wells are required for this assay.
3. Vortex the bead stock for 15 s, and dilute beads in assay buffer. The use of 5,000 beads per target per well is recommended (see Note 13). If more than one analyte is measured, combine the beads against the different analytes at this point. For example, if the concentration of beads is 5×10^6/mL, for 40 wells, the requirement is (50,000 beads)/(5,000 beads/μL) = 40 μL.

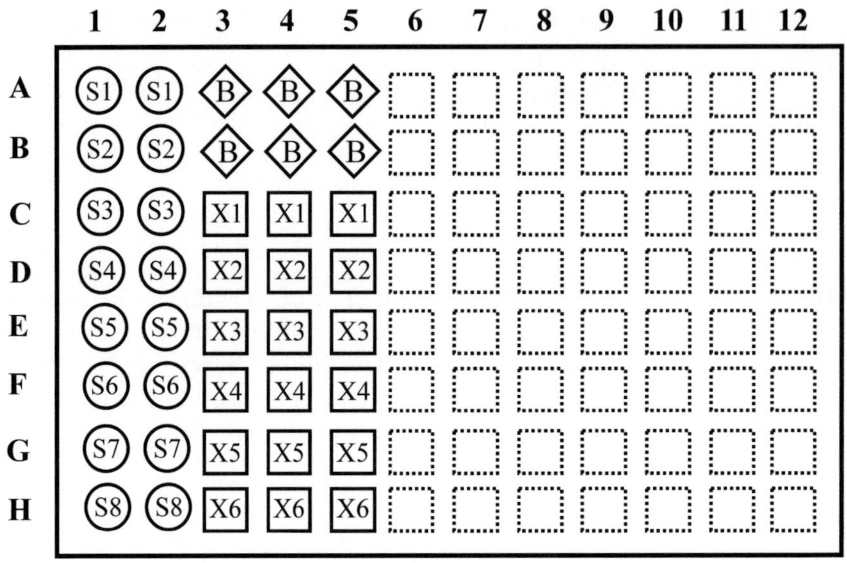

Fig. 2. Plate layout for an assay measuring analyte concentration in six plasma samples. The eight calibration standard sample (S) dilution series are indicated as *circles* and measured in duplicate. The six blank samples (B) are shown as *diamonds*. The six plasma samples (X) are depicted as *squares*, and measured in triplicate.

Hence add 40 μL bead suspension to 1.96 mL assay buffer. It is advisable to prepare stock solution for few more wells than required.

4. Add 150 μL assay buffer to each well.
5. Vortex bead suspension and add 50 μL to each well.
6. Filter wash the plate by placing it into the vacuum manifold and applying a gentle vacuum to pass the solution in the wells through the membrane into the reservoir (see Note 14).
7. Add 100 μL assay buffer to each well.
8. Repeat steps 6 and 7, and filtration (step 6) again.
9. Add 20 μL of the calibration standards, blanks, or plasma samples to the respective wells as outlined in Fig. 2 (see Note 15).
10. Seal the plate, wrap in aluminium foil, and shake for 30 s (high speed oribital) and for 30 min (low speed orbital).
11. Dilute the detection antibody in assay buffer. Based on a stock concentration of 0.1 mg/mL, for 40 wells, dilute 10 μL antibody solution with 9.9 mL assay buffer to give a final concentration of 1 μg/mL.
12. Repeat steps 6 and 7 three times, and step 6 again.
13. Vortex detection antibody. Add 25 μL of detection antibody solution to each well.
14. Seal the plate, wrap in aluminium foil, and shake for 30 s (high speed oribital) and for 30 min (low speed orbital) (see Note 16).

15. Dilute SA–PE in assay buffer: For 40 wells, use 8 μL of SA–PE stock (0.5 mg/mL) and 1.992 mL of assay buffer to give a final concentration of 2 μg/mL.
16. Repeat steps 6 and 7 three times, and step 6 again.
17. Vortex the SA–PE solution and add 50 μL to each well.
18. Seal the plate, wrap in aluminium foil, and shake for 30 s (high speed oribital) and for 10 min (low speed orbital).
19. Repeat steps 6 and 7 three times, and step 6 again.
20. Add 100 μL of fixation solution to each well (see Note 17).
21. Incubate for 30 min.
22. Repeat step 6, and add 125 μL assay buffer to each well.

3.5. Data Acquisition and Analysis

1. Turn the instrument on at least 1 h prior to data acquisition. Start the Bio-Plex Manager and select instrument warm-up. This will take approximately 30 min. See the Bio-Plex Manager software user guide for details on the different steps.
2. Perform instrument start-up.
3. If necessary, perform a calibration or validation (see Note 18).
4. Select "New Protocol". Enter name and description.
5. Select "Select Analyte". If you are running an analyte for the first time, then create a new analyte panel.
6. Format the plate according to the sample layout (Fig. 2).
7. Enter the concentrations for the calibration standards for every analyte. Ensure that the most dilute standard concentration has three significant digits (i.e., enter the concentration of the first standard as 1,000,000 pg/mL) (see Note 19).
8. Enter the details for controls and samples. Samples are of unknown concentration, whereas controls are samples of known concentration, which allows the evaluation of assay accuracy.
9. Load the plate and begin acquisition (see Note 20).

The Bio-Plex Manager 5.0 software comes with a comprehensive data analysis software. A series of fitting models for the calibration standards is available (including linear fit). The recommended model is the five-parameter logistic (5PL) model which takes into account the behaviour of the curve at both low and high analyte concentration, hence covering a larger concentration range.

10. Open the acquired data file.
11. Select the desired curve fitting model. The 5PL model is recommended.
12. Check the measured values for outliers which may have occurred due to experimental errors. To set outliers, open the

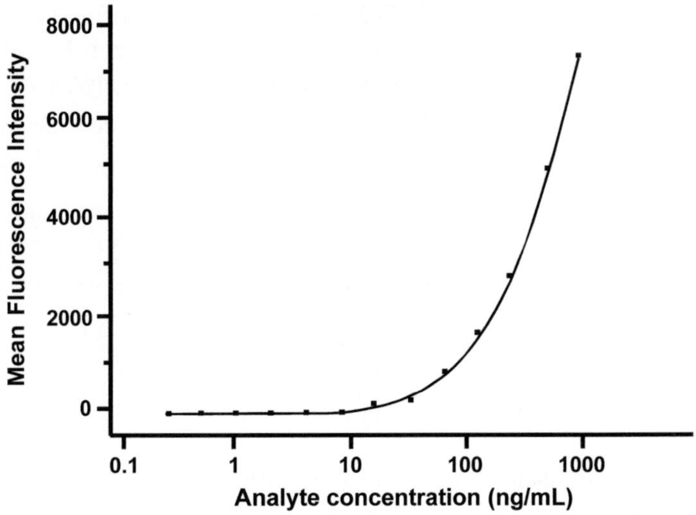

Fig. 3. A typical calibration curve for a standard. The *curve* is automatically calculated by the Bio-Plex Manager 5.0 software.

report table and select "Show replicates". Right-click on the sample number on the left side and select "Set outlier". This sample will no longer be included in the calculation.

13. View the standard curve to evaluate the fit (see Fig. 3).
14. Add the desired parameters to the report table by right-clicking anywhere on the top row of the results table and selecting "Show column". Usually, the observed concentration and the accompanying standard deviation or coefficient of variation (CV) are selected. For hints on troubleshooting, see Note 21.

4. Notes

1. We found R&D Systems (Minneapolis, Minnesota) as a useful source for recombinant proteins and antibodies.
2. When selecting beads for analytes which will be measured simultaneously (multiplexed), bead regions that are not adjoining should be selected. If the regions are next to each other on the bead map, bead signals that are just outside one region (for example, due to bead aggregation) may appear inside the region immediately next to it, compromising the accuracy of the reading in this region.
3. The capture antibody to be coupled to beads must be in a buffer free of sodium azide, BSA, glycine, Tris, or any other amine-containing compound because the coupling chemistry immobilises the proteins via their amine groups. If the capture

antibody is in a buffer containing any such compounds, the buffer must be exchanged with a suitable coupling buffer such as PBS, pH 7.4. A protocol for buffer exchange using Micro Bio-Spin™ 6 Tris columns is provided in the manual of the Bio-Plex amine coupling kit. Alternatively, the protein can be dialyzed against PBS, pH 7.4, overnight or buffer-exchanged using a centrifugal ultrafiltration device.

4. The PE conjugate is required to detect the capture antibody. As most capture antibodies are mouse monoclonal, an anti-mouse Ig–PE conjugate is usually required.
5. Samples can be either prepared fresh on the day of the analysis or prepared as frozen stocks and thawed immediately prior to analysis. We usually prepare a large number of frozen stocks for each sample. Blood and blood-related products may contain harmful pathogens such as HIV or hepatitis viruses. Strict adherence to biosafety rules when working with blood and blood products is, therefore, mandatory. If the contamination status of a sample is not clear, it must be treated as if it was infectious. Waste must be disposed of safely.
6. The analyte concentration can also be measured in tissue culture media or detergent cell lysates:

   Tissue culture medium: Cells must be plated at an optimal seeding density. Density and time of cultivation must be kept constant when different states (e.g. drug treated vs. control) are investigated. We plate cells at 2×10^6 cells per 150-mm dish in 25 mL of media (RPMI with 10% FCS) and incubate for up to 96 h. The supernatant is then collected and spun for 5 min at $480 \times g$ to collect dead cells and for 10 min at $2{,}000 \times g$ to collect other debris. The supernatant should be filtered (0.22-μm syringe-driven filter unit) and stored at −20°C. Tissue culture medium can be used undiluted; however, if it is anticipated to contain a high concentration of analyte, we recommend diluting it to a concentration that lies within the measurable range of the assay.

   Cell lysate: (used to measure cell-associated biomarkers): Prepare cell lysate using a suitable protocol (for example, see ref. (11) and pass the lysate through a 0.22-μm syringe-driven filter unit. The use of detergents such as SDS should be avoided as they may interfere with antibody–analyte binding. They may also promote bead aggregation during the acquisition, which interferes with the measurements. We have used samples containing 0.5% sodium deoxycholate and 0.1% SDS successfully, but we suspect that these detergents promote bead aggregation. The use of milder detergents, such as Triton X-100 and NP-40, and dilution of the samples of at least 20-fold is, therefore, recommended.

7. The filter plates should be low protein binding to avoid loss of analyte and have a pore size well below the diameter of the microspheres (i.e. 0.45 μm).
8. When a monoclonal antibody is used as the detection antibody, it needs to bind the analyte on an independent epitope in relation to the epitope for the capture antibody. If a polyclonal antibody is used, there is an increased chance of cross-reactivity with other analytes when the assay is multiplexed. For multiplex assays, antibodies should be highly specific with no cross-reactivity.
9. We have found that the quality of pipette tips used during the assay (e.g. dilution of the calibration standard and addition of beads, samples, and detection reagents to the wells) is crucial as they influence measurement precision. Therefore, high-quality tips (e.g. Gilson) should be used. We found long-nosed gel-loading tips Cat. #223-9915 (Bio-Rad) to be a suitable and less expensive alternative.
10. The concentration of the analyte in the calibration standard should be around the upper limit of quantification for the assay. This value is different for every combination of an antigen with two antibodies and must be determined first by testing the response for different concentrations of the antigen. We found it suitable to prepare a master stock of 100 μg/mL solution and generate calibration standards at 1 μg/mL from this. Recombinant proteins can be stably stored for at least 3 months at –80°C. The actual assay range is determined by the quality of the antibodies used for capture and detection. Depending on the antibodies, it may be necessary to use a higher concentration than 1 μg/mL as a calibration standard. If the antibodies are of high quality and the assay is very sensitive, a calibration standard with a lower concentration needs to be used and the samples may have to be further diluted.
11. Because of the moisture-sensitive nature of the EDC and Sulfo-NHS required for coupling, it is recommended to perform several coupling reactions in parallel. Alternative protocols for coupling of proteins to COOH beads without the need for purchasing a kit can be found on the Luminex website: http://www.luminexcorp.com/uploads/data/Protein%20Protocols%20FAQs/Protein%20Coupling%20Protocol%200407%2010207.pdf.
12. The beads are sensitive to photobleaching from light sources. Avoid prolonged exposure to light sources by wrapping the tubes and plates in aluminium foil.
13. We have used as few as 2,500 beads per well in our experiments. While reducing experimental costs, using fewer beads may also increase the sensitivity of the assay by increasing the number of analyte molecules captured on each bead. However, using fewer beads can lead to prolonged instrument run

times, while saturation of the beads at comparatively low analyte concentrations may also occur.

14. It is important to use only a low vacuum pressure, since a vacuum that is too high may compromise the filter membrane and render the plate inoperative. Only a very low vacuum pressure should initially be applied and carefully increased until the fluid in the well (~100 μL) drains through the membrane in about 2–3 s.

15. The samples should be vortexed prior to use. To reduce detection of debris by the instrument, the samples should be filtered through a 0.22-μm filter or centrifuged for 10 min in a microcentrifuge at maximum speed.

16. It is recommended to turn the instrument on at this point and start the warm-up routine (see Bio-Plex Manager software user guide for details).

17. This step is required only if biohazardous samples (e.g. human plasma) are used. If the status of a sample is not known, it should always be treated as biohazardous.

18. We do not perform a daily calibration routine. We restrict calibrations to once a month unless required by the instrument (e.g. if a change in temperature was detected). We also found that four drops of the calibration beads plus 80 μL of PBS are sufficient for a successful calibration, which increases the number of calibrations per kit by 25%.

19. The Bio-Plex Manager software contains a calculator function at the right side of the window for the calibration of samples. It will calculate all the concentrations of the different dilutions by simply entering the concentration of the first calibration standard and the dilution factor.

20. If bead aggregation occurs, the signal will appear at the edge or outside of the acceptable bead region of the map towards the top right side. This effect leads to incorrect fluorescent readings for the reporter dye. If bead aggregation occurs, abort the run by clicking "stop" and perform a "wash between plates" routine (see the Bio-Plex Manager user guide for details). The assay plate can then be re-inserted and the run resumed by running the remainder of the plate in "recovery mode".

21. Detailed troubleshooting information for the Bio-Plex system can be found in the Bio-Plex Manager Software user guide, on the Bio-Rad website (http://www.bio-rad.com/cmc_upload/Literature/39187/4006205D.PDF; refer to Section 5) or the Luminex website (http://www.luminexcorp.com/support/tech_tips.html). Our major problems with development of novel bead-based assays were bead aggregation (see Note 20), cross-reactivity when multiplexing, and detection limits too high for plasma-based detection of the analyte. Cross-reactivity

occurs, for example, when the detection antibody for analyte 1 also binds to an epitope on analyte 2 resulting in a higher reading for analyte 2 in a multiplexed experiment compared to that in a monoplex experiment. This is especially prevalent when polyclonal antibodies are used for detection since they are not specific to one epitope. Cross-reactivity can be eliminated only by using specific antibodies and must be evaluated every time two analytes are multiplexed.

The detection limit of bead-based assays is determined predominantly by the quality of the antibodies. The sensitivity may be increased by the use of fewer beads, higher sample volumes, optimization of wash conditions, incubation times, mixing order, and quantities of capture and detection antibodies used. For significant improvement of the sensitivity, however, it is necessary to replace one or both of the antibodies with ones that have a higher affinity for the analyte.

Acknowledgements

We acknowledge the support from the National Health and Medical Research Council (NH & MRC) through Program Grant #487922, and a Peter Doherty Fellowship to O.K.B. (#381411).

References

1. Leng S.X., McElhaney J.E., Walston J.D., et al. (2008) ELISA and multiplex technologies for cytokine measurement in inflammation and aging research. *J. Gerontol.* **63**, 879–884.
2. Elshal M.F., McCoy J.P. (2006) Multiplex bead array assays: performance evaluation and comparison of sensitivity to ELISA. *Methods (San Diego, Calif.)* **38**, 317–323.
3. Horan P.K., Wheeless L.L., Jr. (1977) Quantitative single cell analysis and sorting. *Science (New York, N.Y.)* **198**, 149–157.
4. Vignali D.A. (2000) Multiplexed particle-based flow cytometric assays. *J. Immunol. Methods* **243**, 243–255.
5. Jani I.V., Janossy G., Brown D.W., Mandy F. (2002) Multiplexed immunoassays by flow cytometry for diagnosis and surveillance of infectious diseases in resource-poor settings. *Lancet Infect. Dis.* **2**, 243–250.
6. de Jager W., te Velthuis H., Prakken B.J., Kuis W., Rijkers G.T. (2003) Simultaneousdetection of 15 human cytokines in a single sample of stimulated peripheral blood mononuclear cells. *Clin. Diagn. Lab. Immunol.* **10**, 133–139
7. Liu M.Y., Xydakis A.M., Hoogeveen R.C., et al. (2005) Multiplexed analysis of biomarkers related to obesity and the metabolic syndrome in human plasma, using the Luminex-100 system. *Clin. Chem.* **51**, 1102–1109.
8. Kettman J.R., Davies T., Chandler D., Oliver K.G., Fulton R.J. (1998) Classification and properties of 64 multiplexed microsphere sets. *Cytometry* **33**, 234–243.
9. Fulton R.J., McDade R.L., Smith P.L., Kienker L.J., Kettman J.R., Jr. (1997) Advanced multiplexed analysis with the FlowMetrix system. *Clin. Chem.* **43**, 1749–1756.
10. Chesterton K.A., Pretl K., Sholander J.T., et al. (2003) Rapid and reliable detection of HLA-specific antibodies with the luminex platform. *Human Immunol.* **64**, S108–S108.
11. Kurzinger K., Springer T.A. (1982) Purification and structural characterization of LFA-1, a lymphocyte function-associated antigen, and Mac-1, a related macrophage differentiation antigen associated with the type three complement receptor. *J. Biol. Chem.* **257**, 2412–12418.

Chapter 13

Immuno-Mass Spectrometry: Quantification of Low-Abundance Proteins in Biological Fluids

Vathany Kulasingam, Christopher R. Smith, Ihor Batruch, and Eleftherios P. Diamandis

Abstract

Mass spectrometry is emerging as one of the most promising analytical techniques to examine simultaneously hundreds of analytes quickly, precisely, and accurately, using minute sample volumes. Currently, a major bottleneck in the verification phase of putative biomarkers is the lack of methods/reagents to quantify low levels of analytes in biological fluids. A major objective is to establish a high-throughput multiple reaction monitoring (MRM) assay capable of quantifying low-abundance proteins or peptides in biological fluids (low μg/L range) using mass spectrometry. The experimental procedure we propose, called immuno-mass spectrometry, consists of immuno-capturing analytes of interest from relevant biological fluids in 96-well microtiter plates and performing in-well tryptic digestion, with subsequent MRM of digested peptides on a triple quadrupole mass spectrometer. With such a strategy, limits of detection of 0.1–1 μg/L proteins in serum with a coefficient of variation of <20% can be obtained. This methodology could be adapted quickly and easily to potential candidates of interest, thus providing a much needed technology to bridge the gap between discovery and validation platforms.

Key words: Biomarkers, Multiple reaction monitoring, Quantification, Targeted proteomics, Verification phase, Immuno-mass spectrometry

1. Introduction

There is an emerging interest to quantify proteins and peptides in biological fluids using mass spectrometry (MS) since it provides a rapid and specific assay platform for analyte quantification. The premise is that a peptide that is cleaved from a protein can serve as a representation of the concentration of that intact protein (1). This peptide, referred to as a signature or proteotypic peptide, must be unique to the protein of interest (2). Classically, these experiments are performed on a triple quadrupole mass spectrometer

Richard J. Simpson and David W. Greening (eds.), *Serum/Plasma Proteomics: Methods and Protocols*,
Methods in Molecular Biology, vol. 728, DOI 10.1007/978-1-61779-068-3_13, © Springer Science+Business Media, LLC 2011

in multiple reaction monitoring (MRM) mode. In this approach, the discriminating power of mass analyzers to select a specific analyte is utilized to transmit only ions of a specific mass-to-charge ratio (m/z), which are fragmented and some of the resulting product ions monitored in a two-stage mass selection process. To achieve absolute protein quantification, an internal standard peptide is required to spike into a sample so that the ratio between the synthetic and endogenous peptide can be determined (3–5).

However, the sensitivity of MS-based serum protein quantification assays is limited by the dynamic range of most mass spectrometers (6). Therefore, an enrichment strategy involving immunodepletion, gel-based, or chromatographic techniques is often coupled to MRM assays to increase the sensitivity of detection. The studies published thus far utilizing mass spectrometry as a quantification tool have successfully achieved detection limits of 300–500 μg/L for proteins in serum with minimal or no purification steps (7, 8). However, for an assay to be clinically relevant, the detection limits should be in the low μg/L range.

To achieve this level of sensitivity with MS, a purification/enrichment step is essential. But it is also equally important for these MRM assays to be relatively simple to perform, robust, and high-throughput. Notably, antibody capture of the analyte is one of the most sensitive purification methodologies (9). Combining antibody immunoextraction with MRM assays may achieve low μg/L sensitivity. In a recent study, using an immuno-MS (immunoextraction in combination with mass spectrometry) approach, with minimal prefractionation steps and potentially preparing up to 96 samples simultaneously on a microtiter plate, we were able to demonstrate that MRM can be performed on an ion-trap mass analyzer using a purified recombinant protein digest to achieve a limit of detection (LOD) of 10 attomoles with a coefficient of variation (CV) of <20% (10). Furthermore, we successfully quantified the protein of interest in serum down to less than 1 μg/L, with a detection limit 0.1 μg/L.

2. Materials

2.1. Recombinant Proteins and Antibodies

1. If needed, reconstitute the carrier-free human recombinant protein of interest according to the manufacturer's package insert, ideally in a solvent that is compatible with chromatography/mass spectrometry.
2. If needed, reconstitute the monoclonal or polyclonal antibody of interest according to the manufacturer's package insert.

2.2. Reagents for Immunoextraction on 96-Well Microtiter Plate (See Note 1)

1. 200 mM ammonium bicarbonate (Fisher Scientific, #A643-500).
2. Reagent-grade methanol and deionized water.
3. 20 mM dithiothreitol (DTT) (Sigma–Aldrich, #D9779).
4. 100 mM iodoacetamide (Sigma–Aldrich, #I1149).
5. Sequencing-grade modified trypsin (Promega, #V5111). Dissolve 20 μg of trypsin in 160 μL of 200 mM ammonium bicarbonate (stock concentration of 0.125 μg/μL trypsin solution stored at –20°C). Check the pH of the freshly made solution and adjust accordingly to pH 8.
6. 96-Well white polysterene microplate; high binding, flat bottom (Greiner Bio One, #655074).
7. Omix pipette C18 tip (Varian, Inc. #A57003MB).
8. Coating buffer: 50 mM Tris, pH 7.8.
9. Washing solution: 0.9% NaCl (Fisher Scientific, #S271).
10. Automated 96-well plate washer (Bio-Tek Instruments, ELx405).
11. C18 extraction (Omix pipette tip):
    (a) Buffer A: 0.1% formic acid in water.
    (b) Buffer B: 65% ACN and 0.1% formic acid in water.

2.3. HPLC and Mobile Phases

1. HPLC instrument: Nano-LC system (Proxeon Easy-nLC or equivalent) (capillary liquid chromatography).
2. Columns: 2.5-cm C18 trap column trimmed from a 5-cm IntegraFrit C18 column (inner diameter 150 μm, 300 Å pore size, 5 μm, New Objective #IF150-50-50 H002) and a 5-cm analytical PicoFrit C18 column (inner diameter 75 μm, 90 Å pore size, 2.7 μm, New Objective #PF7515-H035) with a 15-μm tip.
3. Mobile phases: Phase A: 0.1% (v/v) formic acid in an aqueous water solution,and phase B: 0.1% (v/v) formic acid in ACN.

2.4. Mass Spectrometer and Related Items

1. Mass spectrometer: A triple quadrupole mass analyzer (Quantum Ultra, Thermo Fisher).
2. Ionization source: Nano-electrospray ionization.
3. Database search software (Mascot, Matrix Science).

2.5. Synthetic Peptides

1. Heavy-labeled (^{13}C or ^{15}N) synthetic peptides of interest [internal standard, Thermo Scientific, Custom HeavyPeptides AQUA Kit (BSD-10137166, http://www.thermo.com/heavypeptides)].
2. Synthetic peptides (for transition optimization).

3. Methods

For simplicity, the protocol described below will be for one protein of interest (Protein X). However, this method can be expanded to include simultaneous quantification of multiple proteins (3, 4). The experimental procedure requires having one antibody towards Protein (3,4). The choice of a monoclonal antibody or polyclonal antibody will vary depending on the analyte of interest. The overall objective is to establish a high-throughput MRM assay capable of quantifying low-abundance proteins in biological fluids (low ng/L–μg/L range) using mass spectrometry.

3.1. Proteotypic Peptide Selection

The objective of this step is to select two to three proteotypic peptides that display the most intense signal to represent Protein X for quantification purposes using mass spectrometry. We recommend that steps 2–5 be performed on an ion trap mass analyzer, rather than on a triple quadrupole mass analyzer. This information can be obtained by in silico analysis or empirically derived (see Notes 2–5). Below is a detailed protocol for empirically derived proteotypic peptide selection.

1. Digest 10 μg of purified human recombinant Protein X:
   - (a) To a microcentrifuge tube containing 10 μg of purified recombinant protein, add 48 μL of freshly made ammonium bicarbonate (200 mM).
   - (b) Add 40 μL water, 10 μL methanol, and 2 μL DTT (20 mM).
   - (c) Incubate the sample at room temperature for 30 min with shaking.
   - (d) Add 2 μL iodoacetamide (100 mM) and incubate at room temperature for 1 h, shaking in the dark (cover with aluminum foil).
   - (e) Add 0.25 μg trypsin and incubate the sample at 37°C overnight (~16 h).
   - (f) Stop the trypsin digestion process by the addition of 5 μL of 10% (v/v) aqueous formic acid.
2. Inject the digested peptide solution (quantity or volume) into the LC–MS system, operating in data-dependent mode. Refer to Subheading 3.3 for LC gradient and instrument settings.
3. Search the raw data generated with a search algorithm (Mascot, Matrix Science; version 2.1.03) containing appropriate search parameters and database.
   - (a) Analyze the resulting Mascot Generic Files with a forward and reverse concatenated International Protein

Index human database (version 3.62; 1,67,894 entries forward and reverse combined).

(b) Select trypsin as the digestion enzyme.

(c) Perform search with fixed carbamidomethylation of cysteines and variable oxidation of methionine residues selected.

(d) Indicate a fragment tolerance of 0.4 Da and a precursor tolerance of 7 ppm.

(e) Parameters for DTA creation: minimum mass, 300 Da; maximum mass, 4,000 Da; automatic precursor charge selection; minimum peaks, ten per MS/MS scan for acquisition; and minimum scans per group, 1.

4. Inspect the list of identified peptides for peptides containing methionine and cysteine and for miscleavages, and remove these peptides from further consideration (see Note 3 for exceptions to this rule). Additional information regarding guidelines for selection of proteotypic peptides can be found in the literature (11–13).
5. Generate an extracted ion chromatogram (XIC) for each of the peptides identified to inspect which peptides have the highest signal intensity.
6. On the Quantum Ultra (or an equivalent triple quadrupole mass analyzer), perform a selected ion monitoring (SIM) scan on five to seven highest intensity peptides observed for Protein X. It should be noted that some proteins may yield fewer overall peptides and after removal of low-intensity proteotypic peptides, you may be left with a much smaller pool of peptides to work with.
7. From the SIM analysis, select the top two to three proteotypic peptides with the highest signal intensity and chromatographic peak shape for further analysis (12).

3.2. Transitions Selection

1. Use the digested recombinant protein (or alternatively, use the two to three light-labeled synthetic proteotypic peptides) for this step.
2. Using a static spray tip, directly infuse the peptide mixture into the mass spectrometer (triple quadrupole).
3. In tune mode, tune on the peptide of interest in MS mode (optimize capillary temperature, tube lens offset, and spray voltage):

   (a) If the final assay will measure more than one peptide at a time, a specific temperature that will satisfy all peptides must be selected.

4. In MS/MS mode, optimize at least five product ions (transitions) of highest relative intensity (define optimal collision

energies for each). Select product ions with *m/z* values greater than the *m/z* values of the precursor ion (see Notes 6–8):

   (a) After this is completed, the optimized products must correspond to b- or y-type ions.

5. After the transitions have been identified, create a method that contains the transition information.
   (a) Run the method.
   (b) Select the top three to four transitions with the highest signal intensity (good signal-to-noise ratio) for each proteotypic peptide. If all transitions have similar intensity, then select those with *m/z* values greater than precursor *m/z* values.
   (c) Note that all the transitions should have the same retention time for each proteotypic peptide.

At the end of this step, two to three proteotypic peptides to represent Protein X should have been selected and three to four product ions should have been identified for each peptide. Select only the top performing proteotypic peptide (consistently identified, intense signal, reproducible fragmentation, etc.) for further MRM method development.

3.3. MRM Method Development and Verification Studies

The objective of this step is to develop an MRM method, including defining HPLC conditions for the proteotypic peptide and its associated three to four product ions selected from above to monitor Protein X. Once the method is set up, verification studies are needed to characterize the performance of the method using synthetic peptides:

1. Create an MRM method in the Xcalibur software method editor for the Quantum Ultra (or equivalent software method for another triple quadrupole mass analyzer). Some of the typical MS settings are as follows:
   (a) Scan width (*m/z*): 0.002.
   (b) Scan time (s): 0.1.
   (c) Resolution (peak width at full-width half-maximum): Q1: 0.2 and Q3: 0.7.
   (d) Microscans: 1.
   (e) Skimmer offset (V): 7.
   (f) Data type: Centroid.
   (g) Polarity: Positive.
2. Nano-LC conditions: A 40-min dual column (precolumn and analytical column) run consisting of a 7.5-min sample load, 23-min discontinuous gradient, and 5-min analytical and 4.5-min precolumn equilibration step. The flow rate should be

8 μL/min for sample loading/precolumn equilibration, 0.8 μL/min for analytical column equilibration, and 0.4 μL/min for the gradient.

3. Linearity study:
   (a) Using synthetic light-labeled peptides, inject increasing concentrations of the purified peptide (or amounts on column). This should be performed in triplicate. For example, one can inject 0, 0.1, 0.5, 1, 10, and 100 fmol on a column, or equivalent concentrations in μg/L.
4. Determining the LOD and limit of quantification (LOQ):
   (a) LOD is defined as the lowest concentration of analyte that can be discriminated from zero with 95% confidence (analyte concentration at signal of background plus two standard deviations). The background signal can be obtained via injection of the mobile buffers as unknown specimens or by immuno-purification of a complex matrix which does not contain Protein X. The "background" can be determined by calculating the level of noise on both sides of the chromatographic peak. The Xcalibur software will determine the background and integrate the peaks.
   (b) LOQ is defined as the concentration of analyte that can be measured with <20% CV.
5. Precision study:
   (a) Within-day precision analysis: measure ten replicate analysis of two different levels of peptide mixture spanning the dynamic range of the method (as established in the linearity study). For example, inject 0.5 and 10 fmol of peptide mixture.
   (b) Between-day precision analysis: measure the two peptide levels two times per day for 5 days.
   (c) Calculate the imprecision of the method at the two levels.

3.4. Immunoextraction Coupled to MRM (Immuno-MS Approach)

1. Add 100 ng/well of antibody in coating buffer to a 96-well microplate (200 μL), seal the microplate tightly so that no evaporation occurs, and leave the plate overnight at room temperature (see Notes 9–12) (Fig. 1).
2. Wash the plate five times with 0.9% NaCl washing solution (200 μL each); this can be performed using the automated washing machine or manually (see Note 13). Gently pat down the plate on a paper cloth to ensure that residual salt solution is removed.
3. Load 100 μL of serum/plasma/standard to each well (it is recommended to perform all sample analyses in triplicate). To generate a standard curve, use the recombinant protein to

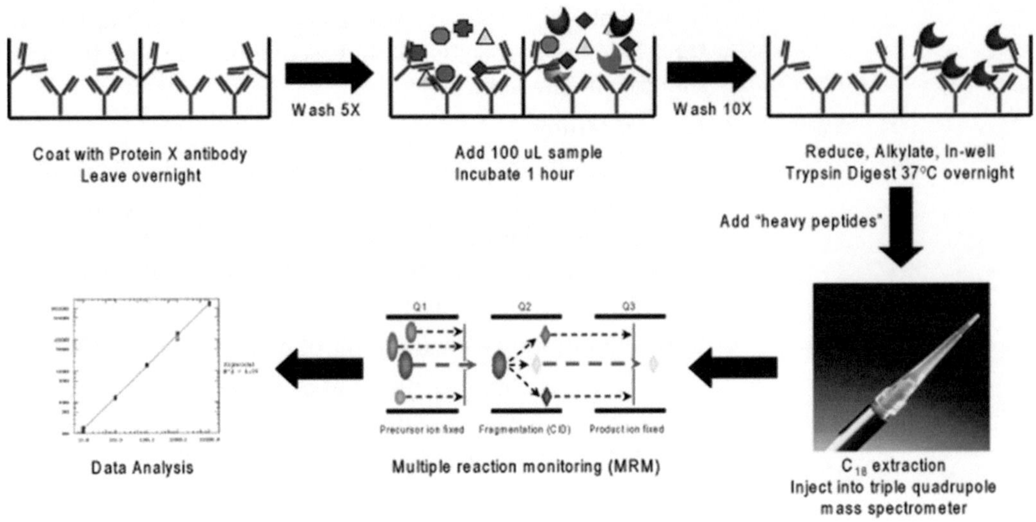

Fig. 1. Strategy for immunoextraction coupled-MRM.

spike into a complex matrix such as 6% bovine serum albumin (BSA) of increasing concentration. Load the unknown samples after the standards (see Notes 9 and 14).

4. Incubate the plate for 1 h at room temperature, with shaking.
5. Wash the plate ten times with 0.9% NaCl washing solution and gently pat down the plate after the last wash.
6. Add 48 μL ammonium bicarbonate (200 mM), 10 μL methanol, 40 μL water, and 2 μL DTT (20 mM) to each well and incubate at room temperature for 30 min on a shaking device.
7. Add 2 μL iodoacetamide (100 mM) and incubate in the dark for 1 h at room temperature.
8. Add 0.25 μg of trypsin to each well and incubate the plate in a 37°C incubator (sealed) overnight (~16 h).
9. Transfer the contents from each well to microcentrifuge tubes.
10. Spike in appropriate concentration of synthetic heavy-labeled peptide (fixed concentration) to serve as the internal standard (e.g., 10 fmol of a heavy-labeled peptide mixture).

3.5. Sample Analysis

1. Acidify the samples by adding 5 μL of 10% formic acid (if not already performed).
2. Perform a C18 extraction using the Omix pipette tips:
   (a) Condition the tip in Buffer B (aspirate and dispense five times).
   (b) Wash the tip in Buffer A.
   (c) Aspirate and dispense the sample ten times.

   (d) Wash the tip in Buffer A.
   (e) Elute the peptides in 5 μL of Buffer B.
   (f) Add 80 μL of Buffer A.
3. Deposit the sample into a 96-well plate.
4. Run the developed LC–MRM method (Subheading 3.3).
5. Run a quality control (QC) sample before and after the analytical run to ensure that the retention time and peak area have not changed. Refer to Note 15 for more detail.

3.6. Data Analysis

1. Manually verify that all transitions for a given peptide elute at the same time and at the appropriate retention time for the QC sample and check for acceptability.
2. Refer to the precision analysis study to determine the variability in retention time for each peptide:
   (a) Peaks should only be considered "real" if they are within the determined ± retention time, and if the signal-to-noise (S/N) ratio is ≥3:1 (14, 15).
3. Export the integrated peak areas for light and heavy proteotypic peptides.
4. Calculate the ratio of the light to heavy peptide peak area for each sample (standard and unknowns):
   (a) Use the peak area of one transition for quantification purposes. The remaining transitions for the proteotypic peptide are used for confirmation of the identification of the peptide.
5. Generate a calibration curve using the standard's light:heavy peptide ratio and create a "line of best fit" with a $y = mx + b$ equation:
   (a) Ensure that it is linear and passes through all points on the curve.
6. Apply the unknown sample's ratios (y) to the equation and solve for x (concentration).

4. Notes

1. All reagents for sample preparation should be made on the day of the experiment.
2. As mentioned, proteotypic peptide selection can be performed using in silico analysis or taken from previously performed experimental data:
   (a) Search the literature for studied proteotypic peptide and transitions for Protein X.

(b) Check various online open-source databases such as GPMDB (http://gpmdb.thegpm.org) or MRM Atlas (http://www.mrmatlas.org/) to see if there is any MS/MS information for Protein X.

(c) Even if information is available, it is recommended to digest the recombinant protein and perform empirical analysis via LC–MS on Protein X since different instruments may yield different fragmentation patterns.

(d) Selecting transitions and/or peptides via in silico means may not always yield the "correct answers" as there is no way to predict ionization efficiency, making it difficult to predict which peptide will yield the best signal intensity.

3. While we do not recommend the use of peptides containing modified amino acids, sometimes modified amino acids are acceptable and may be the peptide for that protein with excellent $S{:}N$ ratio (>3). Trial and error are required at this point to select the proteotypic peptide.

4. In the selection of proteotypic peptides, small peptides (six to seven amino acids) are not ideal candidates since they tend to be nonspecific (11).

5. Generally, selecting more transitions per peptide can increase selectivity but potentially at the risk of decreased sensitivity:

   (a) If adding the peak areas for the three transitions gives better signal-to-noise ratio, then it is recommended to use this consistently for all samples analyzed.

6. It is important to ensure that adequate scans across a peak is performed. For example, a good rule of thumb is to have more than ten scans per peak. A method quantifying many proteins requires monitoring a lot of transitions in a single experiment. In such a case, the dwell time of each individual transition is reduced and hence will vary with the number of peptides monitored. Practical dwell-time settings range between 10 ms for good sensitivity and 100 ms for excellent sensitivity. To ensure precise LC–MS quantification, at least ten data points should be acquired across the chromatographic elution profile of a peptide (12).

7. As a cautionary note, some peptides are not stable for long periods of time. For example, hydrophobic peptides can adhere to plastic, while aspartic and glutamic acid can undergo deamidation to asparagine and glutamine. In addition, methionine can become oxidized over time. To deal with this problem, perform the following: The stock solutions made on day

1 should be run on the mass spectrometer and the resulting file kept as a "reference file." With each analytical run or day, the stock solutions can be run on the mass spectrometer and the resulting file checked against the initial "reference file." A decrease in signal intensity should flag the need to make a new stock solution.

8. It is recommended to analyze samples on the MS immediately after sample preparation to avoid sample degeneration over time.
9. If specimen volume is an issue, one can load 50 μL of specimen and 50 μL of water. It is important to ensure that the final specimen volume does not exceed 150 μL.
10. The amount of capture antibody coated per well may vary for each protein. Test 100, 200, and 500 ng of antibody per well to see which yields the greatest capture. Note that the actual binding capacity of the wells is ~20–50 ng.
11. Do not exceed the stated coating solution volume as the plates are high protein/peptide binding and will mop up any peptides that adhere to it. For example, if 200 μL of coating antibody is used, do not exceed 150 μL of sample volume or digestion volume.
12. It is not recommended to block the wells with agents such as BSA as this will introduce unnecessary peptides into the final sample. In general, we have not experienced any issues with peptides originating from the antibody interfering with our MRM assay.
13. Do not let the plate wells dry up in the process of washing the wells. It is recommended to wash using an automatic plate washer or if performing the washing manually, dispense and discard the solution one strip of wells at a time. The plate can be gently tapped (face down) on a paper cloth to remove excess salt/buffer.
14. It is recommended to have a variety of negative controls in the run. For example, coat the well with the antibody but do not add the sample. Alternatively, coat the well with a different antibody (not against Protein X), or do not coat the well with the antibody but add the 100 μL of sample.
15. A QC sample should be assayed with each analytical run. The purpose of the QC is to ensure that the instrument is performing within the required specifications (accuracy and precision). The QC sample can be BSA, commercially purchased (Michrom Bioresources, Inc., #PTD-00001-15).

References

1. Barnidge, D.R., Goodmanson, M.K., Klee, G.G., and Muddiman, D.C. (2004) Absolute quantification of the model biomarker prostate-specific antigen in serum by LC-Ms/MS using protein cleavage and isotope dilution mass spectrometry. *J.Proteome Res.*, **3**, 644–652.
2. Mallick, P., Schirle, M., Chen, S.S., et al. (2007) Computational prediction of proteotypic peptides for quantitative proteomics. *Nat.Biotechnol.,* **25**, 125–131.
3. Domon, B., Aebersold, R. (2006) Mass spectrometry and protein analysis. *Science*, **312**, 212–217.
4. Barnidge, D.R., Dratz, E.A., Martin, T., et al. (2003) Absolute quantification of the G protein-coupled receptor rhodopsin by LC/MS/MS using proteolysis product peptides and synthetic peptide standards. *Anal.Chem.*, **75**, 445–451.
5. Kuhn, E., Wu, J., Karl, J., Liao, H., Zolg, W., and Guild,B. (2004) Quantification of C-reactive protein in the serum of patients with rheumatoid arthritis using multiple reaction monitoring mass spectrometry and 13C-labeled peptide standards. *Proteomics.*, **4**, 1175–1186.
6. Anderson, L., Hunter, C.L. (2006) Quantitative mass spectrometric multiple reaction monitoring assays for major plasma proteins. *Mol.Cell Proteomics.*, **5**, 573–588.
7. Kuzyk, M.A., Smith, D., Yang, J., et al. (2009) Multiple reaction monitoring-based, multiplexed, absolute quantitation of 45 proteins in human plasma. *Mol.Cell Proteomics.*, **8**, 1860–1877.
8. Addona, T.A., Abbatiello, S.E., Schilling, B., et al. (2009) Multi-site assessment of the precision and reproducibility of multiple reaction monitoring-based measurements of proteins in plasma. *Nat.Biotechnol.*, **27**, 633–641.
9. Rifai, N., Gillette, M.A., and Carr, S.A. (2006) Protein biomarker discovery and validation: the long and uncertain path to clinical utility. *Nat.Biotechnol.,* **24**, 971–983.
10. Kulasingam, V., Smith, C.R., Batruch, I., et al. (2008) "Product ion monitoring" assay for prostate-specific antigen in serum using a linear ion-trap. *J.Proteome Res.,* 7, 640–647.
11. Han, B., Higgs, R.E. (2008) Proteomics: from hypothesis to quantitative assay on a single platform. Guidelines for developing MRM assays using ion trap mass spectrometers. *Brief. Funct.Genomic.Proteomic.*, 7, 340–354.
12. Lange, V., Picotti, P., Domon, B., and Aebersold, R. (2008) Selected reaction monitoring for quantitative proteomics: a tutorial. *Mol.Syst.Biol.*, **4**, 222.
13. Fusaro, V.A., Mani, D.R., Mesirov, J.P., and Carr, S.A. (2009) Prediction of high-responding peptides for targeted protein assays by mass spectrometry. *Nat.Biotechnol.*, **27**, 190–198.
14. Banks, J.F., Gulcicek, E.E. (1997) Rapid peptide mapping by reversed-phase liquid chromatography on nonporous silica with on-line electrospray time-of-flight mass spectrometry. *Anal.Chem.,* **69**, 3973–3978.
15. Stahl-Zeng, J., Lange, V., Ossola, R., et al. (2007) High sensitivity detection of plasma proteins by multiple reaction monitoring of N-glycosites. *Mol.Cell Proteomics.*, **6**, 1809–1817.

Chapter 14

Qualitative and Quantitative Analysis of Glycated Proteins in Human Plasma by Glucose Isotopic Labeling with $^{13}C_6$-Reducing Sugars

Feliciano Priego-Capote, Maria Ramírez-Boo, Denis Hochstrasser, and Jean-Charles Sanchez

Abstract

Glucose is the predominant source of energy in cells. However, a chronic high glucose exposure of proteins modifies a number of biological pathways, known as glucotoxicity. Several studies have suggested that this impaired protein function is associated in part to protein glycation. However, despite the evidence of this glucotoxicity on tissues and cells, the exact mechanisms underlying the loss of protein function by glycation are not well understood. Strategies that will allow the discovery of the identity and function of the glycated plasma proteins generated by chronic hyperglycemia would be of a considerable help to further understand the underlying mechanisms implicated in protein dysfunction associated with glucotoxicity. The present chapter describes an innovative and interdisciplinary proteomics strategy to achieve a comprehensive identification and quantification of glycated proteins in plasma. This should provide new molecular insights into protein glycation mechanisms and identify new targets to improve the subsequent defective protein function.

Key words: Protein glycation, Glucotoxicity, Mass spectrometry, Plasma proteomics, Isotopic labeling

1. Introduction

Glycated proteins are formed by nonenzymatic reactions between reducing carbohydrates (e.g., glucose, fructose, ribose, or derivatives such as ascorbic acid) with amino groups predominantly located in lysine and arginine residues or, also in N-terminal positions. It is worth mentioning the differences between glycation and glycosylation, the latter being enzymatically catalyzed by glycosyl transferase and occurring in specific protein side chains such as asparagine (N-linked), serine, and threonine (O-linked), and the C-termini of cell surface proteins (1). Glycosylation is involved

Richard J. Simpson and David W. Greening (eds.), *Serum/Plasma Proteomics: Methods and Protocols*,
Methods in Molecular Biology, vol. 728, DOI 10.1007/978-1-61779-068-3_14,

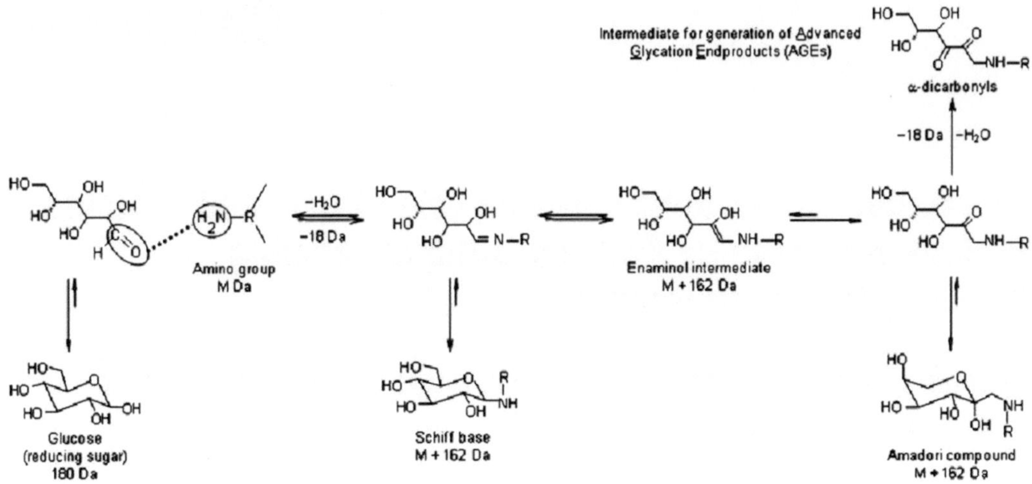

Fig. 1. Mechanism of the glycation process (Maillard reaction)

in many biological processes in contrast to glycation, which is a completely undesired modification from a patho-physiological point of view.

The scheme of the glycation process for glucose is illustrated in Fig. 1. It can be extended to any reducing sugar (2). The process starts with the formation of Schiff base by a condensation reaction between the carbonyl group of the reducing sugar and the amino group of the protein. Then, the Schiff base is rapidly converted into the Amadori compound in a step considered as the first glycation level. The Amadori compound can undergo a series of dehydration steps and intermolecular fragmentations to generate a variety of carbonyl compounds such as methylglyoxal, glyoxal, glucosones, deoxyglucosones, and dehydroascorbate (3). These carbonyl compounds are generally more reactive than the original carbohydrate and act as propagators by reactions with free amino groups, leading to the formation of a variety of heterogeneous structures irreversibly formed and commonly known as advanced glycation end-products (AGEs). Evidently, the glycation effects on biological function will depend on the extent of the process encompassing alterations of the structure, function, and turnover of proteins (4). From a clinical point of view, the detection of this PTM at the initial stage would be helpful for both prognostic and diagnostic purposes.

Glucose plays a crucial role as an energy source in humans, as the main circulating sugar, and, thus, the most relevant molecule in terms of protein glycation. The kinetics of the initial glycation process is governed by the formation of the Amadori compound, a slow process under human physiological conditions (37°C and ~5 mM blood glucose concentration in healthy subjects) (4).

However, reaction kinetics is enhanced under prolonged high glycemic exposure. This chronic supra-physiological glucose concentration (>10 mM) negatively affects a large number of organs and tissues, such as the pancreas, eyes, liver, muscles, adipose tissues, brain, heart, kidneys, and nerves (5). In fact, glucose toxicity is the main cause of diabetic complications, which are often observed only several years after the development of the disease (6). Thus, glycation has often been related to chronic complications of diabetes mellitus, renal failure, and degenerative changes occurring in the course of aging (7–9).

Glycation of proteins is one of the abnormalities involved in glucotoxicity. Calvo et al. evaluated the nonenzymatic glycation rate of high-density lipoproteins (HDL) in type 1 and 2 diabetic patients (10). They isolated glycated apolipoprotein A-I (ApoA-I) from diabetic patients and compared its lipid-binding properties to those of ApoA-I from healthy subjects. They found that ApoA-I glycation promotes a decrease in the stability of the lipid–apolipoprotein interaction and also in its self-association. Therefore, the structural cohesion of HDL molecules is seriously affected by glycation of ApoA-I (10–12). Similarly, in vivo studies in mice proved that glycated insulin exhibits a reduced ability to stimulate glucose oxidation by the isolated mouse diaphragm muscle. This observation is in agreement with previous studies suggesting that glycation of insulin decreases its potency to stimulate lipogenesis in isolated rat adipocytes. This is consistent with the observation that glycated insulin displays a significantly reduced ability to lower plasma glucose concentrations in mice. These and other studies clearly indicate that glycation results in an important impairment of insulin action to regulate plasma glucose homeostasis (13).

The glycemic control of patients is currently assessed indirectly with the reference test of glycated hemoglobin (HbA1c). HbA1c is a long-term indicator of the patient glycemic state because of the erythrocyte lifespan (~120 days). HbA1c concentration represents the memory effect of blood glucose concentrations over the previous 8–12 weeks (14–17). Due to the continuous exposition to glucose, the concentration of HbA1c and glycated human serum albumin (HSA) in blood from healthy subjects is between 5 and 7% and 13 and 15%, respectively (18, 19). Other measurements indicative of short-term glucose perturbation are needed in order to understand its potential biological effect. It should also be taken into account that any protein could be potentially glycated. Therefore, the development of methods for the identification and quantification of glycated proteins is crucial to elucidate their biological impact on plasma protein function. Preliminary experiments have provided the identification of 50 glycated proteins in a reference human plasma (Sigma–Aldrich®) corresponding to a normal glycemic state (20).

2. Materials

2.1. Plasma Collection

1. Evacuated sterile tubes for whole blood hematology determination (BD Vacutainer® EDTA, 6 mL), plastic lavender BD Hemogard™ safety closure, and K_2EDTA spray.
2. Evacuated sterile tubes for plasma determinations (BD Vacutainer® plasma tubes, 6 mL), medical grade PET green BD Hemogard™ safety closure, and lithium heparin (17 IU/mL).

2.2. Isotopic Labeling with $^{12}C_6$-Glucose/$^{13}C_6$-Glucose

1. D-(+)-Glucose (lyophilized powder, ≥99.5%, #G7528, Sigma–Aldrich®) is dissolved in Dulbecco's phosphate-buffered saline 1× (DPBS; #14190-094, Invitrogen™) at 25 mg/mL and stored at 2–8°C.
2. D-Glucose-$^{13}C_6$ (lyophilized powder, 99 atom % ^{13}C; #389374-250 mg, Sigma–Aldrich®) is dissolved in DPBS at 25 mg/mL and stored at 2–8°C.

2.3. Measurement of Protein Concentration and In-Solution Digestion by Enzymatic Hydrolysis

1. Amicon Ultra-0.5 mL, Ultracel® 3K membrane (#UFC500396, 96PK, Millipore™).
2. 1.0 M, pH 8.5 ± 0.1, triethylammonium hydrogen carbonate buffer (TEAB; #90360, Fluka) is used to prepare a 0.5 M aqueous solution (see Note 1).
3. Bovine serum albumin (BSA, lyophilized powder, ≥96%; #05488, Fluka) is dissolved in 0.1 mM DPBS (5 mg/mL concentration). Store in aliquots at −80°C.
4. Protein assay dye reagent concentrate (liquid; #500-0006, Bio-Rad).
5. Spectrophotometer for absorbance measurement at 595 nm.
6. 0.5 M, pH 7.0, Tris-(2-carboxyethyl) phosphine hydrochloride (TCEP; #646547, Sigma–Aldrich®) is diluted in water at 0.05 M (2 μL of TCEP in 20 μL of water).
7. Iodoacetamide (IAA, crystalline, ≥99%; # I-6125, Sigma–Aldrich®) is dissolved in water at 400 mM (37 mg of IAA in 500 μL of water) and stored at 2–8°C.
8. Endoproteinase Glu-C from *Staphylococcus aureus V8* (lyophilized powder; #P2922-500U, Sigma–Aldrich®) is reconstituted in 0.5 M, pH 8.5 ± 0.1, TEAB at 1 mg/mL. Store in aliquots and freeze at −80°C.
9. Shaker

2.4. Isolation of Glycated Peptides by Boronate Affinity Chromatography

1. Speed-Vac.
2. Mobile phase A (1×): 0.2 M ammonium acetate (NH_4Ac, solid, ³98.0%; #09690, Fluka), 0.05 M magnesium chloride ($MgCl_2$, solid, ≥99.0%; #63068, Fluka), pH 8.1 (adjusted with sodium hydroxide pellets, 99%; #102524X, AnalaR®).

Prepare mobile phase A by adding solution of 15.56 g of NH_4Ac and 10.16 g of $MgCl_2$ in 1 L of water, and adjust the pH. Store at room temperature.

3. Mobile phase B (1×): 0.1 M acetic acid (HAc, 99.5%; #45730, Fluka). Prepare mobile phase B by adding solution of 6 mL of HAc in 1 L of water. Store at room temperature.
4. Waters 600E HPLC system (Millipore Corporation, USA) equipped with a TSK gel boronate-5PW matrix chromatographic column (7.5 cm length × 7.5 mm inner diameter, ID; 10-μm particle size; Tosoh Biosciences) at room temperature.

2.5. Desalting and Buffer Exchange by Centrifugal Filtration

1. Speed-Vac.
2. Set-up/reconstitution solvent for isolated peptides: 5% acetonitrile (ACN, CHROMASOLV® gradient grade, for HPLC, ≥99.9%; #3485, Sigma–Aldrich®) and 0.1% formic acid (FA, ~98%; #56302, Fluka). Prepare 50 mL by diluting 2.5 mL ACN and 0.05 mL FA in water. Store at room temperature.
3. 10% FA (prepare 50 mL by diluting 5 mL of FA in water). Store at room temperature.
4. MacroSpin Column™ C18 (Harvard Apparatus, Holliston, MA, USA).
5. Wash/elution solvent: 50% ACN and 0.1% FA. Prepare 50 mL by diluting 25 mL of ACN and 0.05 mL of FA in water. Store at room temperature.

2.6. Chromatographic Analysis and Mass Spectrometry Detection of Glycated Peptides

1. Waters NanoAquity HPLC system (Milford, MA) equipped with a helium degasser and a homemade 100 μm ID × 18 mm long precolumn packed with 200 Å (5 μm) Magic C18 particles (C18AQ; Michrom) and a homemade gravity-pulled 75 μm ID × 150 mm long analytical column packed with 100 Å (5 μm) C18AQ particles (Michrom).
2. Set-up solvent on precolumn: water/ACN (95/5 v/v) with 0.1% FA (v/v).
3. Mobile phase A: 5% ACN and 0.1% FA.
4. Mobile phase B: 95% ACN and 0.1% FA.
5. Microtee union (Upchurch Scientific, Oak Harbor, WA).
6. Hybrid linear ion trap-Orbitrap mass spectrometer (Thermo Fisher, San Jose, CA).

3. Methods

3.1. Plasma Collection and Preparation

1. Venipuncture blood is collected in a syringe or tube containing a suitable anticoagulant such as heparin or EDTA. Capillary

blood from a finger prick can be used, provided that the skin is first cleaned thoroughly to avoid contamination with sweat.

2. After blood collection, the syringe or tube should immediately be placed in an ice bath and brought to the laboratory for analysis.
3. Upon arrival at the laboratory, the blood specimen must be centrifuged immediately at 2,000 × *g* for 10 min at 5°C to avoid hemolysis, and decanted.
4. The sample can be either processed immediately or stored at −80°C until analysis.
5. Alternatively, human reference plasma can be purchased for further experiments (see Note 2).

3.2. Glucose Isotopic Labeling of Plasma Proteins

1. Two aliquots of the reconstituted plasma (50 μL each) in 0.5 mL DPBS are independently incubated with 30 mM ($^{12}C_6$]-glucose and [$^{13}C_6$]-glucose for 24 h at 37°C.
2. Aliquots are separately analyzed or pooled, 1:1 ratio, depending on the purpose, for subsequent analysis.
3. Amicon Ultra-0.5 devices similarly remove glucose and other salts from the isolated proteins that are reconstituted in 0.5 M, pH 8.5 ± 0.1, TEAB (see Note 3).
4. Protein concentration is subsequently measured using the Bradford assay with BSA as calibration protein.

3.3. Enzymatic Digestion with Endoproteinase Glu-C

1. Dilute 1 mg of plasma proteins determined using Bradford assay (21) in 400 μL TEAB.
2. Cysteine groups are reduced with 50 mM TCEP in water (20 μL) by incubation of the reaction mixtures for 60 min at 60°C.
3. Cysteine residues are then alkylated with 400 mM IAA in water (10 μL) shaking for 30 min in the dark at room temperature.
4. Freshly prepared endoproteinase Glu-C is added (100 μL to obtain a ratio of 1:10 w/w) and the digestion performed overnight at 37°C (see Note 4).
5. The digestion mixtures are evaporated under speed vacuum and reconstituted in 50 μL mobile phase A for isolation of glycated peptides.

3.4. Isolation of Plasma-Glycated Peptides by Boronate Affinity Chromatography

1. Reconstituted peptides are fractioned by boronate affinity chromatography (BAC) by interaction between boronic acids and *cis*-diol groups of glycated peptides present at low concentration (see Fig. 2). For this, the target sample (50 μL) is injected in a high pressure liquid chromatograph equipped with a boronate affinity column at room temperature (see Note 5).

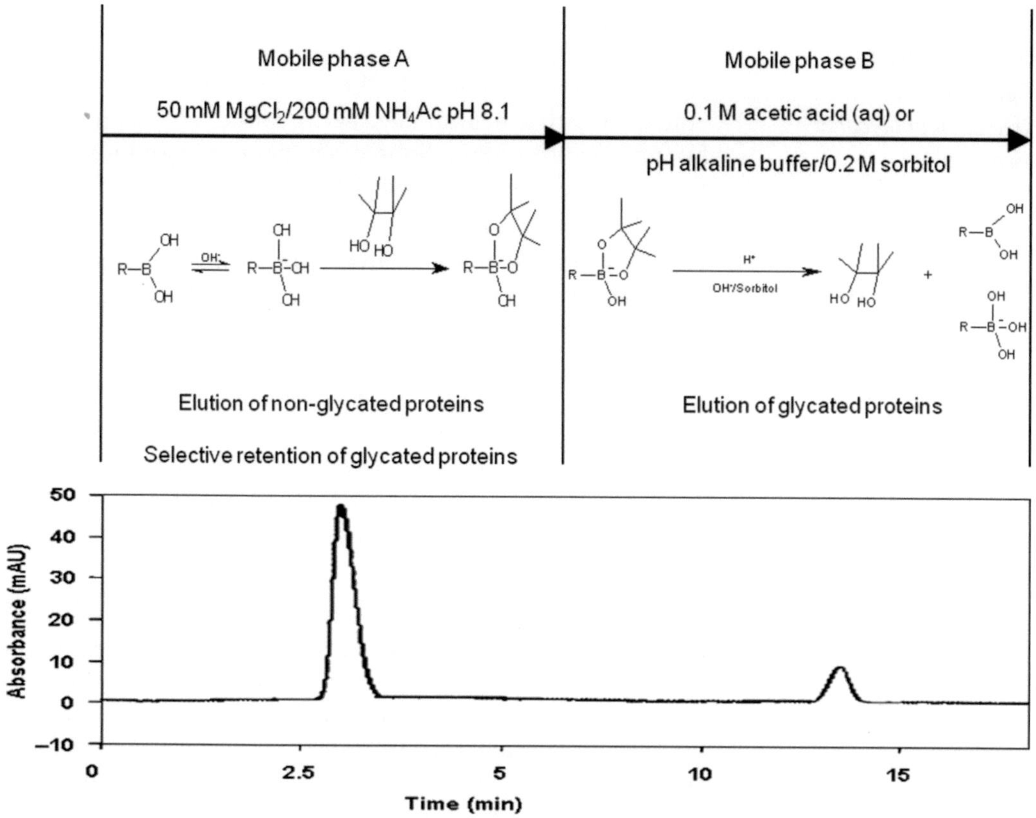

Fig. 2. Mechanism of boronate affinity chromatography (BAC) for separation of glycated and nonglycated proteins.

2. An isocratic chromatographic method is used for affinity separation that consists of the following steps: (1) 0–10 min, 100% mobile phase A for retention of glycated peptides by esterification between boronate ligands and 1,2 *cis*-diol groups of glucose moieties under alkaline conditions, with elution of nonglycated peptides; (2) 10–20 min, 100% mobile phase B for elution of glycated peptides; and (3) 20–30 min, 100% mobile phase A for the equilibration of the column to the initial conditions.
3. Both the nonglycated and the glycated fractions are collected for subsequent evaporation and reconstitution in 5% ACN/0.1% formic acid.
4. The peptides are desalted and preconcentrated prior to LC–MS/MS analysis by preconditioning C18 Macrospin columns with 800 μL 50% ACN/0.1% formic acid, equilibrating with 800 μL 5% ACN/0.1% formic acid, passing sample through the column, eluting peptide fractions in 400 μL 5% ACN/0.1% formic acid, washing with 800 μL 5% ACN/0.1% formic acid, and eluting peptides with 400 μL 50% ACN/0.1% formic acid. This solution was evaporated to dryness for reconstitution with 50 μL 5% ACN/0.1% formic acid (see Note 6).

3.5. Chromatographic Analysis and Mass Spectrometry Detection of Glycated Peptides

1. Peptide digests are analyzed by electrospray ionization in positive ion mode on a hybrid linear ion trap-Orbitrap mass spectrometer. Nanoflow HPLC is performed on a system equipped with a helium degasser. Peptides are trapped on a homemade precolumn. Subsequent peptide separation is performed on a homemade analytical column and directly interfaced to the mass spectrometer (see Note 7).
2. For each LC–MS/MS analysis, an estimated amount of 0.05 mg of glycated peptides (3 μg/μL) is loaded on the precolumn at 3 mL/min in water/ACN (95/5 v/v) with 0.1% formic acid (v/v).
3. After retention, peptides are eluted using an ACN gradient at 220 nL/min with mobile phase A, water, and 0.1% formic acid; and mobile phase B, ACN and 0.1% formic acid. The gradient program is as follows: 0 min, A (95%), B (5%); 55 min, A (65%), B (35%); 60 min, A (15%), B (85%); 65 min, A (85%), B (15%); and 75–90 min, A (95%), B (5%).
4. The electrospray ionization voltage is applied via a liquid junction using a platinum wire inserted into a microtee union located between the precolumn and the analytical column. Ion source conditions are optimized using the tuning and calibration solution recommended by the instrument provider. Two complementary data-dependent tandem mass spectrometry methods are used for analysis of glycated proteins: MS2 with high-energy collisional dissociation (HCD, Fig. 3) as activation mode and MS3 by neutral loss scan with CID as activation mode (Fig. 4).
5. In data-dependent HCD-MS2 analysis, fragmentation of the three most abundant precursor ions is carried out in the octopole collision cell attached to the C-trap (normalized collision energy 50%), followed by Orbitrap detection. The precursor ion isolation window is set to 4 m/z units. MS survey scans are acquired at $R=60{,}000$ resolution in profile mode, while MS2 spectra are acquired at $R=7{,}500$ resolution. Precursor ions of charge state +2 and higher are included for data-dependent selection. In cases where the charge state cannot be identified, the most abundant ion is selected for HCD.
6. Data-dependent acquisition is then performed over the entire chromatographic cycle. Data-dependent CID-MS3 neutral loss scan is entirely carried out in the linear trap (except MS survey scan which is performed with Orbitrap accuracy) with three steps: (1) first fragmentation of medium collision energy (35%) to promote the cleavage of the glucose moiety (−162.05 Da, which corresponds to −81.02 and −54.01 m/z units for doubly and triply charged peptides, respectively) or

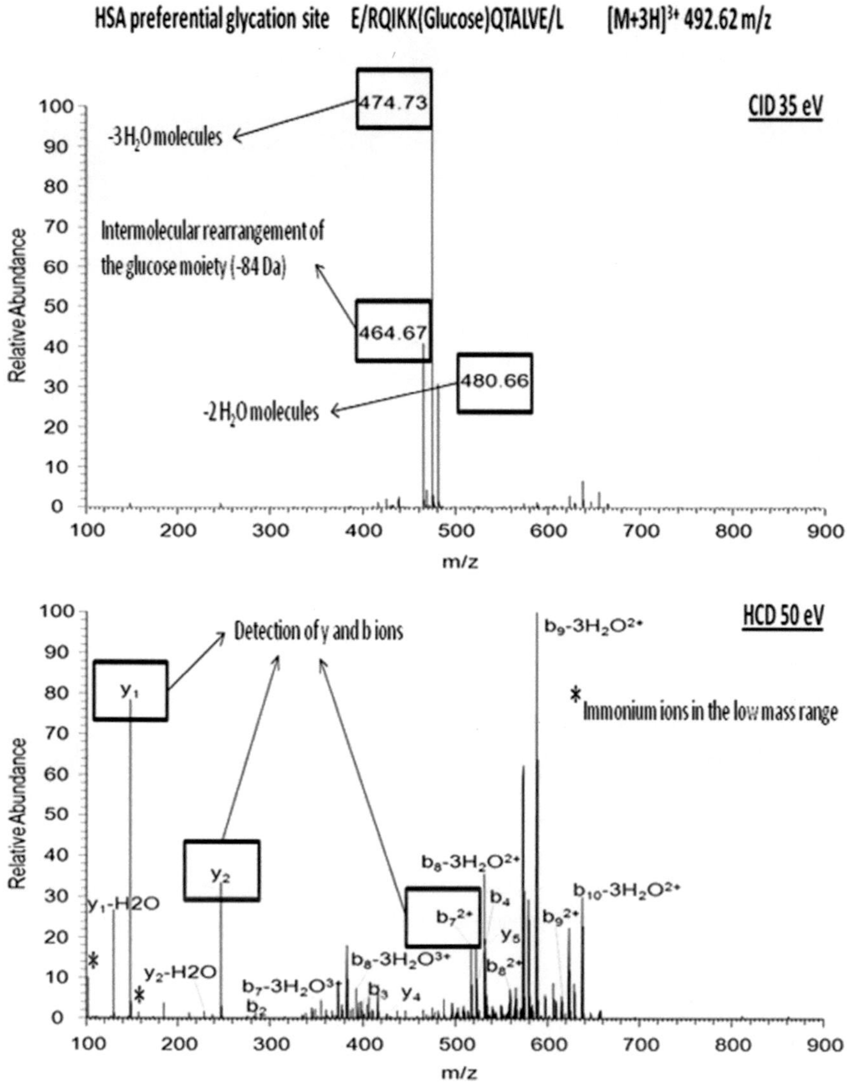

Fig. 3. Comparison of CID and HCD activation modes in MS2 analysis of a glycated peptide from human serum albumin (HSA) identified in plasma. Optimum collision energies for identification were used for each case (35 and 50% for CID and HCD, respectively).

an intermediate fragmentation of the glucose molecule (−84.04 Da, which corresponds to −42.02 and −28.01 m/z units for doubly and triply charged peptides, respectively); (2) isolation of ions in which one of the neutral losses is detected; and (3) fragmentation of the isolated peptide (35%) followed by ion-trap detection. The precursor ion isolation window is set to 2 m/z units (see Note 8).

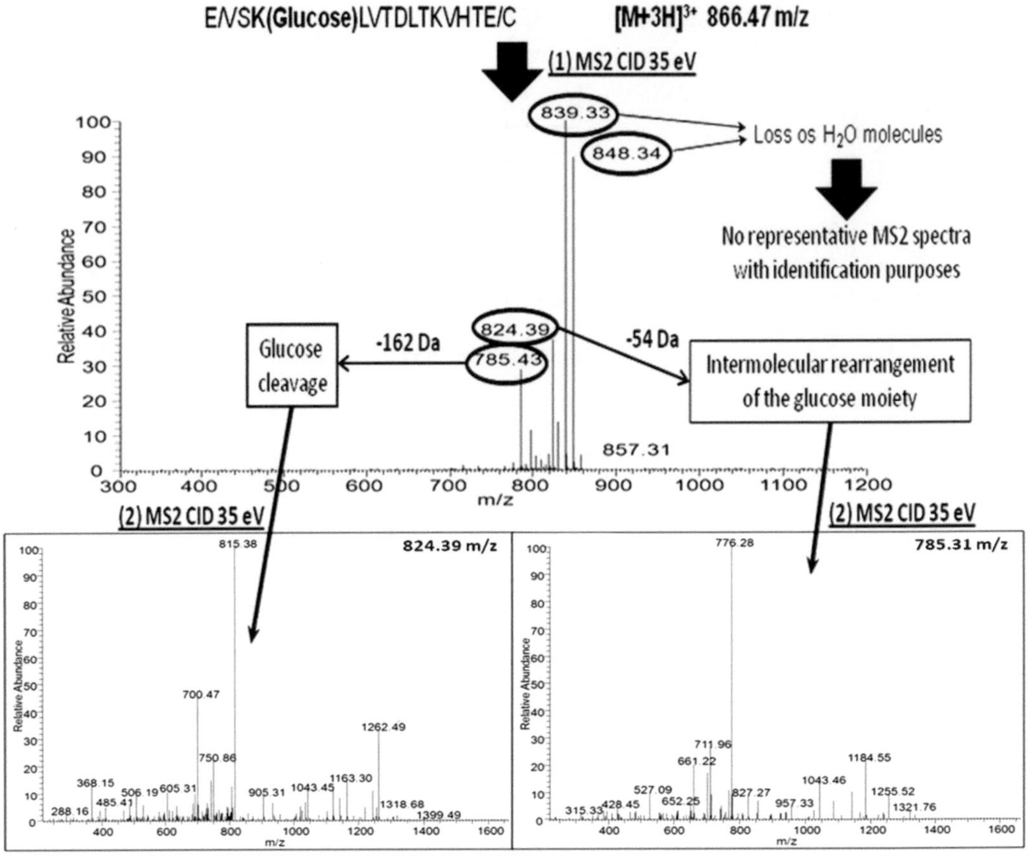

Fig. 4. CID-MS3 analysis in neutral loss scan mode for a HSA glycated peptide identified in plasma. Two neutral losses were monitored: 162.05 and 84.04 Da corresponding to glucose cleavage and an intermolecular rearrangement of the glucose moiety, respectively.

3.6. Identification of Glycated Peptides by Analysis of MS/MS Spectra

After data-dependent acquisition, post-acquisition workflow is initiated specifically for each MS operation mode:

1. For HCD-MS2 experiments, the workflow is based on the detection of precursor ions in an accurate way and correction for misassigned precursor ion isotopes.
2. The resulting data files for both MS operation modes are searched against UniProt-Swiss-Prot/TrEMBL using Phenyx 2.6 (GeneBio, Geneva, Switzerland). *Homo sapiens* is specified as the species for database searching. Common amino acid modifications for both MS operation modes are carbamidomethylation of cysteines and oxidation of methionine, which are set as fixed and variable modifications, respectively.
3. For HCD-MS2 experiments, glycation of lysine and arginine residues or on N-terminal positions (162.052 and 168.072 Da for glycated peptides with [$^{12}C_6$]- or [$^{13}C_6$]-glucose) is selected as a variable modification.

4. For MS3 neutral loss experiments, a variable modification as a consequence of glucose fragmentation after neutral loss of 84.04 Da (78.01 Da for K, R, and on N-terminal positions) is additionally specified.
5. Endoproteinase Glu-C is selected as enzyme, with three potential missed cleavages as maximum (see Note 9).
6. The peptide and fragment ion tolerance depend on the MS operation mode. For HCD-MS2, peptide and fragment ion tolerance is 6 ppm. This tolerance is set to 0.8 Da for fragment ions in MS3 neutral loss.
7. In both modes, two sequential search rounds are used. In the first round, two missed cleavages are allowed. This round is selected in "turbo" search mode. In the second round, three missed cleavages are allowed in half-cleaved mode.
8. The minimum peptide length allowed is 6 for both rounds.
9. The acceptance criteria are slightly lowered in the second-round search. These are as follows for HCD-MS2 experiments: AC score 9.7, peptide *Z*-score 9.7, and peptide *p* value 1×10^{-7} for round 1; AC score 9.5, peptide *Z*-score 9.5, and peptide *p* value 1×10^{-6} for round 2, corresponding to an estimated false-positive ratio of less than 1%.
10. For MS3 in neutral loss experiments, these parameters are changed to AC score 7.0, peptide *Z*-score 7.0, and peptide *p* value 1×10^{-6} for round 1; AC score 6.5, peptide *Z*-score 6.5, and peptide *p* value 1×10^{-5} for round 2, corresponding to an estimated false-positive ratio of less than 1%.
11. False-positive ratios are estimated using a reverse decoy database. This estimation is performed using separate searches in the reverse database to keep the database size constant (22). This involves a slight underestimation of the estimated false-positive ratio. In case of several matching entries, Swiss-Prot entries are preferred to TrEMBL entries. All data are acquired in triplicate (three analytical injections of the same sample) and analyzed in an independent manner.

3.7. Quantitative Analysis of Glycated Proteins by Mass Precursor Scan

1. Quantitation of glycated proteins is possible as after enzymatic digestion, the resulting glycated peptides (with addition of 162 or 168 mass units) produce doublet signals in precursor MS1 scans (labeling with light and heavy glucose). The mass shift of the doublet signals depends on the peptide charge and the number of glycation sites. Peptide quantification is carried out by calculating the ratio between peak areas from extracted ion chromatograms corresponding to both isotopic forms of each glycated peptide.
2. The ratios of [$^{12}C_6$]-glucose peptide/[$^{13}C_6$]-glucose peptide are obtained from the average values of intra-run triplicates (see Notes 10 and 11).

3. Data treatment is automated using the SuperHirn software (version 1.0), which is freely available together with detailed documentation material on http://tools.proteomecenter.org/SuperHirn.php.
4. The *.raw* data files are converted to mzXML file format in profile mode and SuperHirn performs the feature extraction and alignment of the replicate runs (SuperHirn uses standard Orbitrap settings).
5. The post-processing of the feature list is performed in the R statistical programming environment (http://www.r-project.org). Briefly, the SuperHirn result files are parsed in order to find all heavy–light pairs (within a mass tolerance of 0.01 Da and retention time tolerance of 20 s) that appear in at least two replicates.
6. All accepted identifications from the Phenyx Excel export are attributed to a heavy–light pair, if such a pair could be detected (~80% of the cases).

4. Notes

1. Unless stated otherwise, all solutions should be prepared in water with a resistance of 18.2 MΩ cm and total organic content of less than five parts per billion. This standard is referred to as "water" in this text.
2. Human reference plasma containing 3.8% trisodium citrate as anticoagulant was purchased from Sigma (#P9523). Plasma was tested and found negative for antibody to HIV-1/HIV-2, HCV, and HbSAg. According to the manufacturer's instructions, whole blood was collected with anticoagulants (9:1), pooled, and centrifuged. The resulting plasma was filtered (0.45 μm) and lyophilized.
3. After incubation, glucose and other salts were removed with Amicon ultrafiltration devices with an Ultracel® regenerated cellulose membrane with 3K nominal molecular weight cut-off (Millipore). Proteins were recovered in 0.5 M, pH 8.5, TEAB in the same unit according to the manufacturer's instructions.
4. Endoproteinase Glu-C was prepared daily. As glucose attachment is selective for Lys and Arg residues, trypsin digestion pattern was affected with an increase in the number of missed cleaved sites. A high proportion of half-cleaved peptides were also detected. The influence of glucose attachment was less dramatic for endoproteinase Glu-C and identifications of missed cleaved sites and, particularly, the proportion of half-cleaved peptides was considerably reduced.

5. The BAC column was preconditioned daily by injection of mobile phases A and B subsequently in order to regenerate its optimum efficiency.
6. Both the nonglycated and the glycated fractions were desalted with C18 macrospin columns according to the protocol recommended by the manufacturer.
7. Peptides were trapped on a homemade 100 μm ID × 18 mm long precolumn packed with 200 Å (5 μm) Magic C18 particles (C18AQ; Michrom). Peptide separation was performed on a homemade gravity-pulled 75 μm ID × 150 mm long analytical column packed with 100 Å (5 μm) C18AQ particles (C18AQ; Michrom) and directly interfaced to the mass spectrometer.
8. The CID-MS3 mode is a complementary approach to HCD-MS2 as the former is particularly useful for the identification of glycated peptides with charge states +2 and +3.
9. A maximum number of 3 missed cleavage sites should be set. This is an important parameter to be considered in the analysis of MS data for the identification of glycated proteins. Similarly, search should be extended to half-cleaved peptides to increase the number of identifications.
10. The ratio between peak areas corresponding to the peptides labeled with [$^{12}C_6$]- and [$^{13}C_6$]-glucose provides quantitative information in relative terms. Peptides labeled with "heavy" glucose are considered as internal standards, and these isotopic forms specifically mimic physiological conditions. Therefore, in vitro labeling with [$^{13}C_6$]-glucose depends on sample properties such as protein content or pathological factors affecting glycation. The application of this approach is useful for the relative estimation of the extent of glycation at each potential attachment site. In addition, isotopic glucose labeling is valid as a quantitative approach to compare between two glycation states for the same or different patients.
11. The predictive approach proposed here is based on the differential labeling with [$^{12}C_6$]-glucose and [$^{13}C_6$]-glucose, and considers all possible glycations. As glucose labeling is performed by incubation under physiological conditions, glycation of proteins is mimicked in natural terms. As proven above, this favors the evaluation of the impact of glucose concentrations on identified sites.

Acknowledgments

This work was supported by the Spanish Ministry of Science and Innovation, the Spanish Foundation for Science and Technology (Grant with reference 2007-0398), and the Swiss SystemsX.ch initiative (Grant IPP-200 N/ONN).

References

1. Costello CE, Contado-Miller J M, Cipollo JF (2007) A glycomics platform for the analysis of permethylated oligosaccharide alditols. *J Am Soc Mass Spectrom* **18**, 1799–1812.
2. Maillard L C (1912) Action des acides amines sur le sucres: formation des mélanoidines per voie méthodique. *C R Acad Sci* **154**, 66–68.
3. Thornalley P J, Langborg A, Minhas H S (1999) Formation of glyoxal, methylglyoxal and 3-deoxyglucosone in the glycation of proteins by glucose. *Biochem J* **344**, 109–116.
4. Brock J W C, Hinton D J S, Cotham W E, et al. (2003) Proteomic analysis of the site specificity of glycation and carboxymethylation of ribonuclease. *J Proteome Res* **2**, 506–513.
5. Brunner Y, Schvartz D, Priego-Capote F, et al. (2009) Glucotoxicity and pancreatic proteomics. *J Proteomics* **71**, 576–591.
6. Reusch J E (2003) Diabetes, microvascular complications, and cardiovascular complications: what is it about glucose? *J Clin Invest* **112**, 986–988.
7. Baynes J W (2001) The role of AGEs in aging: causation or correlation. *Exp Gerontol* **36**, 1527–1537.
8. Brownlee M (2001) Biochemistry and molecular cell biology of diabetic complications. *Nature* **14**, 813–820.
9. Hipkiss A R (2006) Accumulation of altered proteins and ageing: causes and effects. *Exper Gerontol* **41**, 464–473.
10. Calvo C, Ponsin G, Berthezene F (1998) Characterization of the nonenzymatic glycation of high-density lipoprotein in diabetic patients. *Diabete Metab* **14**, 264–269.
11. Calvo C, Talussot C, Ponsin G, Berthezene F (1988) Nonenzymatic glycation of apolipoprotein A-I. Effects on its self-association and lipid binding properties. *Biochem Biophys Res Comm* **153**, 1060–1067.
12. Calvo C, Verdugo C (1992) Association in vivo of glycated apolipoprotein A-I with high density lipoproteins. *Eur J Clin Chem Clin Biochem* **30**, 3–5.
13. Abdel-Wahab Y H A, O'Harte F P M, Boyd A C, et al. (1997) Glycation of insulin results in reduced biological activity in mice. *Acta Diabetol* **34**, 265–270.
14. Miedema K. (2004) Towards worldwide standardisation of HbA_1c determination. *Diabetologia* **47**, 1143–1148.
15. Ladyzynski P, Wojcicki J M, Bak M. et al. (2008) Validation of haemoglobin glycation models using glycemia monitoring in vivo and culturing of erythrocytes in vitro. *Ann Biomed Eng* **36**, 1188–1202.
16. Saudek C D, Derr R L, Kalyani R R (2006) Assessing glycemica in diabetes using self-monitoring blood glucose and hemoglobin A1c. *J Am Med Assoc* **295**, 1688–1697.
17. Brownlee M, Hirsch I B (2006) Glycemic variability: a hemoglobin A1c-independent risk factor for diabetic complications. *J Am Med Assoc* **295**, 1707–1708.
18. Kisugi R, Kouzuma T, Yamamoto T. et al. (2007) Structural and glycation site changes of albumin in diabetic patient with very high glycated albumin. *Clin Chim Acta* **382**, 59–64.
19. Yoshiuchi K, Matsuhisa M, Katakami N. et al. (2008) Glycated albumin is a better indicator for glucose excursion than glycated hemoglobin in type 1 and type 2 diabetes. *Endocrine J* **55**, 503–507.
20. Priego-Capote F, Scherl A, Müller M, et al. (2010) Glycation isotopic labeling with ^{13}C-reducing sugars for quantitative analysis of glycated proteins in human plasma. Mol Cell *Proteomics* **9**, 579–592.
21. Bradford M (1976) A rapid and sensitive method for the quantitation of microgram quantities of protein utilizing the principle of protein-dye binding. *Anal Biochem* **72**, 248–254.
22. Huttlin E L, Hegeman A D, Harms A C, Sussman M R (2007) Prediction of error associated with false-positive rate determination for peptide identification in large-scale proteomics experiments using a combined reverse and forward peptide sequence database strategy. *J Prot Res* **6**, 392–398.

Part IV

Proteome Analysis of Blood Cell Components, Vesicles, and Blood-Related Fluids

Chapter 15

Exosome Isolation for Proteomic Analyses and RNA Profiling

Douglas D. Taylor, Wolfgang Zacharias, and Cicek Gercel-Taylor

Abstract

While the existence of exosomes has been known for over three decades, they have garnered recent interest due to their potential diagnostic and therapeutic relevance. The expression and release of specific tumor-derived proteins into the peripheral circulation has served as the centerpiece of cancer screening and diagnosis. Recently, tissue-associated microRNA (miRNA) has been shown to be characteristic of tumor type and developmental origin, as well as exhibit diagnostic potential. Tumors actively release exosomes, exhibiting proteins and RNAs derived from the originating cell, into the peripheral circulation and other biologic fluids. Recently, we have demonstrated the presence of miRNAs within the RNA fraction of circulating tumor-derived exosomes. Currently, in over 75 investigations compiled in ExoCarta, over 2,300 proteins and 270 miRNAs have been linked with exosomes derived from biologic fluids. Our previous work has indicated that these circulating exosomal proteins and miRNAs can serve as surrogates for the tumor cell-associated counterparts, extending their diagnostic potential to asymptomatic individuals. In this chapter, we compare currently utilized methods for purifying exosomes for postisolation analyses. The exosomes derived from these approaches were assessed for quantity and quality of specific RNA populations and specific marker proteins. These results suggest that, while each method purifies exosomal material, circulating exosomes isolated by ExoQuick precipitation produces exosomal RNA and protein with greater purity and quantity than chromatography, ultracentrifugation, and DynaBeads. While this precipitation approach isolates exosomes in general and does not exhibit specificity for the originating cell, the increased quantity and quality of exosomal proteins and RNA should enhance the sensitivity and accuracy of down-stream analyses, such as qRT-PCR profiling of miRNA and mass spectrometric and electrophoretic analyses of exosomal proteins.

Key words: Exosomes, Biofluids, Proteomics, miRNA profiling

1. Introduction

The release of small vesicles by cancer and embryonic cell types was initially described by our group three decades ago (1). Since that time, the release of membranous vesicles (exosomes), ranging from 50 to 100 nm, has been demonstrated in multiple cell

Richard J. Simpson and David W. Greening (eds.), *Serum/Plasma Proteomics: Methods and Protocols*,
Methods in Molecular Biology, vol. 728, DOI 10.1007/978-1-61779-068-3_15, © Springer Science+Business Media, LLC 2011

types and systems (2) (see Note 1). Although we demonstrated the presence of microvesicles in the peripheral circulation of patients with cancer and pregnant women in the late 1970s (3), they have received renewed interest due to the recognition that they serve as essential intercellular communication vehicles, are key determinants of the immunosuppressive microenvironment observed in cancer and pregnancy (4, 5) and provide stability to tumor-derived proteins that can serve as diagnostic biomarkers for cancer. During the 1970s, following the development of monoclonal antibody technology, significant work was directed at the identification of new biomarkers for the development of antigen-based assays for cancer. Our original work focused on the identification of specific isoforms of placental-type alkaline phosphatase (PLAP) associated with ovarian cancer (1, 3). We defined the presence of heat-stable PLAP in the sera and ascites of patients with ovarian cancer. Using size exclusion chromatography, we attempted to purify PLAP from the blood and ascites of cancer patients. While the expected molecular weight of PLAP was approximately 70,000 Da, the enzyme activity appeared in the void volume of Sephadex G200 columns. Subsequent separations using high exclusion limit, agarose-based gels revealed a molecular weight between 100 and 150 million Daltons (1). Further analyses of a large group of female patients in a blinded study found that these high molecular weight complexes were observed in all women with cancer and in women who were pregnant; however, the 100–150 million Dalton fractions were not observed in any other groups. We now recognize that the apparent absence of exosomes in normal individuals was the consequence of the limited detection of the methods used, with pathologic conditions, such as cancer and pregnancy, linked with elevated production and release of exosomes.

Analyses of these "high molecular weight" complexes with various extraction approaches indicated that they consisted of lipids and proteins (6). Using specific markers of intracellular membranous components, these complexes expressed components associated with the plasma membrane, but failed to exhibit markers of nuclear or mitochondrial membranes. This suggested that the presence of this membranous material was not the result of cell death. The membranous complexes were shown to possess markers associated with tumors in cancer patients and with placenta in pregnant patients (6, 7). Further, analyses using metabolic inhibitors demonstrated that the release of these membranous structures was energy requiring, further indicating that the release of this material was an active cellular process and not merely the consequence of cell death (6). Our initial electron micrographs of these membranous complexes revealed that they were vesicular (6). Subsequently, Pan and Johnstone (8) described the role of these vesicles from sheep reticulocytes in the recycling of transferrin receptors (TfR) during maturation and that these vesicles of endocytic origin were secreted extracellularly. Thus, this secreted vesicular material became identified as "exosomes."

In addition to recycling "unwanted" proteins during maturation, we now recognize that exosomes play a significant role in intracellular communication. With current mass spectrometry-based proteomic technologies, it is clear that exosomes are comprised of distinct subpopulations of macromolecules, including proteins and RNAs, associated with cell type-linked functions. Recent data indicate that these macromolecules can be transferred to target cells and can mediate intercellular interactions, nonclassical protein secretion, signaling between neighboring cells resulting in pathologic conditions by becoming functional in their new microenvironment (9, 10). While tumor-derived exosomes can induce events associated with the pathology of cancer, the renewed interest in exosomes has been their potential diagnostic utility. Mathivanan and Simpson (11) have established a compendium of exosomal proteins and RNA (ExoCarta), since numerous studies have demonstrated the presence of tissue/cell type-specific proteins associated with exosomes. The presence of tissue/cell type-specific marker proteins associated with specific exosome populations can serve as surrogates, identifying the presence of the originating cell. Similarly, in 2008, we published the initial description of microRNA (miRNA) associated with circulating exosomes derived from ovarian cancer patients and demonstrating their diagnostic use (12), which would extend its utility to screening asymptomatic individuals.

While currently data from 75 studies have been cataloged in ExoCarta, the various groups investigating exosomal components lack consensus on the methods for isolating exosomes from biologic fluids of cancer patients, which include ultracentrifugation, chromatography, and magnetic beads. Include this chapter, we compare these methods along with a new precipitation procedure, for the isolation of exosomes from biologic fluids for subsequent analyses, including RNA profiling or protein analyses using mass spectrometry and electrophoresis-based proteomic techniques.

2. Materials

2.1. Patient Samples

To compare methods for the isolation of exosomes from biologic fluids derived from cancer patients, three ascites specimens were obtained from women diagnosed with stage III serous adenocarcinoma of the ovary.

2.2. Exosome Isolation Methods

1. Size exclusion chromatography. Size exclusion chromatography gel comprised of 2% agarose (Agarose Bead Technologies, Tampa, FL).

2. Magnetic beads. Anti-EpCAM antibodies coupled to magnetic microbeads (#161.02, DynaBeads, Invitrogen, Carlsbad, CA).
3. Ultracentrifugation. Beckman Optima MAX-XP centrifuge with a TLA-120.2 rotor (Beckman Coulter, Fullerton, CA).
4. ExoQuick precipitation. ExoQuick precipitation reagent #EXOQ5A-1 (System Biosciences, Mountain View, CA).

2.3. RNA Isolation and Analysis

1. Trizol (#15596-028, Invitrogen).
2. Agilent small RNA chip and reagent kit (#5067-1548).

2.4. Exosome Protein Analysis by SDS-PAGE and Western Immunoblotting

1. Separating buffer (4× solution): 1.5 M Tris(hydroxymethyl) aminomethane (Tris–HCl), pH 8.7, 0.4% SDS (Bio-Rad Laboratories, Hercules, CA).
2. Stacking buffer (4× solution): 0.5 M Tris–HCl, pH 6.8, 0.4% SDS (Bio-Rad Laboratories).
3. 30% Acrylamide/Bis solution (37.5:1 with 2.6% C) and *N*,*N*,*N*,*N'*-tetramethyl-ethylenediamine (TEMED, Bio-Rad Laboratories).
4. Ammonium persulfate: 10% solution in water prepared immediately prior to use.
5. Running buffer (10× solution): 250 mM Tris, 1,920 mM glycine, 1.0% (w/v) SDS, pH 8.3 (#161-0772, Bio-Rad Laboratories).
6. Prestained molecular weight markers: Kaleidoscope markers (Bio-Rad Laboratories).
7. Transfer buffer (Tris–Glycine): 25 mM Tris, 190 mM glycine, 20% (v/v) methanol plus 0.05% (w/v) SDS.
8. Nitrocellulose membrane (0.45 μm, #162-0117, Bio-Rad Laboratories) and thick filter paper (#170-5040, Bio-Rad Laboratories).
9. Tris-buffered saline, as a 10× solution (#170-6435, Bio-Rad Laboratories) with 0.1% Tween-20 (TBS-T).
10. Blocking buffer: 5% (w/v) nonfat dry milk (#170-6404, Bio-Rad Laboratories) in TBS-T.
11. Bradford protein microassay kit with BSA standard (#500-0203, Bio-Rad Laboratories).
12. Primary antibody dilution buffer: Super-Block solution (#37537, Pierce Chemical Co.).
13. Primary antibody: mouse antihuman placental alkaline phosphatase (sc-47691, Santa Cruz Biotechnology, Santa Cruz, CA).
14. Secondary antibody: Antimouse IgG conjugated to horseradish peroxidase (#170-5047, Bio-Rad Laboratories).

15. Immun-Star HRP chemiluminescent reagents (#170-5040 from Bio-Rad) and Bio-Max Light-2 film (Kodak, Rochester, NY).
16. Visualized bands were digitized and quantitated using Un-Scan-it Software (Silk Scientific Corp., Orem, UT).

3. Methods

Exosomes can be isolated from blood (plasma or serum) or ascites (see Note 1). For plasma and ascites, sodium heparin (1,000 IU/L) is added prior to isolation. To isolate exosomes from biologic fluids, the ascites (2 mL) was initially centrifuged at 12,000 ×*g* for 15 min at 4°C to remove any cellular debris.

3.1. Exosome Isolation by Ultracentrifugation

1. The cell-free ascites specimens were centrifuged at 100,000 ×*g* for 1 h at 4°C.
2. The pellet containing exosomes was resuspended in PBS and then recentrifuged at 100,000 ×*g* for 1 h at 4°C.
3. The resulting pellet was used for Trizol extraction for RNA and protein analyses.

3.2. Exosome Isolation by Size Exclusion Chromatography

1. For isolation by size exclusion chromatography, 2 mL aliquots of patient-derived cell-free ascites were applied to a 2% agarose-based gel column (2.5 × 16 cm). For optimal separation, the sample volume should be 1/20 of the total column volume (as defined by $\pi r^2 h$).
2. The column was eluted isocratically with PBS at a flow rate of 1 mL/min, monitoring at 280 nm and fractions (2 mL) were collected.
3. The void volume fractions (based on absorbance at 280 nm) were pooled and centrifuged at 100,000 ×*g* for 1 h at 4°C.
4. The resulting pellet was used for Trizol extraction for RNA and protein analyses.

3.3. Exosome Isolation by Magnetic Beads

1. For isolation of specific subsets of exosomes, exosomes were selectively isolated from ascites specimens by absorption to anti-EpCAM antibodies coupled to magnetic microbeads.
2. Anti-EpCAM coupled to microbeads (50 μL) were added to the ascites specimens (2 mL), mixed, and incubated on a shaker for 2 h at room temperature.
3. Each tube was placed in the magnetic separator and fluid removed, leaving the magnetic beads and bound exosomes attached to the side of the tube.
4. The tube was removed from the magnetic separator and the beads rinsed with 500 μL TBS and the separation repeated.

5. After the wash step, the tube was removed from the magnetic holder and the bead/exosome complex was used for Trizol extraction for RNA and protein analyses.

3.4. Exosome Isolation by ExoQuick Precipitation

1. To isolate exosomes by the precipitation approach, the ascites specimen (2 mL) was transferred to a sterile tube and 0.5 mL of ExoQuick exosome precipitation solution was added and mixed.
2. The mixture was incubated overnight (at least 12 h) at 4°C and then the mixture was centrifuged at 10,000 × *g* in a microfuge for 5 min at 4°C.
3. The supernatant was aspirated and the exosome pellet was extracted using the Trizol extraction procedures for RNA and protein analyses.

3.5. RNA Isolation and Analysis (See Notes 2–4)

1. Total RNA was isolated from exosomes derived from each isolation method using Trizol according to manufacturer's instructions (Invitrogen), except with the isopropanol precipitation step extended to overnight.
2. The RNA quality and yield was accessed using a GeneQuant II.
3. Small RNAs were analyzed with the Agilent 2100 Bioanalyzer Lab-on-a-Chip instrument system (Agilent Technologies, Santa Clara, CA), using the Agilent Small RNA chip and reagent kit (see Note 3).
4. Approximately 100 ng of isolated total RNA in 1 μL was applied to each run.
5. The manufacturer's recommended protocol was strictly followed to obtain Bioanalyzer profiles for the size range of 6–150 nucleotides (nt).
6. The profiles were calibrated for size (nt) using the small RNA ladder supplied with the kit, containing markers of 20, 40, 60, 80, and 150 nt in size, as reference.
7. The instrument software quantitated the peak area between 0 and 150 nt as small RNA region, the area within 10–40 nt as miRNA region, and provides percentages of miRNA detected for each sample (see Note 5).

3.6. Exosome Protein Analysis by Western Immunoblotting (See Note 6)

1. Exosomal protein isolation was performed by continuing the Trizol isolation procedure, as described by the manufacturer.
2. The quantity of protein was determined by the Bradford microassay method, using BSA as a standard.
3. SDS-PAGE was performed by the method of Laemmli (13). Proteins from each exosome isolate were standardized to the original sample volume and equal volumes were applied per lane of a 12.5% SDS-PAGE gel.

4. Western immunoblotting was performed to analyze the presence of the specific marker protein, PLAP as described by Brown (14).
5. The SDS-PAGE gel was transferred to a nitrocellulose membrane, the membrane blocked for 1 h at room temperature with nonfat dried milk, and probed overnight at 4°C with primary antibody.
6. The bound immune complexes were visualized by ECL onto X-ray film and the resulting bands were quantified by digitizing the X-ray film image.

4. Notes

In the late 1970s, biofluids (ascites and blood) were fractionated by size exclusion chromatography to identify and isolate circulating tumor-derived antigens as potential diagnostic biomarkers (see Notes 7–13). Based on these initial studies, we identified a "high molecular weight" form of these ovarian tumor antigens (in excess of 50 million Daltons). Due to their size, we attempted to sediment these high molecular weight antigens by ultracentrifugation prior to chromatographic separation from other blood components; however, little of this material could be pelleted (also see Notes 2 and 6). By the early 1980s, we referred to these vesicular structures as "membrane vesicles," which were subsequently termed exosomes. A major obstacle to the study of circulating tumor proteins is their relative instability. The association of tumor associated proteins with exosomes, in contrast, creates a stable environment for these proteins. This enhanced stability has renewed interest in exosomes and their associated components as diagnostic biomarkers. The current use of mass spectrometry for proteomics on circulating biomarkers focuses on proteins in the peripheral circulation. The association of tumor-derived proteins with exosomes allows the proteins to be studied in their intact forms and with appropriate conformations:

1. Exosomes have been isolated from most biofluids (blood, ascites, milk, urine) (10, 15–19). While exosomes have been isolated from both plasma and serum with similar results, our experience is that better yields of protein and RNA are obtained with plasma.
2. For isolation of RNA for further analyses, based on the OD_{260}/OD_{280} ratios, in general, the ExoQuick precipitation (Subheading 3.4) produced an enhanced concentration of RNA and a higher ratio indicating greater purity (Fig. 1). The use of size exclusion chromatography (Subheading 3.2) to

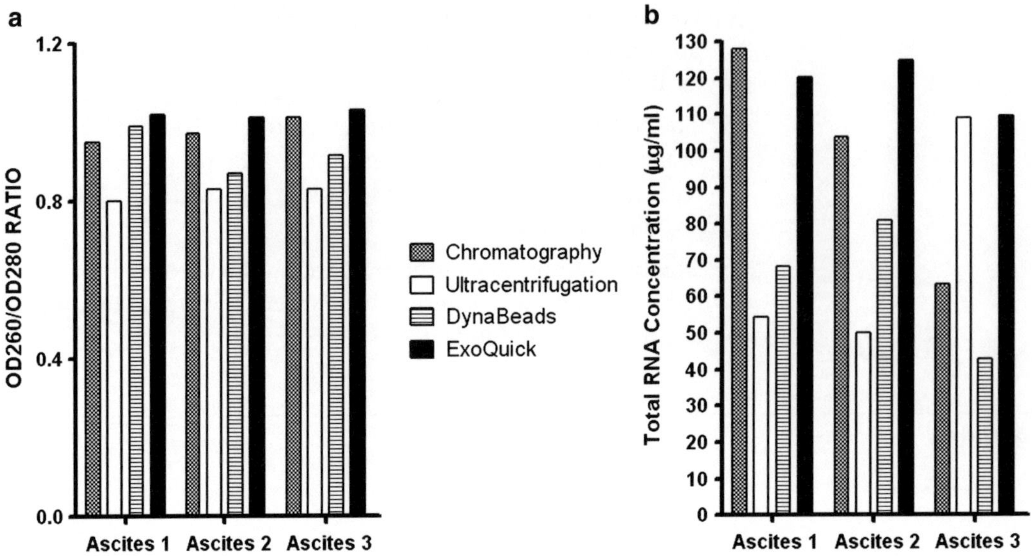

Fig. 1. Total RNA from exosomes isolated from biologic fluids (ascites) of ovarian cancer patients by size exclusion chromatography, ultracentrifugation, anti-EpCAM DynaBeads, and ExoQuick precipitation procedure. Panel (**a**) presents OD_{260}/OD_{280} ratio for RNA purity. Panel (**b**) presents total RNA isolated.

isolate circulating exosomes also yields similar ratios of purity. Ultracentrifugation (Subheading 3.1) produced the lowest ratios for RNA integrity, potentially due to cosedimentation of protein impurities. The use of anti-EpCAM dynabeads (Subheading 3.3) isolated exosomal RNA similar in purity to chromatography, although the quantity was significantly less.

3. In our initial characterization of exosomal RNA, the analysis was limited by the use of the RNA6000 Pico LabChip kit (Agilent), which defined only the absence of 18S and 28S RNA and poorly discriminated small RNAs [12). Thus, for this comparative characterization of small RNA populations derived from circulating exosomes, the small RNA LabChip kit was used.
4. In terms of RNA isolation, the ExoQuick precipitation (Subheading 3.4) produced the greatest quantity of exosomal RNA in all three ascites specimens with the greatest purity. When the isolated exosomal small RNA fraction was analyzed by Bioanalyzer scanning (Fig. 2), it revealed that the exosomal RNA derived by the ExoQuick precipitation approach yielded more small RNA with a greater percent of miRNA than that observed with anti-EpCAM DynaBeads (percent of miRNA was 53% vs. 47%, respectively).
5. Since miRNA signatures have been linked with cancer type, stage and grade, as well as predicting therapeutic responses, the enhanced isolation of miRNA is essential for accurate miRNA profiling.

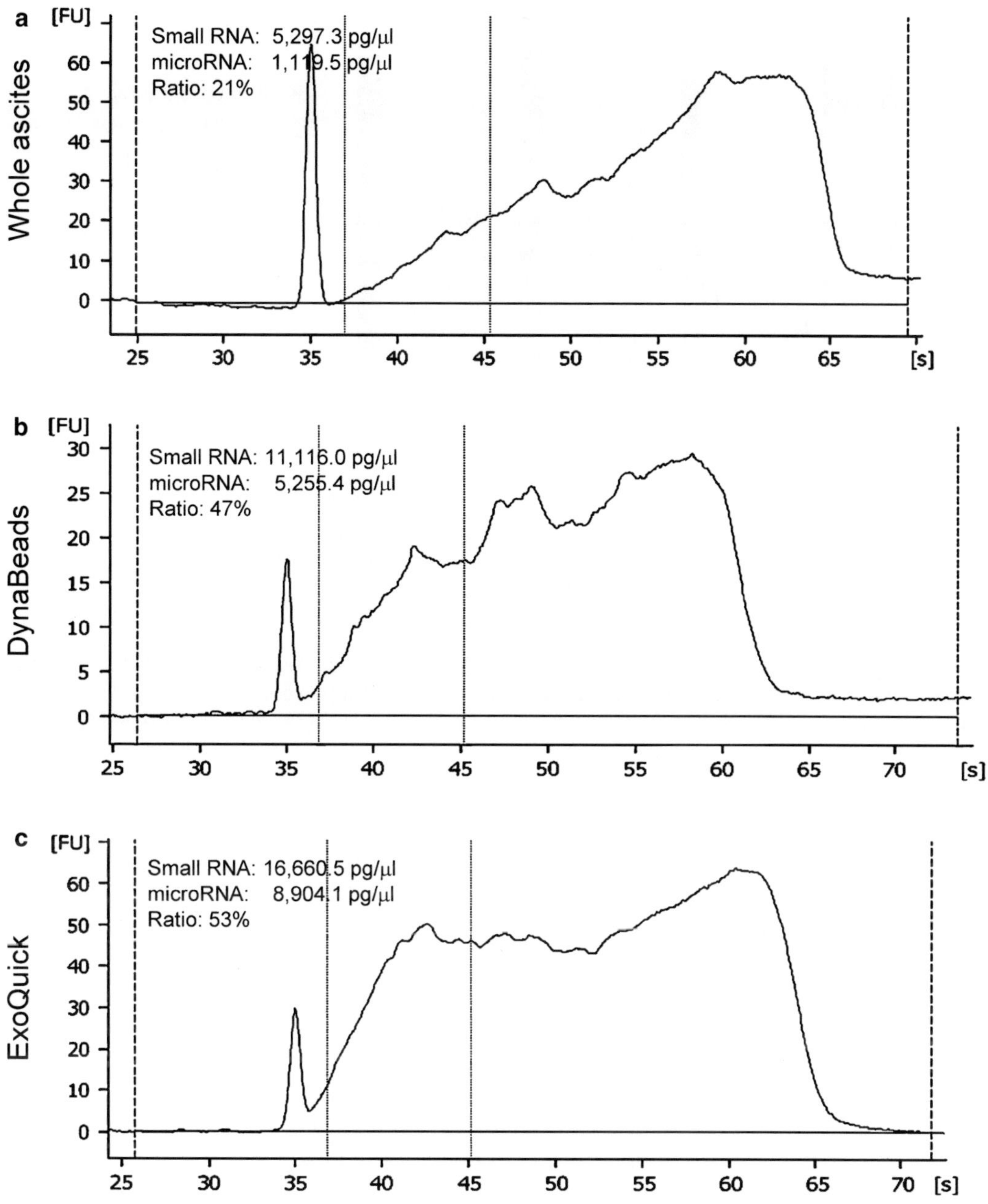

Fig. 2. Bioanalyzer electropherogram characterization of the small RNA fractions from Trizol extraction of unfractionated ovarian cancer ascites (panel **a**) compared with Trizol extraction of exosomes isolated from the same ascites by anti-EpCAM DynaBeads (panel **b**) and ExoQuick precipitation (panel **c**).

6. For isolation of exosomal protein for further analyses, the ExoQuick precipitation yielded an increased concentration of protein from each ascites sample (Fig. 3). The use of size exclusion chromatography (Subheading 3.2) to isolate circulating exosomes resulted in the second highest concentration of exosomal protein. Ultracentrifugation (Subheading 3.1) produced the lowest ratios of exosomal protein. The use of

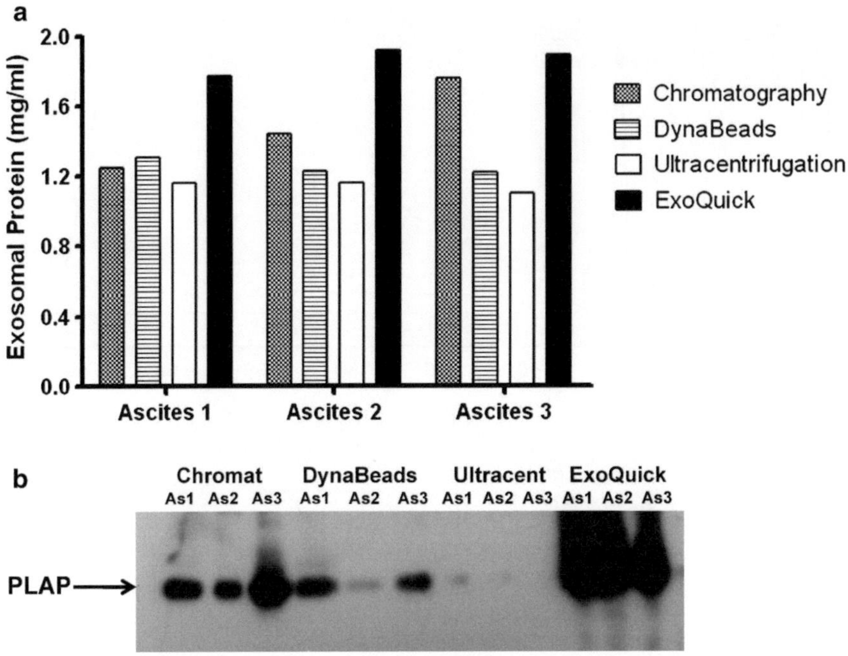

Fig. 3. Exosomal proteins isolated from biologic fluids (ascites) of ovarian cancer patients by size exclusion chromatography, ultracentrifugation, anti-EpCAM DynaBeads, and ExoQuick precipitation. Panel (**a**) presents total exosomal protein concentrations isolated by each technique. Panel (**b**) presents the western immunoblots for PLAP, visualized by ECL, associated with exosomes isolated by each approach.

anti-EpCAM Dynabeads (Subheading 3.3) isolated exosomal proteins at a level intermediate between chromatography and ultracentrifugation.

7. During the past three decades, we evaluated different approaches for exosome isolation from blood and other biofluids that would allow high-throughput isolation of intact exosomal proteins for proteomic analyses, while also retaining their biologic activities. Analysis of these isolated exosomal proteins for the level of the specific protein marker, PLAP, indicated that, when standardized to volume of the initial sample, the ExoQuick-derived exosomes exhibit the highest level of the tumor marker, followed by chromatography and DynaBeads.
8. It should be noted that chromatography, ultracentrifugation, and ExoQuick precipitation cannot preferentially isolate tumor-derived exosomes. These are general approaches that isolate all circulating exosome populations.
9. We have focused on approaches appropriate for the small sample volumes, feasible with clinical specimens. Chromatography is labor-intensive and requires specialized equipment. Chromatographic separation of exosomes exhibits an incomplete recovery of exosomes, due to their dilution

into multiple fractions and nonspecific adherence to the agarose chromatography beads. Sedimentation of exosomes requires an ultracentrifuge and we have previously demonstrated that direct ultracentrifugation of biofluids does not pellet all, or even most, of the exosomes present. This appears to be due to protein composition of the biofluid and the resulting viscosity; exosomes pellet poorly, without prior chromatographic separation (1). The ExoQuick precipitation uses only low-speed microfuge and does not dilute the biofluid-derived exosomes.

10. Of the approaches described here, only immunoaffinity, using the anti-EpCAM, can selectively derive any specific population of exosomes.
11. Since we are investigating cancers of epithelial origin, our immunoaffinity approach is based on anti-EpCAM. Based on our previous work, we are aware that antibody binding to exosomal antigens is distinct from binding their counterparts expressed on cells (1). This may relate to differences in glycosylation of exosomal proteins or other aberrant posttranslational modifications or differences in lipid/protein ratios of the membranes of exosomes versus cells, resulting in differential exposure of antigenic epitopes on integral membrane proteins or changes in protein motility due to the rigidity of exosomal membranes compared to cellular membranes. These may effectively reduce the binding of exosomal target proteins with antibodies linked to beads and reducing the levels of specific exosomes isolated.
12. Other investigators have utilized other surface markers or exosome markers to isolate exosomes from biologic fluids. Selection for A33-positive exosomes has been reported for isolating exosomes from colorectal cancer patients (11). Other investigators have also used anti-CD63 for immunoaffinity selection of circulating exosomes (9); however, as a general exosome marker, the use of CD63 isolates all circulating exosomes, cancerous and noncancerous.
13. Since the level of exosomes in biologic fluids is significantly elevated in pathologic conditions, it may not be essential to separate tumor-derived exosomes from the normal exosome background. The presence of tissue/cell type-specific marker proteins or cell-specific miRNA signatures associated with specific exosome populations can serve as surrogates, identifying the presence of the originating cell. Thus, postisolation analyses by RNA profiling or proteomics should be capable of identifying the presence of specific tissue/cell type components or signatures associated with tumor exosomes above the background of exosomes from normal cells.

References

1. Taylor DD, Doellgast GJ. (1979) Quantitation of peroxidase-antibody binding to membrane fragments using column chromatography. *Anal Biochem* **98**, 53–59.
2. Simpson RJ, Jensen SS, Lim JWE. (2008) Proteomic profiling of exosomes: Current perspectives. *Proteomics* **8**, 4083–4099.
3. Taylor DD, Homesley HD, Doellgast GJ. (1980) Binding of specific peroxidise-labeled antibody to placental-type alkaline phosphatase on tumor-derived membrane fragments. *Cancer Res* **40**, 4964–4969.
4. Raposo G, Tenza D, Mecheri S, et al. (1997) Accumulation of major histocompatibility complex class II molecules in mast cell secretory granules and their release upon degranulation. *Mol Biol Cell* **8**, 2631–2645.
5. Heijnen HFG, Schiel AE, Fijnheer R, et al. (1999) Activation platelets release two types of membrane vesicles: Microvesicles by surface shedding and exosomes derived from exocytosis of multivesicular bodies and alpha granules. *Blood* **94**, 3791–3799.
6. Taylor DD, Black PH. (1986) Shedding of plasma membrane fragments: Neoplastic and developmental importance. In: *Developmental Biology*, (M. Steinberg, ed.) **3**, 33–57.
7. Taylor DD, Bohler HC, Gercel-Taylor C. (2006) Pregnancy-linked suppression of TcR signaling pathways by a circulating factor absent in recurrent spontaneous pregnancy loss. *Mol Immunol*, **43**, 1872–1880.
8. Pan BT, Johnstone RM. (1983) Fate of the transferring receptor during maturation of the sheep reticulocytes in vitro: Selective externalization of the receptor. *Cell* **33**, 967–978.
9. Skog J, Wurdinger T, van Rijn S, et al. (2008) Glioblastoma microvesicles transport RNA and proteins that promote tumour growth and provide diagnostic biomarkers. *Nature Cell Biol* **10**, 1470–1476.
10. Taylor DD, Gercel-Taylor C. (2005) Tumour-derived exosomes as mediates of T-cell signaling defects. *Brit J Cancer* **92**, 305–311.
11. Mathivanan S, Simpson RJ. (2009) ExoCarta: A compendium of exosomal proteins and RNA. *Proteomics* **9**, 4997–5000.
12. Taylor DD, Gercel-Taylor C. (2008) MicroRNA signatures of tumor-derived exosomes as diagnostic biomarkers of ovarian cancer. *Gynecol Oncol* **110**, 13–21.
13. Laemmli U K. (1970) Cleavage of structural proteins during the assembly of the head of bacteriophage T4. *Nature* **227**, 680–685.
14. Brown, R., Clugston, C., Burns, P., et al. (1993) Increased accumulation of p53 protein in cisplatin-resistant ovarian cell lines. *Int J Cancer* **55**, 678–684.
15. Runz S, Keller S, Rupp C, Stoeck A, Issa Y, Koensgen D, Mustea A, Sehouli J, Kristiansen G, Altevogt P. (2007) Malignant ascites-derived exosomes of ovarian carcinoma patients contain CD24 and EpCAM. *Gynecol Oncol* **107**, 563–571.
16. Fernández-Llama P, Khositseth S, Gonzales PA, Star RA, Pisitkun T, Knepper MA. (2010) Tamm-Horsfall protein and urinary exosome isolation. Kidney Int 77, 736–742.
17. Chen C, Skog J, Hsu CH, Lessard RT, Balaj L, Wurdinger T, Carter BS, Breakefield XO, Toner M, Irimia D. (2010) Microfluidic isolation and transcriptome analysis of serum microvesicles. *Lab Chip* **10**, 505–511.
18. Admyre C, Johansson SM, Qazi KR, Filén JJ, Lahesmaa R, Norman M, Neve EP, Scheynius A, Gabrielsson S. (2007) Exosomes with immune modulatory features are present in human breast milk. *J Immunol* **179**, 1969–1978.
19. Palanisamy V, Sharma S, Deshpande A, Zhou H, Gimzewski J, Wong DT. (2010) Nanostructural and transcriptomic analyses of human saliva derived exosomes. *PLoS One* **5**, e8577.

Chapter 16

Extraction and Proteome Analysis of Liver Tissue Interstitial Fluid

Wei Sun, Ying Jiang, and Fuchu He

Abstract

Tissue interstitial fluid (TIF) forms the interface between circulating body fluids and intracellular fluid. Pathological alterations of liver cells could be reflected in TIF, making it a promising source of liver disease biomarkers. Here, we introduce the method of extracting TIF from mouse liver. We confirm the absence of cellular protein contamination by western blot using organelle specific protein markers including lamin B, cytochrome C, and flotillin-1. Using two-dimensional differential in-gel electrophoresis (2D-DIGE), we demonstrate different profiling patterns among TIFs of liver tissue and hepatocytes. The high solubility and even distribution of liver TIF supports its suitability for proteome analysis. Based on the method introduced here, liver TIF in animal models and clinical samples could be extracted and analyzed for further application in liver disease biomarker discovery.

Key words: Liver, Tissue interstitial fluid, Proteome, Extraction, Plasma

1. Introduction

The liver performs essential functions by expressing hepatocyte-specific genes, which encode plasma proteins, enzymes involved in metabolism, and proteins that participate in the detoxification of exogenous chemicals (1). In addition to hepatocytes, which contribute roughly 80% of the liver mass, there are many types of nonparenchymal cells (NPCs), such as sinusoidal endothelial cells, Kupffer cells, hepatic stellate cells, and pit cells. Cross talk occurs between the NPCs and hepatocytes through paracrine secretion, and NPCs form the cellular microenvironment of hepatocytes. Further, blood, bile, and lymph fluid form the fluid microenvironment of the hepatocytes and accommodate substances that play a role in signal exchange within the cellular microenvironment.

Richard J. Simpson and David W. Greening (eds.), *Serum/Plasma Proteomics: Methods and Protocols*,
Methods in Molecular Biology, vol. 728, DOI 10.1007/978-1-61779-068-3_16, © Springer Science+Business Media, LLC 2011

The liver tissue interstitial fluid (TIF), therefore, is the medium between circulating body fluids, such as plasma, and the intracellular fluid. Because pathological changes in liver cells could be reflected in the microenvironment and composition of the TIF, it could provide a promising resource for liver-specific disease biomarker discovery.

Compared with plasma, a commonly used resource for biomarker discovery, TIF has the advantage of not having the disease-associated proteins diluted in a large biofluid volume (5 L of plasma in adult) or being combined with proteins from other organs and tissues. It has been hypothesized that concentrations of potential biomarkers are highest in the tumor and its immediate microenvironment (i.e., TIF), and this will be significantly diluted (~1,500-fold) upon passage to the circulatory system via the lymphatics (2, 3). And only those tissue proteins which could be secreted into body fluids have the potential for clinical application.

At the same time, it is a technical challenge to analyze the plasma proteome directly, due to the high dynamic protein concentration range. Sufficient identification of low-abundant proteins is hampered by existence of high-abundant proteins. Two-dimensional differential in-gel electrophoresis (2D-DIGE) of mouse liver TIF shows the even distribution of proteins in TIF and the absence of obvious high-abundant proteins (4). Comparison between TIF and plasma/serum in their application in disease proteomics is shown in Table 1.

Table 1
Comparison between TIF and plasma/serum in their application in disease proteomics. (Reprinted with permission from (4). Copyright 2010 American Chemical Society)

	TIF	Plasma/serum
Concentration of target proteins	High Especially low-abundant proteins such as regulation proteins	Low Diluted by peripheral plasma (5 L)
Disease specificity	High From local focus	Low Complex and chaoed with unrelated proteins from other parts of the human body
High-abundance proteins	Not so serious	Serious
Degradation	Relatively stable in local focus	Tend to be degraded in blood circulation
Others	Cellular interaction information in the microenvironment is involved	

Celis et al. have profiled the TIF proteome of human breast carcinoma, making a significant contribution to the current body of knowledge in this field (5, 6). However, the absence of cellular protein contamination in TIF has not been demonstrated; such an analysis is critical for determining the appropriate method for extraction of TIF and for data annotation. Here, we evaluate the contamination of TIF extracted from liver tissue and show its different profiling patterns with liver tissue and hepatocytes by 2D-DIGE.

2. Materials

2.1. TIF Collection

1. PBS (Phosphate-buffered solution): 137 mM NaCl, 2.7 mM KCl, 4.3 mM Na_2HPO_4, and 1.5 mM KH_2PO_4, pH 7.3.
2. Protease inhibitor cocktail (Roche, Penzberg, Germany).
3. 15-mL Tubes (BD, San Jose, CA, USA).

2.2. SDS–Polyacrylamide Gel Electrophoresis

1. Separating buffer: 1.5 M Tris–HCl, pH 8.8. Store at room temperature.
2. Stacking buffer: 0.5 M Tris–HCl, pH 6.8. Store at room temperature.
3. 30% (w/v) Acrylamide/bis solution (29:1) and *N,N,N,N′*-Tetramethyl-ethylenediamine (TEMED, Bio-Rad, Hercules, CA).
4. Ammonium persulfate: prepare 10% (w/v) solution in water and immediately freeze in single use (200 μL) aliquots at −20°C.
5. 10% (w/v) Sodium dodecyl sulfate (SDS).
6. Running buffer (5×): 125 mM Tris–HCl, 960 mM glycine, and 0.5% (w/v) SDS, pH 8.3.
7. SDS–PAGE loading buffer (2×): 120 mM Tris–HCl (pH 6.8), 4% (w/v) SDS, 20% (v/v) glycerol, bromophenol blue (trace), and 200 mM dithiothreitol (DTT) (add before use).

2.3. Western Blotting

1. Transfer buffer: 25 mM Tris–HCl (do not adjust pH), 190 mM glycine, 0.05% (w/v) SDS, and 20% (v/v) methanol.
2. Supported polyvinylidene fluoride membranes (Amersham Pharmacia Biotech, Uppsala, Sweden).
3. Tris-buffered saline with Tween (TBS-T) (10×): Prepare stock with 1.37 M NaCl, 27 mM KCl, 250 mM Tris–HCl, pH 7.4, 1% Tween-20. Dilute 100 mL with 900 mL of water for use.
4. Blocking buffer: 5% (w/v) nonfat dry milk in TBS-T.

5. Antibody dilution buffer: TBS-T supplemented with 5% (w/v) nonfat dry milk.
6. Primary antibodies: antilamin B polyclonal antibody (diluted 1:1,000, Santa Cruz, CA, USA), anticyto C (cytochrome C) monoclonal antibody (diluted 1:1,000, Neomarkers, Fremont, CA, USA), antiflotillin-1 monoclonal antibody (diluted 1:500, BD, San Jose, CA, USA), antibeta-actin polyclonal antibody (diluted 1:500, Neomarkers, Fremont, CA, USA).
7. Secondary antibody: anti-mouse IgG conjugated to horse radish peroxidase (Santa Cruz, CA).
8. Enhanced chemiluminescent (ECL) reagents (Pierce, Rockford, IL, USA).

2.4. Two-Dimensional Differential In-Gel Electrophoresis

1. Cy2, Cy3, and Cy5 dyes (GE Healthcare, Piscataway, NJ, USA).
2. Dimethylformamide (DMF), 99.8% (Aldrich, Steinheim, Germany).
3. Thiourea, >99% (Fluka, Buchs, Switzerland).
4. Immobilized pH gradient (IPG) strips (24 cm, pH 3–10 NL) and Pharmalyte (pH 3–10) (GE Healthcare).
5. Lysis buffer: 7 M urea, 2 M thiourea, and 4% (w/v) CHAPS.
6. 2× Sample buffer: 7 M urea, 2 M thiourea, 4% (w/v) CHAPS, 1% (v/v) Pharmalyte pH 3–10, and 20 mg/mL DTT.
7. Rehydration buffer: 7 M urea, 2 M thiourea, 4% (w/v) CHAPS, 0.5% (v/v) Pharmalyte pH 3–10, and 10 mg/mL DTT.
8. Equilibration buffer: 6 M urea, 2% SDS, 50 mM Tris–HCl (pH 8.8), and 30% (v/v) glycerol.
9. All buffers were prepared with Milli-Q water (Millipore, Bedford, MA, USA).

3. Methods

3.1. TIF Collection

C57BJ/6L mice are used in this experiment. The TIF collection protocol is modified based on a method developed by Celis et al (4, 5). Since the liver is one of the most fragile organs and contains an abundance of proteolytic enzymes, we perform additional steps to avoid cell lysis. All the experimental procedures should be performed on ice or at 4°C.

1. Fresh liver tissue (see Note 1) is cut into small pieces (1–3 mm^3) in PBS with Protease inhibitor cocktail (see Note 2) on ice.

2. The tissue is transferred to a 15-mL tube, washed carefully with PBS (without protease inhibitor), and immersed in an equal volume of PBS (see Note 3).
3. Sample is incubated at 37°C in a CO_2 incubator for 1 h (see Note 4).
4. The sample is centrifuged at 1,000 × *g* for 3 min at 4°C and supernatant is transferred to a fresh Eppendorf tube (see Note 5).
5. The sample is further centrifuged at 2,000 × *g* for 8 min at 4°C and supernatant is transferred to a fresh Eppendorf tube (see Note 6).
6. After centrifugation at 20,000 × *g* for 30 min at 4°C, supernatant is concentrated in a vacuum dehydrator for 1 h to obtain the final TIF (see Note 7).
7. After split charging, the TIF sample is snap-frozen in liquid nitrogen and stored at −80°C. No additional freeze–thaw cycles should be performed (see Note 8).

3.2. SDS–Polyacrylamide Gel Electrophoresis

1. The following instructions assume the use of a Mini-PROTEAN 3 system (Bio-Rad, Hercules, CA). They are easily adaptable to other formats, including minigels. It is critical that the glass plates for the gels are scrubbed clean and rinsed extensively with distilled water.
2. Prepare a 0.75-mm thick, 12% gel by mixing 1.3 mL of separating buffer with 2 mL of acrylamide/bis solution, 1.6 mL of water, 50 μL of SDS solution, 50 μL of ammonium persulfate solution, and 2 μL TEMED. Pour the gel, leaving space for a stacking gel, and overlay with water-saturated isobutanol. The gel should polymerize in about 30 min.
3. Pour off the isobutanol and rinse the top of the gel twice with water.
4. Prepare the stacking gel by mixing 250 μL of stacking buffer with 330 μL of acrylamide/bis solution, 1.4 mL of water, 20 μL of SDS solution, 20 μL of ammonium persulfate solution, and 2 μL of TEMED. Pour the stacking gel and insert the comb. The stacking gel should polymerize within 30 min.
5. Prepare the running buffer by diluting 100 mL of the 5× running buffer with 400 mL of water in a measuring cylinder. Cover with Parafilm and invert to mix.
6. Add the running buffer to the upper and lower chambers of the gel unit and carefully remove the comb and use a 3-mL syringe fitted with a 22-gauge needle to wash the wells with running buffer.
7. Add an equal volume of 2× loading buffer to the sample and boil for 5 min in a water bath. Load 30 μL of each

sample in a well. Include one well for prestained molecular weight markers (5 μL).

8. Complete the assembly of the gel unit and connect to a power supply. The gel can be run at 50 V for 30 min through the stacking gel and 200 V through the separating gel. The dye fronts (blue and pink) can be allowed to run off the gel if desired, but if the pink dye (pyronin-Y) is retained, then it will be transferred to the nitrocellulose membrane and identify the positions of the lanes.

3.3. Western Blotting for Organelle Marker Proteins

1. The samples that have been separated by SDS–PAGE are transferred to supported nitrocellulose membranes electrophoretically. These directions assume the use of a Mini-PROTEAN Tetra electrophoresis system (Bio-Rad, Hercules, CA). A tray of transfer buffer is prepared that is large enough to lay out a transfer cassette with its pieces of foam and with two sheets of 3MM paper submerged on one side. A sheet of the nitrocellulose cut just larger than the size of the separating gel is laid on the surface of a separate tray of distilled water to allow the membrane to wet by capillary action. The membrane is then submerged in the transfer buffer on top of the 3MM paper.
2. The gel unit is disconnected from the power supply and disassembled. The stacking gel is removed and discarded and one corner cut from the separating gel to allow its orientation to be tracked. The separating gel is then laid on top of the nitrocellulose membrane.
3. Two further sheets of 3MM paper are wetted in the transfer buffer and carefully laid on top of the gel, ensuring that no bubbles are trapped in the resulting sandwich. The second wet foam sheet is laid on top and the transfer cassette is closed.
4. The cassette is placed into the transfer tank such that the nitrocellulose membrane is between the gel and the anode. It is vitally important to ensure this orientation or else the proteins will be lost from the gel into the buffer rather than being transferred to the nitrocellulose.
5. The ice bath is used to maintain a low temperature, and a magnetic stir bar in the tank is activated.
6. The lid is put on the tank and the power supply is activated. Transfers can be accomplished at either 30 V overnight or 70 V for 2 h.
7. Once the transfer is complete, the cassette is taken out of the tank and carefully disassembled, with the top sponge and sheets of 3MM paper removed. The gel can be discarded. The colored molecular weight markers (and pyronin-Y lane

markers, if the latter are retained on the gel during the PAGE run) should be clearly visible on the membrane.

8. The nitrocellulose is then incubated in 50 mL of blocking buffer for 1 h at room temperature on a rocking platform.
9. The blocking buffer is discarded, and the membrane is quickly rinsed prior to addition of a diluted primary antibody in antibody dilution buffer for overnight incubation at 4°C on a rocking platform. Four primary antibodies are used, respectively, including antilamin B polyclonal antibody (diluted 1:1,000), anticyto C (cytochrome C) monoclonal antibody (diluted 1:1,000), antiflotillin-1 monoclonal antibody (diluted 1:500), and antibeta-actin polyclonal antibody (diluted 1:500).
10. The primary antibody is then removed (see Note 9), and the membrane is washed three times for 5 min, each time with 50 mL of TBS-T.
11. The secondary antibody is freshly prepared for each experiment as 1:10,000-fold dilution in blocking buffer and added to the membrane for 1 h at room temperature on a rocking platform.
12. The secondary antibody is discarded, and the membrane is washed three times for 5 min, each time with TBS-T.
13. During the final wash, 2-mL aliquots of each portion of the ECL reagent are warmed separately to room temperature, and the remaining steps are done in a dark room under safe light conditions. Once the final wash is removed from the blot, the ECL reagents are mixed together and then immediately added to the blot, which is then rotated by hand for 1 min to ensure even coverage.
14. The blot is removed from the ECL reagents after another 4-min reaction and then placed between the leaves of an acetate sheet.
15. The acetate containing the membrane is then placed in an X-ray film cassette with film for a suitable exposure time, typically a few minutes. An example of the results produced is shown in Fig. 1.

3.4. 2D-DIGE of Mouse Liver Tissue, TIF, and Hepatocytes

1. The concentration of the protein is adjusted to 5 mg/mL (see Note 10) with lysis buffer, and the pH is adjusted to 8.5 using 50 mM NaOH (see Note 11).
2. Mouse liver TIF and hepatocyte lysate are labeled with Cy3 or Cy5, while mouse liver tissue lysate is labeled with Cy2 using 400 pmol fluorochrome per 50 μg protein (see Note 12). Labeling is performed for 30 min on ice in the dark. Reactions are then quenched by the addition of 1 μL of lysine (10 mM) for 10 min on ice in the dark.

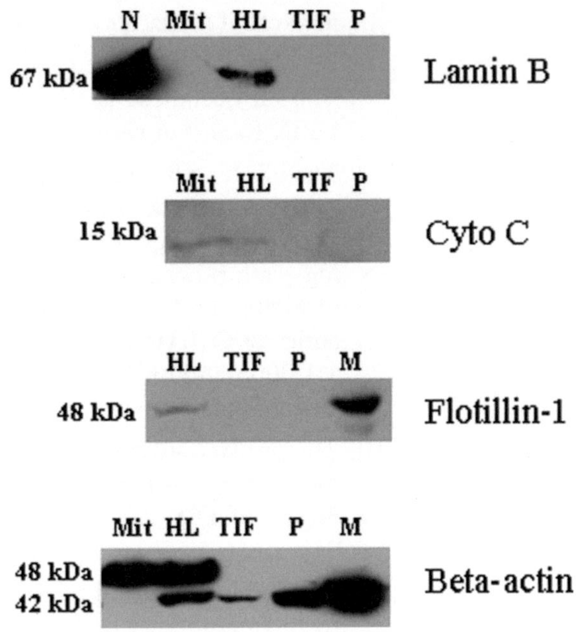

Fig. 1. Western blot of nuclear protein lamin B, mitochondrial proteins cyto C, and plasma membrane protein flotillin-1 for evaluation of contamination in mouse liver TIF. *HL* hepatocyte lysate, *P* perfusate, *N* nuclear protein, *Mit* mitochondrial protein, *M* plasma membrane protein. Twenty-five micrograms of protein were loaded into each lane. Lamin B, cyto C, and flotillin-1 could be detected in the positive controls and the hepatocyte lysate, while none were detected in TIF or liver perfusate. Two β-actin bands were detected in mouse hepatocyte, while only the lower band was detected in mouse TIF and perfusate. (Reprinted with permission from (4). Copyright 2010 American Chemical Society).

3. Fifty micrograms of Cy2-, Cy3-, and Cy5-labeled samples are combined and an equal volume of 2× sample buffer is added. The total volume of the sample is made up to 410 μL with rehydration buffer.
4. The sample is actively rehydrated into 24-cm 3-10 NL IPG strips at 17°C for 12 h using a IPGphor (GE Healthcare), followed by isoelectric focusing for a total of 80 kVh (ramp to 250 V in 30 min, hold at 1,000 V for 1 h, ramp to 10,000 V in 5 h, hold at 10,000 V for 60,000 Vh) (see Note 13).
5. The IPG strips are equilibrated in equilibration buffer supplemented with 0.5% DTT for 15 min at room temperature, followed by 4.5% iodoacetamide in equilibration buffer for 15 min with incubation at room temperature.
6. IPG stripes are placed on the top of 12% homogeneous polyacrylamide gels. The second-dimensional SDS–PAGE is carried out using Ettan DALT Six electrophoresis units (GE Healthcare).
7. After 2DE, gels are scanned on the Typhoon 9410 scanner (GE Healthcare) with Ettan DALT gel alignment guides

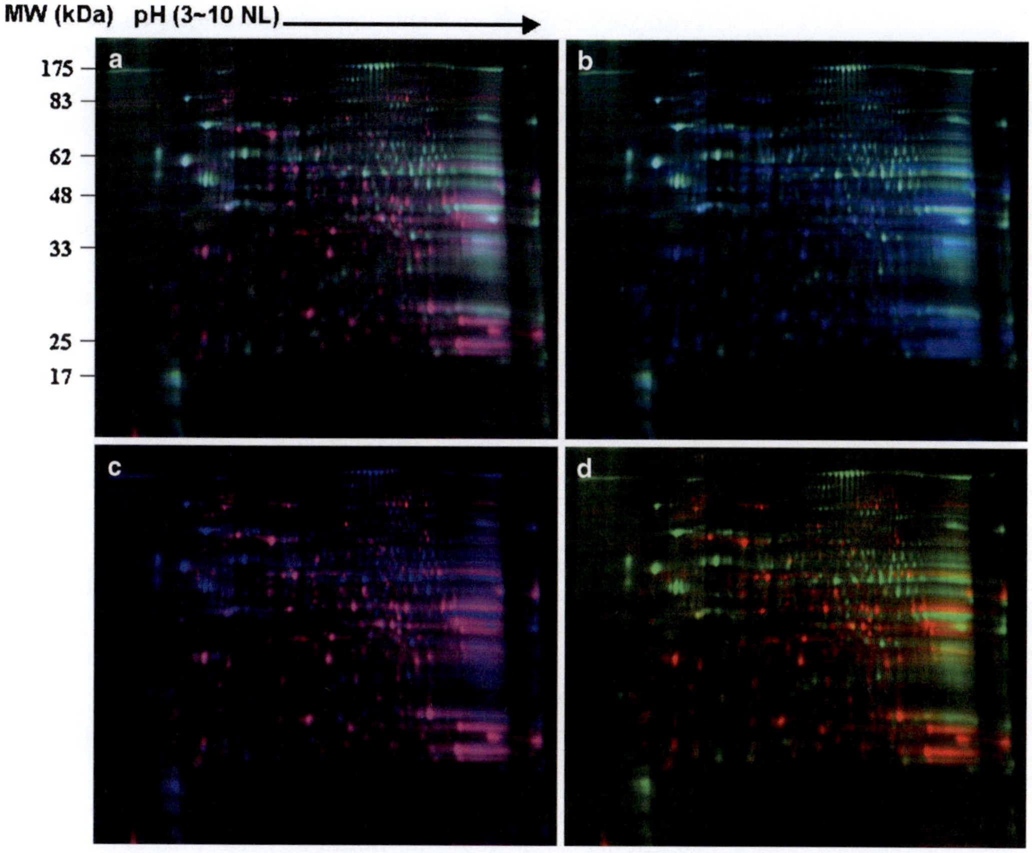

Fig. 2. Two-dimensional differential in-gel electrophoresis (2D-DIGE) maps of proteins from mouse liver tissue (Cy2-labeled, *blue*), hepatocyte (Cy3-labeled, *green*), and TIF (Cy5-labeled, *red*) showed different profiling patterns among these samples. (**a**) Merged 2D-DIGE image of mouse liver tissue, hepatocyte, and TIF. (**b**) Merged image (*cyan*) of mouse liver tissue and hepatocyte. Hepatocyte proteins were included in liver tissue. (**c**) Merged image (*magenta*) of mouse liver tissue and TIF. TIF proteins were included in liver tissue. (**d**) Merged image (*yellow*) of mouse hepatocytes and TIF. (Reprinted with permission from (*4*). Copyright 2010 American Chemical Society).

using excitation/emission wavelengths specific for Cy2 (488/520 nm), Cy3 (532/580 nm), and Cy5 (633/670 nm). The intensity is adjusted to ensure that the maximum pixel volume of each image is within 60,000–90,000. An example of the results produced is shown in Fig. 2.

4. Notes

1. Only fresh liver tissues should be used for TIF extraction. Freezing or long time in vitro exposure will lead to breakdown of cells in the liver. Long-distance transportation duration should be limited within 2 h on ice; otherwise, samples should be treated in a nearby laboratory.

2. Protease inhibitor cocktails are recommended to be added into PBS in which liver tissue is cut into small pieces. Protease inhibitor cocktails could be replaced by 1 mM PMSF, 0.01 mg/mL leupeptin, and 0.01 mg/mL aprotinin.
3. Tissue washing should be performed gently. Three times inversion is recommended. Then, the tube should be set on ice for deposition. The supernatant is discarded prior to adding fresh PBS. The tissue should be washed at least five times, until the supernatant is colorless and transparent.
4. The cap of the tube is screwed slightly as required for cell culture bottles. Then, the tube is placed in a CO_2 incubator as flatly as possible to maximize the exposure area.
5. The aim of this step is to remove large cells (e.g., hepatocytes) in the supernatant. Pasteur pipette is used for liquid transfer to keep any remaining cells intact. Pipette tips should not be used before step 6.
6. The aim of this step is to thoroughly remove smaller cells (e.g., white blood cells) remaining in the supernatant. A new Eppendorf tube should be used after each time of centrifugation.
7. This 1-h concentration time is recommended for getting the final sample concentration around 5 mg/mL.
8. TIF samples should be vortexed before split charging. No additional freeze–thawing should be performed before use.
9. The primary antibody can be saved for subsequent experiments by addition of 0.02% final concentration sodium azide (conveniently done by dilution from a 10% stock solution; exercise caution since azide is highly toxic) and storage at 4°C. These primary antibodies could be reused for three to five times in the next several months, with the only adjustment required being increasing length of exposure to film at the ECL step.
10. Ensure that the sample protein concentration exceeds 1 mg/mL. Too low concentration will affect the label efficiency. Samples could be concentrated in the vacuum dehydrator for longer time to increase protein concentration.
11. Test the sample pH by spotting 1 μL on a pH indicator strip.
12. Cy2, Cy3, and Cy5 dyes need to be dissolved in DMF, giving a concentration of 1 nmol/L (stock solution). Stock solution is stable for 2 months at –20°C. Further dilution is needed before use. Pay attention that fluors and opened DMF have a limited life.
13. Steps 4–6 should be performed in the dark.

Acknowledgments

We thank Dr. Celis Julio E. for the constructive discussion with him on the extraction protocol. Also, we thank Drs. Yunwei Hao, Lihai Guo, Ping Wan, Yuanbiao Guo, and Xinyu Deng for their help and advices. This work was partially supported by Chinese State High-tech Program (2006AA02A308), Chinese State Key Projects for Basic Research (Nos. 2006CB910801, 2006CB910401, 2006CB910602, 2010CB912700 and 2006CB503910), National Natural Science Foundation of China (30700988, 30700356, 30972909), Chinese State Key Project Specialized for Infectious Diseases (2008ZX10002-016, 2009ZX10004-103, 2009ZX09301-002), International Scientific Collaboration Program (2009DFB33070, 2008GR0763) and supported by the State Key Laboratory of Proteomics (SKLP-K200801, SKLP-Y200801, SKLP-O200801) in part with permission from Sun et al. (2010) (4). Copyright 2010 American Chemical Society.

References

1. Jungermann, K. and Katz, N. (1989) Functional specialization of different hepatocyte populations. *Physiol Rev* **69**, 708–764.
2. Simpson, R. J., Bernhard, O. K., Greening, D. W., and Moritz, R. L. (2008) Proteomics-driven cancer biomarker discovery: looking to the future, *Curr Opin Chem Biol* **12**, 72–77.
3. Ahn, S.-M., and Simpson, R. J. (2007) Body fluid proteomics: Prospects for biomarkers discovery, *Proteomics Clin Appl* **9**, 1004–1015.
4. Sun, W., Ma, J., Wu, S., et al. (2010) Characterization of the liver tissue interstitial fluid (TIF) proteome indicates potential for application in liver disease biomarker discovery, *J Proteome Res* **9**, 1020–1031.
5. Celis, J. E., Gromov, P., Cabezon, T., et al. (2004) Proteomic characterization of the interstitial fluid perfusing the breast tumor microenvironment: a novel resource for biomarker and therapeutic target discovery. *Mol Cell Proteomics* **3**, 327–344.
6. Celis, J. E., Moreira, J. M., Cabezon, T., et al. (2005) Identification of extracellular and intracellular signaling components of the mammary adipose tissue and its interstitial fluid in high risk breast cancer patients: toward dissecting the molecular circuitry of epithelial-adipocyte stromal cell interactions. *Mol Cell Proteomics* **4**, 492–522.

Chapter 17

A Protocol for the Preparation of Cryoprecipitate and Cryodepleted Plasma

Rosemary L. Sparrow, David W. Greening, and Richard J. Simpson

Abstract

Cryoprecipitate is a concentrate of high-molecular-weight plasma proteins that precipitate when frozen plasma is slowly thawed at 1–6°C. The concentrate contains factor VIII (antihemophilic factor), von Willebrand factor (vWF), fibrinogen, factor XIII, fibronectin, and small amounts of other plasma proteins. Clinical-grade preparations of cryoprecipitate are mainly used to treat fibrinogen deficiency caused by acute bleeding or functional abnormalities of the fibrinogen protein. In the past, cryoprecipitate was used to treat von Willebrand disease and hemophilia A (factor VIII deficiency), but the availability of more highly purified coagulation factor concentrates or recombinant protein preparations has superseded the use of cryoprecipitate for these coagulopathies. Cryodepleted plasma ("cryosupernatant") is the plasma supernatant that remains following removal of the cryoprecipitate from frozen-thawed plasma. It contains all the other plasma proteins and clotting factors present in plasma that remain soluble during cold-temperature thawing of the plasma.

Key words: Plasma, Cryoprecipitate, Antihemophilic factor, Factor VIII, Cryodepleted, Fibrinogen, Proteomics, Cryosupernatant

1. Introduction

Cryoprecipitate (also known as cryoprecipitated antihemophilic factor) was first described in the mid-1960s as a method to concentrate factor VIII (antihemophilic factor) from plasma for use in patients with hemophilia, von Willebrand disease, or hypofibrinogenemia (1, 2). Cryoprecipitate is the insoluble concentrate of high-molecular-weight plasma proteins that precipitate when frozen plasma is slowly thawed at 1–6°C (2). Cryoprecipitate is enriched for plasma coagulation proteins, in particular factor VIII, fibrinogen, von Willebrand factor, factor XIII, and fibronectin. Small amounts of other plasma proteins, such as immunoglobulins, may also be present.

Richard J. Simpson and David W. Greening (eds.), *Serum/Plasma Proteomics: Methods and Protocols*, Methods in Molecular Biology, vol. 728, DOI 10.1007/978-1-61779-068-3_17,

Cryoprecipitate is a standard blood transfusion component manufactured by most blood transfusion services (3). Clinically, the main indications for use of cryoprecipitate is for the treatment of fibrinogen deficiency (hypofibrogenemia) caused by significant bleeding due to trauma, massive transfusion or disseminated intravascular coagulation, or dysfibrogenemia arising from functionally abnormal fibrinogen (3, 4). Previously, cryoprecipitate was used to treat hemophilia and von Willebrand disease, but with the advent of specific coagulation factor concentrates, it has been relegated to second-line therapy for these diseases. Over the years, attempts have been made to improve the yield in cryoprecipitate by the use of techniques such as thaw-siphon technique (5), the effect of temperature on freezing and thawing (6), or by the use of various additives, such as heparin (7) and sodium citrate (8, 9).

Cryodepleted plasma (cryosupernatant) is the plasma supernatant that remains following removal of the precipitated cold insoluble proteins (i.e., cryoprecipitate) from frozen-thawed plasma. It is, therefore, significantly depleted of fibrinogen, factor VIII, von Willebrand factor, factor XIII, and fibronectin, but contains all the other plasma proteins and clotting factors in similar concentrations as the original plasma. Clinically, cryodepleted plasma is used for plasma exchange in thrombotic thrombocytopenic purpura and in situations requiring rapid temporary reversal of warfarin anticoagulant therapy (10).

In the following protocol, the preparation of small research-scale cryoprecipitate and cryodepleted plasma is outlined and is based on the procedures used by blood transfusion services for the preparation of clinical-scale cryoprecipitate (10).

2. Materials

2.1. Blood Collection, Plasma Preparation, and Storage

1. Whole Blood Collection Tubes containing citrate anticoagulant (e.g., BD Vacutainer® with 3.2% sodium citrate anticoagulant, 4.5-mL tube BD Biosciences #366415; BD Vacutainer with acid–citrate–dextrose (ACD) Sol A anticoagulant (22.0 g/L trisodium citrate, 8.0 g/L citric acid, and 24.5 g/L dextrose), 8.5-mL tube, BD Biosciences #364606] (see Note 1).
2. Blood collection needles compatible with the blood collection tubes, e.g., BD Vacutainer® Safety-Lok™ Blood Collection Set #367283; 23 G butterfly needle with attached sterile tubing).
3. Alcohol and swabs for disinfection of the venipuncture site.
4. Gloves, gown, and eye protection.
5. Disposal container for biological hazards.
6. Polypropylene tubes (1.5 mL, 15 mL).

7. Labels for blood sample tubes.
8. Tube storage rack.
9. Centrifugation unit (with swing-bucket rotor, compatible with 1.5/15 mL tubes, programmable temperature setting; range 4–25°C).
10. Pipettes.
11. Freezer (–20°C or lower).

2.2. Cryoprecipitate/Cryodepleted Plasma Preparation

1. Refrigerator or water bath set at 1–6°C.
2. Centrifuge, refrigerated (swing-bucket rotor, compatible with 1.5/15 mL tubes, programmable temperature setting).
3. 0.9% Saline solution.

3. Methods

3.1. Blood Collection/Phlebotomy

1. Blood collection must only be performed by personnel trained in phlebotomy/venipuncture. Safety precautions for the collection and handling of blood must be employed at all times (see Note 2). Particular care must be taken with insertion of the needle into the vein to limit the possibility of activation of the coagulation factors in the blood, which could compromise the quality of the blood sample.
2. It is important to collect the volume of blood specified for the particular type of blood collection tube to ensure that the correct blood/anticoagulant ratio is achieved (see Note 3).
3. After blood collection, gently mix the blood by inverting the tube several times to ensure thorough mixing with the anticoagulant. For thorough mixing of blood collected into citrate tubes, it is recommended to invert the tube 3 4 times, while ACD tubes should be inverted eight times.
4. Blood samples should be maintained at temperate conditions (i.e., 20–24°C) and centrifuged within 4 h of blood collection. Superior factor VIII yields are obtained from blood that is maintained at 20–24°C before processing.
5. To separate the plasma, centrifuge the blood samples at $1,200 \times g$ for 10 min at 22°C. If needed, RCF for a centrifuge can be calculated. For an online calculator tool, refer: http://www.currentprotocols.com/tools/g-forcerpm-conversion-tool.
6. After centrifugation, the plasma layer will be the upper layer of the separated blood. The plasma should be a clear, straw-yellow colored fluid. Mononuclear cells and platelets form a thin whitish mid-layer (buffy coat) that settles directly on top of the red blood cell layer.

7. Carefully, collect the plasma layer with an appropriate transfer pipette without disturbing the buffy coat layer. Do not attempt to collect all of the plasma. Do not allow the tip of the transfer pipette within 5 mm of the buffy coat layer and avoid touching the wall of the tube with the pipette. This helps to avoid inadvertent contamination of the plasma with cells that may only be softly sedimented in the buffy coat–plasma interface (see Note 4). If more than one tube is collected from the same donor, pool the plasma samples from both tubes into a 15-mL polypropylene tube. If necessary, aliquot plasma into smaller volumes. A practical minimum volume is 1–1.5 mL.
8. Close the tube tightly and place on ice or immediately freeze by placing in the freezer. This process should be completed within 30 min of centrifugation. Plasma should be frozen as quickly as possible to minimize loss of labile coagulation factors, such as factor VIII. Frozen plasma should be stored at below –20°C.

3.2. Cryoprecipitation

1. For the preparation of cryoprecipitate, remove tube(s) of frozen plasma from the freezer and immediately place in a thermostatically controlled water bath or refrigerator set at 1–6°C. Slowly, thaw the plasma until it becomes "slushy" (the required time will depend on the volume of plasma being thawed) (see Note 5). Optimum temperature for cryoprecipitate formation is 3°C.
2. Immediately, sediment precipitated proteins in a refrigerated centrifuge (1–6°C) at 5,000 × *g* for 15 min. A white precipitate should be evident at the bottom of the tube.
3. Carefully, remove the supernatant (Note: this is the cryodepleted plasma). If this cryodepleted plasma is required, aliquot into separate polypropylene tube(s). Leave a small amount of plasma above the deposited cryoprecipitate (5–10% v/v, 50–100 μL for 9–10 mL of blood collection volume). If the cryoprecipitate or cryodepleted plasma is not required immediately, freeze at –20°C (see Note 6).

3.3. Thawing of Cryoprecipitate/Cryodepleted Plasma

1. Thaw cryoprecipitate or cryodepleted plasma in a water bath at 30–37°C. The cryoprecipitate should be evenly dissolved at warming temperature.
2. Cryoprecipitate can be suspended in diluent, such as 0.9% saline, at 20–24°C. Cryoprecipitate should be maintained at 20–24°C and used within 6 h of thawing. Progressive loss of labile proteins, including Factor VIII and fibrinogen, occurs following thawing (see Notes 7–10).
3. Thawed cryodepleted plasma can be maintained at 2–6°C.

4. Notes

1. *Anticoagulant.* For all physiological coagulation studies, sodium citrate or ACD is the anticoagulant of choice. Cryoprecipitate prepared by blood transfusion services for clinical use is prepared from whole blood collected into citrate–phosphate–dextrose (CPD) anticoagulant or from plasma collected into ACD anticoagulant by apheresis (11). Cryoprecipitate prepared from CPD anticoagulated plasma has been shown to give a higher yield of Factor VIII compared to ACD anticoagulated plasma (11). The preparation of cryoprecipitate from plasma anticoagulated with noncitrate anticoagulants (e.g., heparin) may yield a different profile of precipitated proteins (12).
2. *Safety.* All blood and biological specimens and materials coming in contact are considered biohazards. Use gloves, gowns, eye protection, other personal protective equipment, and controls to protect from blood splatter, blood leakage, and potential exposure to blood-borne pathogens. Use aseptic technique and sterile disposables (tubes, pipettes) throughout to prevent contamination of the blood. Risk factors for possible transfusion transmissible infections should be rigorously screened for prior to blood collection. Handle as if capable of transmitting infection and dispose of with proper precautions in accordance with federal, state, and local regulations. Refer to your institutional regulations regarding the screening of blood for specific infectious disease markers (i.e., HIV, Hepatitis B, Hepatitis C, etc.). Discard all blood collection materials in biohazard containers approved for their disposal.
3. *Evacuated blood collection tubes* (e.g., BD Vacutainers) are manufactured to draw the blood volume specified for the particular tube. Filling is complete when vacuum no longer continues to draw blood into the tube. Partially filled tubes will not have the correct blood/anticoagulant ratio and should not be used for the purpose of plasma sample preparation. If a citrated blood sample is the first sample to be collected from the donor, it is important to first collect a small volume of blood into a discard tube. This ensures that the "dead" volume in the needle/tubing set is filled with blood prior to the citrated tube being connected, thus ensuring that the correct blood volume is drawn into the tube.
4. *Quality control of plasma sample.* It is recommended that following the preparation of the plasma sample, the sample is monitored for hemolysis and cell contamination. This should be validated by your institution or testing laboratory. The plasma sample should also be observed with respect to its

color (normally straw-yellow) and clarity. The presence of high lipid content in a plasma sample can be observed by the opaqueness of the plasma. Pink or red coloration of the plasma is indicative of hemolysis of red blood cells. Plasma samples with high lipid content or hemolysis should be avoided for protein chemistry studies of normal plasma. Plasma that has been previously thawed and refrozen will give inferior yields if used for the preparation of cryoprecipitate.

5. *How to prevent cryoprecipitation when thawing plasma.* If plasma is to be used for coagulation studies, it is important to prevent cryoprecipitation from occurring. To avoid cryoprecipitation of the cold-insoluble proteins when thawing replete plasma, the plasma must be thawed quickly at 37°C. This can be achieved by placing the frozen plasma samples in a 37°C water bath or dry heating system set at 37°C. Such equipment must be maintained and kept clean to avoid inadvertent bacterial contamination of plasma samples.
6. *Cryoprecipitate and cryodepleted plasma storage.* The prepared cryoprecipitate and cryodepleted plasma have a shelf life of 12 months when stored at or below –18°C (10).
7. *Cryoprecipitate content and specifications.* A cryoprecipitate unit prepared from plasma obtained from a standard 450–500 mL CPD-anticoagulated whole blood donation and processed according to blood banking procedures should contain at least 150 mg of fibrinogen and a minimum of 80 international units (IU) of Factor VIII (10, 13). This equates to 30–70% of the factor VIII/vWF and fibrinogen content of the original plasma. Proportionally similar yields should be achieved when cryoprecipitate is prepared from smaller starting volumes of plasma, provided care is taken with the processing, freezing, and thawing of the plasma samples.
8. *Effect of ABO blood group.* Plasma/cryoprecipitate from blood group O individuals has lower levels of Factor VIII and vWF than A, B, or AB blood groups.
9. *Effect of pathogen reduction treatment on coagulation factor proteins.* Some blood transfusion services supply plasma units that have undergone an additional pathogen reduction process to further minimize the already very low risk of infectious disease transmission by blood transfusion. Various pathogen reduction technologies have been developed, including treatments with solvent/detergent, or photoactivating agents such as methylene blue, riboflavin, or psoralen (14). A trade-off for the increased safety rendered by pathogen reduction treatment is a reduced yield and activity of plasma coagulation factors, including fibrinogen and factor VIII (14–16). The nature of the biochemical changes that occur to the coagulation proteins following pathogen reduction treatment is yet to be fully determined.

10. *Platelet microparticle concentration.* The platelet membrane microparticle concentration of the cryoprecipitate has been shown to be 29-fold greater than that of the cryosupernatant plasma and 265-fold greater than that of the original plasma sample (17). There is a need to further clarify whether such microparticles have a role in hemostasis, vascular function, inflammation, or alloimmunoreactivity (17, 18). It is unknown whether the platelet membrane microparticles retain hemostatic function after processing and freezing of cryoprecipitate.

References

1. Pool, J.G., E.J. Gershgold, and A.R. Pappenhagen (1964), High-Potency Antihaemophilic Factor Concentrate Prepared from Cryoglobulin Precipitate. *Nature.* **203**, 312.
2. Pool, J.G. (1965), Preparation and testing of antihemophilic globulin (Factor 8) sources for transfusion therapy in hemophilia. Description of a new sterile concentrate process for blood banks. *Scand J Clin Lab Invest.* **17**, p. Suppl 84, 70–7.
3. Callum, J.L., K. Karkouti, and Y. Lin (2009), Cryoprecipitate: the current state of knowledge. *Transfus Med Rev.* **23**(3), 177–88.
4. Erber, W.N. and D.J. Perry (2006), Plasma and plasma products in the treatment of massive haemorrhage. *Best Pract Res Clin Haematol.* **19**(1), 97–112.
5. Mason, E.C. (1978), Thaw-siphon technique for production of cryoprecipitate concentrate of factor VIII. *Lancet.* **2**(8079), 15–7.
6. Rock, G.A. and P. Tittley (1979), The effects of temperature variations on cryoprecipitate. *Transfusion.* **19**(1), 86–9.
7. Rock, G.A., W.H. Cruickshank, E.S. Tackaberry, et al. (1979), Improved yields of factor VIII from heparinized plasma. *Vox Sang.* **36**(5), 294–300.
8. Shanbrom, E. and W.J. Owens (2001), A novel method to enhance the formation of cryoprecipitate from fresh plasma. *Blood.* **98**, 60a.
9. Yousef, H., D. Neurath, M. Freedman, et al. (2006), Cryoprecipitate production: the use of additives to enhance the yield. *Clin Lab Haematol.* **28**(4), 237–40.
10. (2009), American Association of Blood Banks. Technical Manual. 23rd ed, ed. M. Bethesda.
11. Wensley, R.T. and T.J. Snape (1980), Preparation of improved cryoprecipitated factor VIII concentrate. A controlled study of three variables affecting the yield. *Vox Sang.* **38**(4), 222–8.
12. Meilin, E., S. Sela, and B. Kristal (2008), Heparin cryoprecipitation reduces plasma levels of non-traditional risk factors for atherosclerosis in vitro. *Blood Purif.* **26**(3), 238–48.
13. (2004), Menitove J. 5.7.5.14 Standards of Blood Banks and Transfusion Services. American Association of Blood Banks. .ed 23rd, ed. M.D. Bethesda. U.S.A. 30.
14. Prowse, C. (2009), Properties of pathogen-inactivated plasma components. *Transfus Med Rev.* **23**(2), 124–33.
15. Cardigan, R., K. Philpot, P. Cookson, et al. (2009), Thrombin generation and clot formation in methylene blue-treated plasma and cryoprecipitate. *Transfusion.* **49**(4), 696–703.
16. Hornsey, V.S., O. Drummond, A. Morrison, et al. (2009), Pathogen reduction of fresh plasma using riboflavin and ultraviolet light: effects on plasma coagulation proteins. *Transfusion.* **49**(10), 2167–72.
17. George, J.N., E.B. Pickett, and R. Heinz (1986), Platelet membrane microparticles in blood bank fresh frozen plasma and cryoprecipitate. *Blood.* **68**(1), 307–9.
18. McVerry, B.A. and S.J. Machin (1979), Incidence of allo-immunization and allergic reactions to cryoprecipitate in haemophilia. *Vox Sang.* **36**(2), 77–80.

Chapter 18

Preparation of Platelet Concentrates

David W. Greening, Rosemary L. Sparrow, and Richard J. Simpson

Abstract

Platelets are specialized blood cells that play central roles in physiologic and pathologic processes of hemostasis, wound healing, host defense, thrombosis, inflammation, and tumor metastasis. Activation of platelets is crucial for platelet function that includes a complex interplay of adhesion, signaling molecules, and release of bioactive factors. Transfusion of platelet concentrates is an important treatment component for thrombocytopenia and bleeding. Recent progress in high-throughput mRNA and protein profiling techniques has advanced the understanding of platelet biological functions toward identifying novel platelet-expressed and secreted proteins, analyzing functional changes between normal and pathologic states, and determining the effects of processing and storage on platelet concentrates for transfusion. It is important to understand the different standard methods of platelet preparation and how they differ from the perspective for use as research samples in clinical chemistry. Two simple methods are described here for the preparation of research-scale platelet samples from whole blood, and detailed notes are provided about the methods used for the preparation of platelet concentrates for transfusion.

Key words: Platelet concentrate, Platelet-rich plasma, Buffy coat, Apheresis, Transfusion, Proteomics, Research, Protocol, SOP

1. Introduction

Platelets, the smallest of the human blood cells (~3.6×0.7 μm), are central players in processes of hemostasis and thrombosis. In addition, platelets are involved in clot retraction, vessel constriction and repair, inflammation, including promotion of atherosclerosis, host defense, and even tumor growth/metastasis (1). They are produced by differentiation of the bone-marrow-derived megakaryocytes (2) and are released as anucleated fragments into the circulation. Circulating platelets have a discoid form, are the second most numerous corpuscle, are normally circulating between 150 and 450×10^9/L, and have the lowest specific gravity of all blood cells (2). Their shape and small size enables them to

Richard J. Simpson and David W. Greening (eds.), *Serum/Plasma Proteomics: Methods and Protocols*,
Methods in Molecular Biology, vol. 728, DOI 10.1007/978-1-61779-068-3_18,

access the edge of blood vessels, thereby enabling them constantly to survey vascular integrity. Complex, regulated reactions occur between platelets, von Willebrand factor, collagen, and soluble coagulation factors in regions of disturbed vasculature. These changes induce platelet adherence to vessel walls and platelet activation, which leads to platelet aggregation, procoagulant activity, spreading, microparticle release, and formation of a primary hemostatic plug (3).

Since platelets lack nuclear DNA and their genome consists of a subset of megakaryocyte-derived mRNA transcripts, they represent a simplified biological model when investigating cell function (4). While valuable information may be gathered from studies of mRNA (5, 6), the rapid signaling and regulatory events in platelets are not governed by, or dependent on, alterations in gene expression (7). In contrast, proteomics can provide the necessary tools for characterizing, at the protein level, the vicissitudes of platelet function (reviewed in (8, 9)). Most proteomic efforts can be grouped into several distinct categories: (1) cataloging the spectrum of proteins that comprise the normal, resting (quiescent) platelet proteome, (2) characterizing proteins released from activated (functional) platelets, and (3) identifying specific platelet subproteomes (i.e., membrane (10–12), phosphorylation (13–16)), or functional end points (i.e., microparticles (17–19)) in response to external stimuli, as well as in applications for transfusion (20–22) and disease (23, 24).

The separation of platelets from whole blood is based on the differential densities of various cellular components when blood is subjected to variable centrifugation forces. Platelets, being the smallest and lightest cellular elements of blood, remain suspended in the liquid plasma when whole blood is centrifuged at a low centrifugal force. Protocols for the preparation of platelets rely on this characteristic, including those used by blood transfusion services for the preparation of platelet concentrates (PCs) for clinical use. Transfusion of PCs is indicated for the treatment of thrombocytopenia and bleeding, caused by hematological disease, the effects of chemotherapy, and post-operative bleeding.

PCs for transfusion can be prepared by three different methods including (i) platelet-rich plasma-platelet concentrates (PRP PC), (ii) buffy coat-platelet concentrates (BC-PC), and (iii) apheresis-platelet concentrates (apheresis-PC). In the preparation of PRP-PC, an initial soft centrifugation produces PRP, which is separated from white and red cells, and subsequently platelets are pelleted at a higher centrifugation force. This methodology is typically used for preparing research platelet samples from smaller volumes of whole blood. Disadvantages of this method are that it is difficult to avoid aspirating some white and red cells with the PRP (11) and that the platelets are hard spun against the surface of the container, which can cause increased platelet activation and/or damage. In contrast, platelets prepared by the BC-PC method are not subjected to being pelleted, but rather are cushioned by the red cells and then centrifuged at much lower centrifugation force to suspend the platelets in the supernatant, consequently limiting the platelet pelleting and reducing the level of cellular contaminants (25, 26). For apheresis-PC, a specialized automated blood cell separator is used where blood is immediately mixed with anticoagulant and separated into components; the target component is collected in a separate bag, while the other components are returned to the donor.

In developed countries, blood for transfusion is collected, processed, and distributed by licensed blood transfusion services that operate in a highly regulated environment governed by strict codes of practice similar to those that apply to the manufacturers of medicinal products [i.e., the code of good manufacturing practice (cGMP)]. These regulations are designed to ensure standardization of all procedures to maximize quality and safety of blood transfusion components, such as PCs.

In the following protocol, two research-scale methods for the preparation of platelets from whole blood are described that yield a minimally manipulated platelet specimen suitable for use in proteomics analysis and other research applications. An overview of the methods used by blood transfusion services to prepare PCs from whole blood donations and by single-donor apheresis collection is detailed in Subheading 4.

2. Materials

2.1. Blood Collection and Platelet Sample Preparation

1. Acid-citrate dextrose (ACD) Blood Collection Tubes (e.g., Whole Blood Glass Vacutainer® Tube with Anticoagulant, ACD Sol A, 8.5 mL, BD Biosciences #364606) (see Note 1).
2. Alcohol and swabs for disinfection of the venipuncture site.
3. Gloves, gown, and eye protection.
4. Disposal container for biological hazards.
5. Polypropylene tubes for processing and storage (15 mL).
6. Tube rack.
7. Labels for blood sample tubes.
8. Transfer pipettes.
9. Centrifugation unit (swing-bucket rotor, compatible with 15-mL tubes, programmable temperature setting; set at 22°C).
10. Platelet wash buffer (see Note 2).

3. Methods

3.1. Blood Collection

1. For the preparation of normal, resting platelets, the blood donor must be healthy, with no signs of infection or inflammation, and must not have taken medication that have antithrombotic effects, such as aspirin.
2. Blood collection must only be performed by personnel trained in phlebotomy/venipuncture. Safety precautions for the collection and handling of blood must be employed at all times (see Note 3). Particular care must be taken with insertion of the needle into the vein to limit the possibility of activation of the hemostasis/coagulation system, which could compromise the quality of the blood sample. For routine venipuncture procedures, a 21-gauge needle is recommended to minimize hemolysis.
3. It is important to collect the volume of blood specified for the particular blood collection tube to ensure that the correct blood/anticoagulant ratio is achieved (see Note 4).
4. After blood collection, gently mix the tube by inverting the tube 3–4 times to ensure thorough mixing with the ACD anticoagulant.
5. Blood samples must be maintained at temperate conditions (i.e., 20–24°C), and platelet preparations are prepared within 4 h of blood collection.

3.2. Platelet-Rich Plasma Preparation

1. Platelets are extremely labile and are very easily activated during sample preparation. It is important to limit the extent of manipulation of the blood sample to avoid unintentional activation of the platelets. All procedures must be performed at 20–24°C to maintain platelet quality and viability. Platelets undergo cold-storage-induced activation if subjected to temperatures below 20°C (27, 28). The temperature of all equipment (e.g., centrifuge) and wash buffers (if used) must be between 20 and 24°C prior to use.
2. The PRP is separated from whole blood (typical volume 8.5 mL) by light spin centrifugation at $110 \times g$ for 15 min at 22°C. Under these centrifugation conditions, platelets will remain suspended in the plasma (upper fraction, yellow-colored fluid), while the white and red cells will be softly sedimented in the lower fraction. Using a pipette, carefully collect the upper 25% of the PRP, to limit contamination by white and red cells, and place into a fresh tube. Take a small aliquot of the collected PRP sample to use for quality assessment, which should be performed as quickly as possible (see Note 5).

3. To sediment the platelets in the PRP sample, centrifuge at a higher centrifugation rate (1,000 × *g* for 15 min, 22°C). Carefully remove the plasma supernatant and discard, if not required (alternatively, the plasma can be stored frozen at <–20°C). If necessary, the platelets can be washed with an appropriate physiological buffer to remove residual plasma proteins (see Note 2), using the same centrifugation conditions, but care must be taken with this step to limit activation of the platelets.
4. The platelets should be used immediately upon isolation. For example, prepare a platelet lysate using a relevant protocol suitable for subsequent proteomic analysis. One 8.5-mL tube of whole blood from a donor with a normal platelet count (i.e., 150–450 × 10^9 platelets/L) should be sufficient to yield enough platelets to prepare a suitable sample for proteomic analysis.
5. For an overview of the procedures used by blood transfusion services to prepare platelet concentrates, see Note 6 (PRP-PC and BC-PC), Note 7 (Apheresis-PC), and Notes 8–13 for details of additional processes.

3.3. Buffy Coat Platelet Preparation

1. Fractionate whole blood by centrifuging at 2,000 × *g* for 10 min at 22°C. This will separate the blood into an upper plasma layer, a lower red cell layer, and a thin white-colored interface called the BC, containing the majority of the platelets and white cells.
2. Carefully aspirate the plasma with a transfer pipette, being careful not to disturb the BC layer, and place into a separate clean tube. Plasma may be stored frozen at <–20°C or discarded if not required.
3. Using a wide-bore transfer pipette, carefully aspirate the BC layer. Aspirate slowly, using a circular motion, to pull all the BC layer into the transfer pipette. Some contamination with the underlying red cells is unavoidable, although care should be taken to minimize the amount of red cells collected.
4. Suspend the BC in at least 5 mL of platelet wash buffer (see Note 2).
5. Centrifuge the BC suspension using a light centrifugation at 520 × *g* for 6 min at 22°C. This sediments the white and red cells and leaves behind the platelets in suspension.
6. Collect the platelet-rich supernatant with a clean transfer pipette. Be careful in not disturbing the cell pellet, which contains the white and red cells.
7. Take a small aliquot of the collected platelet sample to use for quality assessment, which should be performed as quickly as possible (see Note 5).

8. To sediment the platelets, if necessary, centrifuge at a higher centrifugation rate (1,000×*g* for 15 min, 22°C). Carefully remove the supernatant and discard.
9. The platelets should be used immediately upon isolation. For example, prepare a platelet lysate using a relevant protocol suitable for subsequent proteomic analysis. One 8.5-mL tube of whole blood from a donor with a normal platelet count (i.e., $150–450 \times 10^9$ platelets/L) should be sufficient to yield enough platelets to prepare a suitable sample for proteomics analyses.

4. Notes

1. *Anticoagulant.* ACD is the preferred anticoagulant for preparation of platelets, as it preserves cell quality and function. The ACD used in the blood collection tubes suggested here (i.e., #364606, 8.5 mL, BD Biosciences) contains 22.0 g/L trisodium citrate, 8.0 g/L citric acid, and 24.5 g/L dextrose (known as ACD Solution A), which is similar to the ACD-A formulation used by blood transfusion services for the preparation of apheresis-PCs (10). CTAD (citrate, theophylline, adenosine, and dipyridamole) is an alternative citrate-based anticoagulant that contains inhibitors of platelet activation (theophylline and dipyridamole) and may be useful if platelet activation must be minimized (29). CTAD Vacutainer® tubes are available from BD Biosciences (#367947, 4.5 mL).
2. *Platelet wash buffer.* Tyrode's salt solution is a suitable physiological buffer for washing platelets. 0.265 g/L $CaCl_2·2H_2O$, 0.214 g/L $MgCl_2·6H_2O$, 0.2 g/L KCl, 1.0 g/L $NaHCO_3$, 8.0 g/L NaCl, and 1.0 g/L glucose, pH 7.2.
3. *Safety.* All blood and biological specimens and materials coming in contact are considered as biohazards. Use gloves, gowns, eye protection, other personal protective equipment, and controls to protect from blood splatter, blood leakage, and potential exposure to blood-borne pathogens. Use aseptic technique and sterile disposables (tubes, pipettes) throughout to prevent contamination of the blood. Risk factors for possible transfusion transmissible infections should be rigorously screened for prior to blood collection. Handle as if capable of transmitting infection and dispose of with proper precautions in accordance with federal, state, and local regulations. Refer to your institutional regulations regarding the screening of human blood for specific infectious disease markers (i.e., HIV, hepatitis B, Hepatitis C, etc.). Discard all blood collection materials in biohazard containers approved for this purpose.

4. *Evacuated blood collection tubes* (e.g., BD Vacutainers®) are manufactured to draw the blood volume specified for the particular tube. Filling is complete when vacuum no longer continues to draw blood into the tube. Partially filled tubes will not have the correct blood/anticoagulant ratio and should be avoided for the purpose of platelet sample preparation.
5. *Platelet sample quality assessment.* The PRP sample should be assessed to determine the platelet count as well as the level of contaminating white and red cells. This is best performed on an automated hematology analyzer. If access to a hematology analyzer is not possible, a qualitative assessment can be performed by examining the sample by light microscopy for the presence of contaminating white and red cells. Other platelet quality parameters that could be considered include assessment of the expression of CD62P as a marker of platelet activation (29). An important note is to process the platelet sample as quickly as possible and minimize transition and storage time prior to quality assessment. This greatly affects not only platelet yield and viability but also the level of activated platelets. Temperatures below room temperature can result in cold-induced platelet activation (27, 28).
6. *Whole blood-derived, pooled-donor platelet transfusion components.* Whole blood is collected into a sterile soft-plasticized PVC blood collection bag system that consists of a series of interconnected bags (including a gas-permeable bag for the storage of platelets) joined by flexible tubing, which enables closed system processing. Whole blood (450 mL ± 10%) is collected into the primary collection bag containing 63 mL (±10%) of citrate–phosphate–dextrose (CPD) anticoagulant, or alternative formulations such as CPDA-1 and CP2D (10). Blood collection must be completed within 12 min. Longer collection times are indicative of poor venous access, collapsed vein, or insufficient hydration of the donor, all of which can adversely affect the quality of the collected blood. Whole blood for the preparation of platelet components is maintained at 20–24°C and must be processed within 24 h of collection (an 8-h max limit applies in USA). Two different methods can be used to prepare platelet concentrates from whole blood: the PRP method or the BC-PC method. The PRP method is used in USA, while the BC-PC method is used in many countries in Europe, UK, Canada, and Australia. The BC-PC method can be used in conjunction with synthetic protein-free platelet storage solutions as an alternative to plasma as the suspending fluid (30, 31). Further details about methods used for the preparation of platelet concentrates are reviewed by Vassallo and Murphy (32).

   For the PRP method, the whole blood donation is centrifuged by a soft spin (2,000 × *g*, 3 min at 22°C) to sediment the white and red cells, and the PRP is expressed into an attached separate platelet storage bag using a blood bag pressing device. Typically, this process will also include passing the PRP through a leukocyte-depletion filter to remove leukocytes, in a process known as prestorage leukoreduction. The PRP is recentrifuged, but this time using a heavier spin (5,000 × *g*, 5 min at 22°C) to sediment the platelets and the excess plasma is expressed into an attached separate storage bag, leaving behind approximately 40–70 mL of plasma in which the platelets are resuspended. Typically, a unit of platelets from a single whole blood donation should contain approximately 5.5×10^{10} platelets. Individual donor platelet units are stored on a reciprocating platform agitator (50–70 oscillations/min) at 20–24°C. In order to prepare a therapeutic dose of PRP-PC (>3×10^{11} platelets), 4–6 individual donor PRP units must be pooled. Pooling is done either 24 h after preparation of the PRP and subsequently stored for use within 5 days of collection or just prior to issue to the patient.

For the BC-PC method, whole blood is collected into a quadruple "top-bottom" bag system (25, 26). The whole blood unit is centrifuged by a hard spin (5,000 × *g* for 10 min at 22°C) that sediments the platelets into the BC layer, and the red cells and plasma are expressed by a semiautomated blood component separator into separate storage bags, leaving the platelet-rich BC in the original collection bag. BCs are rested for several hours or overnight and then concentrates from four to six preparations are pooled and suspended in either plasma or platelet additive solution (~300 mL) (33) and centrifuged by a soft spin (520 × *g*, 6 min at 22°C), which sediments the white cells and leaves the platelets in suspension. The pooled BC-PC is passed through a leukocyte-reduction filter into a 1.3-L gas-permeable platelet storage bag, which is then placed on a reciprocating platform agitator at 20–24°C and stored for use within 5 days of collection.

7. *Apheresis, single-donor platelet transfusion components.* In contrast to the whole blood methods (described in Note 6), which require pooling of platelets from four to six donors to achieve a therapeutic dose of platelets (>3×10^{11} platelets), platelets collected by apheresis yield a therapeutic dose from a single donor (32). Platelet collection by apheresis uses a specialized programmable instrument that processes over 3.5 L of the donor's blood in a sterile closed system, extracts the platelets into a gas-permeable platelet collection bag, and returns the remainder of the blood to the donor (sometimes with saline fluid replacement). Blood processed by apheresis is anticoagulated with ACD. Leukocyte reduction is incorporated into the automated processing protocol. Typically, apheresis platelets are suspended in plasma, although protocols

that use platelet additive solutions as the storage medium are becoming available. The procedure normally takes 1.5–2 h, with the final volume of apheresis-PC between 100 and 400 mL. Storage and shelf life of apheresis-PCs are the same as PRP and BC-PCs. The majority of the PCs produced in USA are by apheresis collection, and it is increasingly being used in other Western countries.

8. *Storage of platelet transfusion components and storage effects.* PCs for transfusion are stored at 20–24°C with continuous gentle horizontal agitation (reciprocating platform agitator, set at 50–70 oscillations/min) in specifically designed platelet storage bags that permit O_2 and CO_2 exchange to maintain platelet viability and quality (34–36). This combination of storage container, agitation, suspension medium (100% plasma or platelet additive solution/plasma mix), and temperature permits satisfactory preservation of platelets for up to 7 days (37–39). Because PCs are stored at temperate conditions, there is an increased risk of bacterial growth that can occur for various reasons, most often being donor-derived. Consequently, the storage time for PCs is typically limited to up to 5 days from collection (40).

   During storage, platelets undergo numerous physicochemical changes, known as the platelet storage lesion, which reduces the quality, function, and viability of the stored platelets. These changes include platelet activation, shape change, release of bioactive factors and granules, shedding of microparticles, and apoptosis/necrosis (21, 41).

9. *Leukofiltration of PCs.* Removal of white cells from PCs is achieved by filtration using specialized filters that reduce the residual white cells in the order of $>4\log_{10}$ (42). In USA, residual white cells must be $<5 \times 10^6$/PC, while in Europe and other countries the limit is $<1 \times 10^6$/PC. Prestorage leukoreduction of PCs has been widely implemented to reduce the risk of adverse transfusion reactions, alloimmunization, and transmission of white-cell-borne viruses (43).

10. *Pathogen reduction of PCs.* Strategies to minimize the risk of transmission of infectious diseases are a fundamental aspect of the blood banking industry (44, 45). Pathogen reduction technology (PRT) suitable for PCs is a new development in blood banking and is designed to inactivate transfusion transmissible infectious agents, such as viruses and bacteria. For PCs, treatments have been developed with Psoralen/ultraviolet-A light or Riboflavin/ultraviolet-B light (46). Both products are CE marked in Europe but only currently approved for clinical trials in USA. The treatment causes some reduction in platelet function and viability, but PRT-treated PCs meet standard acceptance criteria for clinical use (47, 48).

11. *Irradiation of PCs.* Gamma irradiation is used to inactivate residual white cells present in transfusion components. Irradiated blood components are prescribed for patients at risk of transfusion-associated graft-versus-host disease (TA-GvHD), including immunosuppressed patients or patients that are partially tissue matched (haploidentical) with the transfusion donor. Platelets are relatively unaffected by gamma irradiation at the prescribed dosage of 25–50 Gy routinely used to irradiate blood components. Consequently, the shelf life of PCs is unchanged (49).
12. *Washing and cryopreservation of PCs.* In rare circumstances, PCs have to be washed prior to transfusion to avoid adverse transfusion reactions in hypersensitive recipients. The action of centrifugation and resuspension inevitably results in loss of platelets and some degree of loss of platelet function (50). Platelets can be cryopreserved, typically using 5–6% dimethylsulfoxide (DMSO) and frozen either at –80°C or in liquid nitrogen. Upon thawing and subsequent washing, there is significant loss of platelets and reduction in platelet quality (51, 52). Cryopreserved PCs are a specialized product and are not routinely available.
13. *PC quality assessment.* The standard indicator of PC quality is pH (at expiry, 6.4–7.4). A range of other tests can be used, including functional assays (e.g., extent of shape change, adhesion, aggregation, and thrombus formation/retraction), measures of platelet activation and viability (e.g., CD62P expression, swirling index; release of bioactive factors and granules, procoagulant activity, release of annexin V and microparticles, mitochondrial activity), and metabolic indices (e.g., pH, lactate, and ATP content, hypotonic shock response) (53–55).

References

1. Jurk, K. and B.E. Kehrel (2005), Platelets: physiology and biochemistry. *Semin Thromb Hemost.* **31**(4), 381–92.
2. George, J.N. (2000), Platelets. *Lancet.* **355**(9214), 1531–9.
3. Harrison, P. (2005), Platelet function analysis. *Blood Rev.* **19**(2), 111–23.
4. Italiano, J.E., Jr. and R.A. Shivdasani (2003), Megakaryocytes and beyond: the birth of platelets. *J Thromb Haemost.* **1**(6), 1174–82.
5. McRedmond, J.P., S.D. Park, D.F. Reilly, et al. (2004), Integration of proteomics and genomics in platelets: a profile of platelet proteins and platelet-specific genes. *Mol Cell Proteomics.* **3**(2), 133–44.
6. Harrison, P. and A.H. Goodall (2008), "Message in the platelet"--more than just vestigial mRNA! *Platelets.* **19**(6), 395–404.
7. Garcia, A. (2006), Proteome analysis of signaling cascades in human platelets. *Blood Cells Mol Dis.* **36**(2), 152–156.
8. Gnatenko, D.V., P.L. Perrotta, and W.F. Bahou (2006), Proteomic approaches to dissect platelet function: one-half of the story. *Blood.* **108**(13), 3983–91.
9. Senzel, L., D.V. Gnatenko, and W.F. Bahou (2009), The platelet proteome. *Curr Opin Hematol.* **16**(5), 329–33.
10. Greening, D.W., K. Glenister, R. Sparrow, et al., Enrichment of human platelet

membrane-cytoskeletal proteins for proteomic analysis, in Proteomic analysis of membrane proteins: Methods and protocols. *Methods in Molecular Medicine Series.*, M. Pierce, Editor. 2009, Humana Press. Vol. 528, 245–258.

11. Greening, D.W., K.M. Glenister, E.A. Kapp, et al. (2008), Comparison of human platelet-membrane cytoskeletal proteins with the plasma proteome: Towards understanding the platelet-plasma nexus. *Proteomics Clin. Appl.* **2**, 63–77.
12. Moebius, J., R.P. Zahedi, U.U. Lewandrowski, et al. (2005), The human platelet membrane proteome reveals several new potential membrane proteins. *Mol Cell Proteomics.* **4**(11), 1754–61.
13. Qureshi, A.H., V. Chaoji, D. Maiguel, et al. (2009), Proteomic and phospho-proteomic profile of human platelets in basal, resting state: insights into integrin signaling. *PLoS One.* **4**(10), e7627.
14. Dittrich, M., I. Birschmann, S. Mietner, et al. (2008), Platelet protein interactions: map, signaling components, and phosphorylation groundstate. *Arterioscler Thromb Vasc Biol.* **28**(7), 1326–31.
15. Zahedi, R.P., U. Lewandrowski, J. Wiesner, et al. (2008), Phosphoproteome of resting human platelets. *J Proteome Res.* 7(2), 526–34.
16. Gevaert, K., M. Goethals, L. Martens, et al. (2003), Exploring proteomes and analyzing protein processing by mass spectrometric identification of sorted N-terminal peptides. *Nat Biotechnol.* **21**(5), 566–9.
17. Garcia, B.A., D.M. Smalley, H. Cho, et al. (2005), The platelet microparticle proteome. *J Proteome Res.* **4**(5), 1516–1521.
18. Jin, M., G. Drwal, T. Bourgeois, et al. (2005), Distinct proteome features of plasma microparticles. *Proteomics.* **5**(7), 1940 52.
19. Janowska-Wieczorek, A., M. Wysoczynski, J. Kijowski, et al. (2005), Microvesicles derived from activated platelets induce metastasis and angiogenesis in lung cancer. *Int J Cancer.* **113**(5), 752–60.
20. Thiele, T., L. Steil, S. Gebhard, et al. (2007), Profiling of alterations in platelet proteins during storage of platelet concentrates. *Transfusion.* **47**(7), 1221–33.
21. Glenister, K.M., K.A. Payne, and R.L. Sparrow (2008), Proteomic analysis of supernatant from pooled buffy-coat platelet concentrates throughout 7-day storage. *Transfusion.* **48**(1), 99–107.
22. Thon, J.N., P. Schubert, M. Duguay, et al. (2008), Comprehensive proteomic analysis of protein changes during platelet storage requires complementary proteomic approaches. *Transfusion.* **48**(3), 425–35.
23. Sacristan, D., M. Marques, S. Spinella, et al. (2008), Modifications by Olmesartan medoxomil treatment of the platelet protein profile of moderate hypertensive patients. *Proteomics Clin. Appl.* **2**(9), 1300–1312.
24. Springer, D.L., J.H. Miller, S.L. Spinelli, et al. (2009), Platelet proteome changes associated with diabetes and during platelet storage for transfusion. *J Proteome Res.* **8**(5), 2261–72.
25. van Rhenen, D.J., J. Vermeij, J. de Voogt, et al. (1998), Quality and standardization in blood component preparation with an automated blood processing technique. *Transfus Med.* **8**(4), 319–24.
26. Murphy, S. (2005), Platelets from pooled buffy coats: an update. *Transfusion.* **45**(4), 634–9.
27. Murphy, S. and F.H. Gardner (1969), Effect of storage temperature on maintenance of platelet viability--deleterious effect of refrigerated storage. *N Engl J Med.* **280**(20), 1094–8.
28. Wandall, H.H., K.M. Hoffmeister, A.L. Sorensen, et al. (2008), Galactosylation does not prevent the rapid clearance of long-term, 4 degrees C-stored platelets. *Blood.* **111**(6), 3249–56.
29. Neufeld, M., U. Nowak-Gottl, and R. Junker (1999), Citrate-theophylline-adenine-dipyridamol buffer is preferable to citrate buffer as an anticoagulant for flow cytometric measurement of platelet activation. *Clin Chem.* **45**(11), 2030–3.
30. Gulliksson, H. (2007), Platelet additive solutions: current status. *Immunohematology.* **23**(1), 14–9.
31. van der Meer, P.F. (2007), Platelet additive solutions: a future perspective. *Transfus Clin Biol.* **14**(6), 522–5.
32. Vassallo, R.R. and S. Murphy (2006), A critical comparison of platelet preparation methods. *Curr Opin Hematol.* **13**(5), 323–30.
33. Greening, D.W., K.M. Glenister, R.L. Sparrow, et al. (2010), International blood collection and storage: Clinical use of blood products. *J Proteomics.* **73**(3), 386–395.
34. Murphy, S. and F.H. Gardner (1975), Platelet storage at 22 degrees C: role of gas transport across plastic containers in maintenance of viability. *Blood.* **46**(2), 209–18.
35. Holme, S., K. Vaidja, and S. Murphy (1978), Platelet storage at 22 degrees C: effect of type of agitation on morphology, viability, and function in vitro. *Blood.* **52**(2), 425–35.
36. Snyder, E.L., M. Ezekowitz, R. Aster, et al. (1985), Extended storage of platelets in a new

plastic container. II. In vivo response to infusion of platelets stored for 5 days. *Transfusion.* **25**(3), 209–14.

37. AuBuchon, J.P., H. Taylor, S. Holme, et al. (2005), In vitro and in vivo evaluation of leukoreduced platelets stored for 7 days in CLX containers. *Transfusion.* **45**(8), 1356–61.
38. Cardigan, R., J. Sutherland, M. Garwood, et al. (2008), In vitro function of buffy coat-derived platelet concentrates stored for 9 days in CompoSol, PASII or 100% plasma in three different storage bags. *Vox Sang.* **94**(2), 103–12.
39. Skripchenko, A., A. Myrup, D. Thompson-Montgomery, et al. (2010), Periods without agitation diminish platelet mitochondrial function during storage. *Transfusion.* **50**(2), 390–9.
40. Mathai, J. (2009), Problem of bacterial contamination in platelet concentrates. *Transfus Apher Sci.* **41**(2), 139–44.
41. Shrivastava, M. (2009), The platelet storage lesion. *Transfus Apher Sci.* **41**(2), 105–13.
42. Dzik, S. (1993), Leukodepletion blood filters: filter design and mechanisms of leukocyte removal. Transfus Med Rev. 7(2), 65–77.
43. Blajchman, M.A. (2006), The clinical benefits of the leukoreduction of blood products. *J Trauma.* **60**(6 Suppl), S83–90.
44. Dodd, R., W. Kurt Roth, P. Ashford, et al. (2009), Transfusion medicine and safety. *Biologicals.* **37**(2), 62–70.
45. Allain, J.P., S.L. Stramer, A.B. Carneiro-Proietti, et al. (2009), Transfusion-transmitted infectious diseases. *Biologicals.* **37**(2), 71–7.
46. Solheim, B.G. (2008), Pathogen reduction of blood components. *Transfus Apher Sci.* **39**(1), 75–82.
47. Picker, S.M., V. Schneider, L. Oustianskaia, et al. (2009), Cell viability during platelet storage in correlation to cellular metabolism after different pathogen reduction technologies. *Transfusion.* **49**(11), 2311–8.
48. Van der Meer, P.F., J.L. Kerkhoffs, J. Curvers, et al. (2009), In vitro comparison of platelet storage in plasma and in four platelet additive solutions, and the effect of pathogen reduction: a proposal for an in vitro rating system. *Vox Sang.* **98**, 517–24.
49. Moroff, G. and N.L. Luban (1997), The irradiation of blood and blood components to prevent graft-versus-host disease: technical issues and guidelines. *Transfus Med Rev.* **11**(1), 15–26.
50. Kelley, W.E., B.B. Edelman, C.B. Drachenberg, et al. (2009), Washing platelets in neutral, calcium-free, Ringer's acetate. *Transfusion.* **49**(9), 1917–23.
51. Balint, B., D. Paunovic, D. Vucetic, et al. (2006), Controlled-rate versus uncontrolled-rate freezing as predictors for platelet cryopreservation efficacy. *Transfusion.* **46**(2), 230–5.
52. Hornsey, V.S., L. McMillan, A. Morrison, et al. (2008), Freezing of buffy coat-derived, leukoreduced platelet concentrates in 6 percent dimethyl sulfoxide. *Transfusion.* **48**(12), 2508–14.
53. Cardigan, R., C. Turner, and P. Harrison (2005), Current methods of assessing platelet function: relevance to transfusion medicine. *Vox Sang.* **88**(3), 153–63.
54. Albanyan, A.M., P. Harrison, and M.F. Murphy (2009), Markers of platelet activation and apoptosis during storage of apheresis- and buffy coat-derived platelet concentrates for 7 days. *Transfusion.* **49**(1), 108–17.
55. Tynngard, N. (2009), Preparation, storage and quality control of platelet concentrates. *Transfus Apher Sci.* **41**(2), 97–104.

Chapter 19

Phosphoproteome Analysis of the Platelet Plasma Membrane

Thomas Premsler, Urs Lewandrowski, Albert Sickmann, and René Peiman Zahedi

Abstract

Blood platelets are key players standing at the crossroads between physiologically occurring hemostasis and pathologic thrombus formation. As these cellular particles lack a nucleus, intra- and intercellular processes involved in platelet activity and function are almost exclusively regulated on the protein level. In particular, posttranslational protein modification by phosphorylation, which allows for a quick and highly dynamic transduction of cellular signals, is discussed in this context. In addition, since platelet activation and aggregation usually require surface contact with the surrounding tissue, special interest focuses on this contacting region, and hence on the subproteome of the platelet plasma membrane. In this chapter, we present a mass spectrometry-driven approach capable of dealing with the task of platelet plasma membrane proteomics and phosphoproteomics. The outlined protocols include strategies for the isolation and purification of plasma membrane proteins by aqueous two-phase partitioning and subsequent enrichment of phosphopeptides via titanium dioxide chromatography.

Key words: Aqueous two-phase partitioning, Mass spectrometry, Metal oxides, Phosphoproteome, Plasma membrane, Platelet

1. Introduction

Blood platelets are small, anucleated cellular particles that originate from fragmentation of precursor megakaryocytes in the bone marrow and circulate in the blood in a quiescent state (usual concentration: $2–3 \times 10^{11}$ platelets/L). Under physiological conditions, they play an essential role in hemostasis, i.e. the arrest of hemorrhage after tissue trauma (1, 2). Accordingly, vascular lesion – and hence exposure of subendothelial matrix proteins – first leads to adhesion and then activation of encompassing platelets. This activation induces a cascade of downstream intracellular signaling processes based on the phosphorylation and/or dephosphorylation of

Richard J. Simpson and David W. Greening (eds.), *Serum/Plasma Proteomics: Methods and Protocols*,
Methods in Molecular Biology, vol. 728, DOI 10.1007/978-1-61779-068-3_19, © Springer Science+Business Media, LLC 2011

specific substrates, and results in characteristic platelet shape change, translocation of receptors to the platelet surface and secretion of vasoactive substances. As a consequence, activated platelets start to aggregate, which finally culminates in the formation of a platelet plug loosely occluding the region of injury. This primary, cellular hemostasis is followed by a series of enzymatic reactions, known as humoral coagulation cascade, which help to further stabilize the platelet plug via generation of a strong fibrin fiber mesh (secondary hemostasis). In addition, activated platelets also release a variety of growth factors, e.g. platelet-derived growth factor (PDGF), transforming growth factor-β (TGF-β), and insulin-like growth factor-1 (IGF-1), shown to be involved in immune response, repair and regeneration of connective tissue in the process of wound healing (3, 4).

In order to fully understand the mechanisms and signaling cascades balancing between physiological and pathological platelet activity, a comprehensive knowledge of the proteome of these anucleated cellular particles is needed. Respective proteomics-based approaches may be performed on a global scale (5), as well as on subproteome level, including analysis of platelet organelles (6), secretome (7), and posttranslational protein modifications (8, 9). Since platelet activation is strongly dependent on interaction and adherence to the vascular surface, this chapter focuses on the field of platelet plasma membrane proteomics and in particular the related phosphoproteome.

Usually, organelle-targeted approaches are limited by the overall issue of cross-contamination by components of other cellular compartments, and thus adequate isolation and enrichment strategies have to be specifically designed. Regarding membranous structures and the respective protein content, so-called aqueous two-phase partitioning, as published by Persson and Jergil in 1995 (10), offers a feasible enrichment technique. Being more effective than density gradient centrifugation-based approaches (11), it will be presented as a method for robust and straight forward isolation of platelet plasma membranes in the following. Membrane-specific subfractionation also helps the analysis of comparably low-abundant phosphorylated membrane protein residues. Of the various techniques capable to enrich for phosphorylated protein/peptide residues – i.e. strong cation-exchange chromatography (SCX) (12), immobilized metal-ion-affinity chromatography (IMAC) (13), and chromatography based on metal oxides (such as titanium dioxide (TiO_2) (14)) prior to nano-LC–MS/MS – we will describe the latter one in detail in the course of this chapter. The method portrayed here utilizes apheresis concentrates; however, the approach can be downscaled for the analysis of platelet membranes from fresh blood samples.

2. Materials

2.1. Platelet Isolation and Purification

1. 500 mL of fresh apheresis concentrates from healthy donors (leukocyte depleted, 2–4 × 10^{11} platelets/250 mL, Department of Transfusion Medicine, University of Wuerzburg, Wuerzburg, Germany).
2. Citric acid buffer (pH 4.0): 300 mM citric acid (Merck, Darmstadt, Germany) (see Note 1).
3. HEPES washing buffer (pH 7.2): 152 mM NaCl, 4 mM KCl, 3 mM EDTA (Merck, Darmstadt, Germany), and 10 mM *N*-2-hydroxyethylpiperazine-*N′*-2-ethanesulfonic acid (HEPES; Applichem, Darmstadt, Germany).
4. Liquid nitrogen.

2.2. Platelet Lysis and Plasma Membrane Purification by Aqueous Two-Phase Partitioning

1. Two-phase system: 6.3% (w/v) poly-ethylene glycol 3350 (PEG; Sigma, Seelze, Germany) and 6.3% (w/v) Dextran T500 (Roth, Karlsruhe, Germany) dissolved in 200 mL of 15 mM Tris–base/H_2SO_4, pH 7.8 (Applichem, Darmstadt, Germany), and 1% Phosphatase Inhibitor Cocktails 1 and 2 (Sigma, Seelze, Germany). The solution is pre-equilibrated over night at 6°C.
2. Carbonate extraction buffer (pH 11.5): 100 mM Na_2CO_3 (Merck, Darmstadt, Germany), at 4°C.
3. Bandelin UW2070 ultrasonicator (Sonoplus, Berlin, Germany).
4. TLA100.4 rotor (Beckmann Coulter, Krefeld, Germany).

2.3. Protein Digestion and Peptide Purification

1. Digestion buffer (pH 7.8): 50 mM NH_4HCO_3 (Merck, Darmstadt, Germany), freshly prepared.
2. Dithiothreitol (DTT; Merck, Darmstadt, Germany).
3. Iodoacetamide (IAA; Merck, Darmstadt, Germany).
4. Lyophilized trypsin from porcine pancreas (Fluka, Steinheim, Germany), stored at −30°C.
5. 20, 40, 60, and 100% (v/v) aqueous acetonitrile (Lichrosolv; Merck, Darmstadt, Germany, HPLC grade).
6. 0.2% (v/v) aqueous formic acid (FA; Merck, Darmstadt, Germany, HPLC grade).
7. Reversed-phase SPEC C18AR columns, volume 3 mL, sorbent weight 30 mg (Varian, Darmstadt, Germany).
8. SpeedVac SPD131DDA (Thermo Scientific, Dreieich, Germany).

2.4. Isolation of Phosphopeptides Using TiO_2 Particles (Batch Protocol)

1. Loading buffer: saturated phthalic acid (PA) solution, in 80% (v/v) aqueous acetonitrile/2.5% (v/v) aqueous trifluoroacetic acid (TFA; Merck, Darmstadt, Germany, HPLC grade).
2. Washing buffer: 80% (v/v) aqueous acetonitrile/0.1% (v/v) aqueous TFA.
3. 0.1% (v/v) aqueous TFA.
4. Elution buffer (pH 10.5): 125 mM NH_4HCO_3.
5. 100% (v/v) aqueous FA.
6. Titansphere TiO_2 particles (GL Science, Tokyo, Japan), particle size 5 μm.
7. SpeedVac SPD131DDA.

2.5. MS Acquisition (Nano-LC–MS/MS)

1. 0.1% (v/v) aqueous TFA.
2. Solvent A: 0.2% (v/v) aqueous FA.
3. Solvent B: 0.2% (v/v) aqueous FA/84% (v/v) aqueous acetonitrile.
4. Qstar Elite mass spectrometer (Applied Biosystems, Darmstadt, Germany).
5. Famos™, Switchos™, Ultimate™ nano-HPLC System (Dionex, Idstein, Germany).
6. C18 trapping column: Synergi Hydro RP, ID 100 μm, length 2 cm, particle size 4 μm, pore size 80 Å (Phenomenex, Aschaffenburg, Germany), self-packed.
7. C18 main column: Synergi Hydro RP, ID 75 μm, length 30 cm, particle size 2.5 μm, pore size 80 Å (Phenomenex, Aschaffenburg, Germany), self-packed.
8. Self-made column oven, 60°C.

2.6. MS Data Evaluation: Software Tools

1. Analyst 2.0 QS software (MDS Sciex, Toronto, Canada), mascot.dll plug-in 1.6b7.
2. Mascot server version 2.2 (Matrix Science, London, UK).
3. Swiss-Prot human database (http://www.uniprot.org).
4. IPI human database (http://www.ebi.ac.uk/IPI/IPIhelp.html).
5. MS-Product (University of California, Mass Spectrometry Facility, San Francisco, CA, http://prospector.ucsf.edu/prospector/4.0.8/html/msprod.htm).

3. Methods

3.1. Blood and Its Components

Vertebrate blood is composed of cells of mesenchymal origin (blood cells) and liquid intercellular substance (blood plasma). Plasma is mainly a clear protein and salt solution (90% water) and usually makes up 55% of the total blood volume (15). It contains sugars, lipids, vitamins, mineral ions, hormones, enzymes, blood clotting factors, antibodies, and other proteins. Blood cells, on the other hand, comprise erythrocytes (which constitute about 99% of all blood cells), leukocytes (part of the immune system, $4–11 \times 10^9$/L blood), and platelets ($2–3 \times 10^{11}$/L blood). Due to their differences in molecular weight/size and density, blood components can be separated by differential centrifugation. Accordingly, platelets are usually enriched in the resulting upper-phase after centrifugation, followed by leukocytes in the interphase, and finally erythrocytes in the bottom-phase. The desired layer can then be extracted for further purification (see Note 2) and analysis.

3.1.1. Platelet Isolation and Purification (8)

1. 500 mL of fresh, leukocyte-depleted apheresis concentrates (see Note 2) from healthy donors are collected in 50 mL tubes and centrifuged for 20 min at $300 \times g$ and room temperature.
2. The supernatant is centrifuged once again to remove remaining leukocytes and erythrocytes.
3. Afterwards, the resulting supernatant is adjusted to pH 6.5 with citric acid buffer and centrifuged for 20 min at $400 \times g$ and room temperature.
4. The supernatant is discarded and the obtained pellet is carefully resuspended twice with 20 mL HEPES washing buffer to remove residual plasma proteins, and centrifuged for 20 min at $400 \times g$ and room temperature for each resuspension step.
5. The final supernatant is discarded and the resulting platelet pellet is directly frozen in liquid nitrogen and stored at –80°C until further usage, if not directly used for aqueous two-phase partitioning.

3.2. Aqueous Two-Phase Partitioning

Generally, aqueous two-phase systems, i.e. two aqueous but immiscible phases, are generated when two different polymers (most commonly PEG 3350 and Dextran T500) are mixed at specific concentrations. According to Persson and Jergil (10), these systems proved to be ideally suited for the separation of both soluble and particulate biological materials, which consequently partition between the two phases and/or the interphase in dependence on their surface properties. Plasma membrane fragments and associated structures, for instance, mainly distribute between the upper PEG-phase and the interface. This enrichment of negatively charged membranes in the top phase can be influenced by the addition of salts containing hydrogen phosphate, sulfate, citrate, and borate, as these anions tend to accumulate in the bottom phase. By including a final carbonate extraction step with 100 mM Na_2CO_3 (pH 11.5), soluble proteins loosely adherent to or trapped inside membrane vesicles are removed (16). So far, this method has been used with several cell types to obtain plasma membranes, and can also cope with low amounts of starting material (17).

3.2.1. Platelet Lysis and Plasma Membrane Purification by Aqueous Two-Phase Partitioning (Based on (17))

1. The 6.3% (w/v) PEG/Dextran solution is equilibrated the day before usage at 6°C. During this time, separation into two saturated phases occurs, with the upper one mainly containing PEG and the lower one Dextran (see Note 3).
2. An aliquot of purified platelets is dissolved in an appropriate volume (e.g. 20 mL for 50–100 μL platelet pellet volume) of a 1:1 mixture of both the PEG- and the Dextran-phase.
3. The chilled suspension is lysed by ultrasonication using a Bandelin UW2070 and six pulses for 15 s at 70% maximum power. Upon mixing by several invertations and vortexing, phase separation is accelerated by 10 min of centrifugation at 500 × *g* and 6°C.
4. The upper PEG-phase (i.e. putative membrane fraction) is collected and re-extracted twice with an equal volume of fresh Dextran-phase by mixing and centrifuging as described above.
5. Subsequently, the resulting upper PEG-phase is diluted 1:1 with water and ultracentrifuged for 1 h at 100,000 × *g* and 4°C using a TLA100.4 rotor.
6. The supernatant is discarded and the remaining membranous pellet is incubated with 3 mL carbonate extraction buffer for 30 min on ice.
7. Finally, the purified platelet plasma membranes are pelleted by ultracentrifugation as described above and stored at –80°C until further usage.

3.3. Protein Digestion and Peptide Purification

1. Purified platelet plasma membranes are resuspended in an appropriate volume (500 μL to 1 mL) of digestion buffer.
2. Prior to protein digestion, carbamidomethylation of the sample is performed as follows: disulfide bonds are reduced by incubating with 10 mM DTT (final concentration) at 56°C for 30 min. Resulting free cysteines are then alkylated using 30 mM IAA (final concentration) at 25°C for 30 min in the dark. The reaction is stopped by the addition of 10 mM DTT (final concentration).
3. Upon adding lyophilized trypsin 1:50 in relation to the protein content of the respective sample, protein digestion is performed for at least 8 h at 37°C.
4. The resulting proteolytic peptide solution is both desalted and purified from contaminating lipids via solid phase

extraction using SPEC C18AR columns. The columns are first preconditioned with 100% acetonitrile, followed by equilibration with 0.2% (v/v) aqueous FA. The sample is acidified to a final concentration of 0.2% (v/v) aqueous FA and loaded on the columns. Upon repeated washing with 0.2% (v/v) aqueous FA, peptides are finally eluted from the solid material using 20, 40, and 60% (v/v) aqueous acetonitrile.

5. Elution fractions are pooled, dried in a SpeedVac, and stored at –80°C until further usage.
6. Optional: Protein digestion can be controlled by comparative 1D-SDS-PAGE of sample aliquots before and after digestion and subsequent silver staining procedure according to Blum et al. (18).

3.4. Phosphopeptide Enrichment by Metal Oxide-Based Liquid Chromatography

Due to the low abundance and stoichiometry of phosphorylated protein residues, phospho-specific enrichment prior to a profound mass spectrometric analysis is inevitable for complex samples. For this purpose, several methods have been applied so far, including immunoprecipitation (in particular with regard to phosphotyrosine (pTyr) (19)), SCX, and IMAC (20). In addition, based on the self-assembling properties of alkyl phosphate monolayers on the surface of metal oxides (21, 22), TiO_2-based enrichment of phosphopeptides was established (23, 24) and finally integrated in an online 2D-nano-LC–MS/MS setup in 2004 (14). Nowadays, it has evolved into the most frequently used technique in the field. As in IMAC, phosphopeptides are bound under acidic and eluted under alkaline conditions. Unwanted copurification of nonphosphorylated peptides with acidic sidechains can, for instance, be reduced upon addition of 2,5-dihydroxybenzoic acid (DHB) or PA to the loading buffer, which presumably have a competitive effect for TiO_2 binding of nonphosphorylated peptides (25). Apart from TiO_2, zirconium dioxide and aluminum hydroxide can also be used, although their specificity was reported to be lower than the one of TiO_2 (26).

3.4.1. Isolation of Phosphopeptides Using TiO_2 Particles (Based on (25, 27), Batch Protocol)

1. 200 µg of purified proteolytic peptides are reconstituted in 80% (v/v) acetonitrile/2.5% (v/v) TFA/saturated PA (see Note 4).
2. Then the peptide solution is added to 2 mg of equilibrated (in loading buffer) TiO_2 particles in a 500-µL reaction tube and incubated for 15 min with end-over-end rotation.
3. After this incubation step, the beads are washed twice with loading buffer, twice with washing buffer, and finally twice with 0.1% (v/v) TFA (1 min incubation time).
4. Peptides are eluted from the TiO_2-material with 125 mM NH_4HCO_3 and acidified immediately with 100% (v/v) FA to pH 3.0 to avoid β-elimination of phosphoamino acids.

5. The eluted peptides are dried in a SpeedVac and reconstituted in 0.1% (v/v) TFA for nano-LC–MS/MS analysis.

3.5. MS Acquisition (Nano-LC–MS/MS) (8)

1. The enriched phosphopeptides (reconstituted in 0.1% (v/v) TFA) are loaded onto a C18 trapping column for further desalting and preconcentration in an Ultimate nano-HPLC system (flow rate: 7 μL/min 0.1% (v/v) TFA, prolonged loading time of 25 min).
2. Peptide separation is then performed on a reversed-phase main column by applying an 80-min binary gradient of solvents A and B ranging from 5 to 50% solvent B at a flow rate of 270 nL/min, followed by column washing at 95% solvent B for 5 min, and finally re-equilibration at 5% solvent B for 15 min.
3. MS analysis is conducted on a Qstar Elite mass analyzer online coupled to the LC-system (see Note 5).
4. Spray voltage is set to 2,200 V.
5. Full 0.5-s MS survey scans from 380 to 1,500 m/z are acquired, and the three most intensive peaks are subjected to MS/MS, taking into account a dynamic exclusion.
6. Dynamic background subtraction is deactivated and linear collision energy equations for 2^+, 3^+, and 4^+ peptides are optimized manually using synthetic standard peptides. Automatic MS/MS accumulation is set to highest quality at a maximum accumulation time of 2 s.

3.6. MS Data Evaluation (8)

1. Qstar Elite MS data files are transformed into MGF format using the Analyst 2.0 QS software with the mascot.dll plug-in 1.6b7.
2. Charge states are determined from survey scans. Spectra with less than ten signals are omitted, precursor mass tolerance for grouping is set to 0.05 Da, and the maximum number of cycles between groups is set to 3. Signals with intensities 0.1% below base peak are removed.
3. Using Mascot (28), the generated peak lists can be searched either against the human Swiss-Prot or the human IPI database (29).
4. For all samples, trypsin is chosen as protease with a maximum of two missed cleavage sites. Peptide- and MS/MS tolerances are set to 0.2/0.5 Da, respectively. Carbamidomethylation of Cys residues (+57.02 Da) is chosen as fixed modification; oxidation of Met (+15.99 Da) and phosphorylation of Ser, Thr, and Tyr (+79.96 Da) as variable ones.
5. All phosphopeptide MS/MS spectra with a minimum Mascot score of 34 (p-value of 0.05 or below) and a length between 7 and 30 amino acid residues are taken into consideration for

further interpretation and are validated manually (see Note 6). This process includes evaluation of spectra quality (signal-to-noise ratio, annotation of prominent signals, sequence coverage in particular of b- and y-ions), interpretation of neutral losses of H_2O, NH_3, or H_3PO_4 (in case of recent and stable mass calibration, generally peptide hits with delta masses >0.02 can be omitted), and in particular verification of phosphorylation site assignment by Mascot. In this context, the software tool MS-Product, with "loss of H_3PO_4" and "multiple losses" enabled as additional options, helps to cope with hitherto unassigned internal fragment ions by comparison to fragment ion *m/z* predictions.

4. Notes

1. All solutions and buffers, unless otherwise specified, are to be prepared with water having a resistivity of 18.2 MΩ cm and a TOC of 1–5 ppb. Except for the digestion buffer, which needs to be freshly prepared, they can be stored at 4°C.
2. In case there are no leukocyte-depleted apheresis concentrates available, it is also possible to purify platelets from common blood donations. The blood is collected into one-fifth volume anticoagulation buffer (pH 6.5) consisting of 100 mM sodium citrate, 7 mM citric acid, 140 mM glucose, and 15 mM ethylenediaminetetraacetic acid (EDTA) and centrifuged for 20 min at 300×*g* at room temperature. Then the platelet-rich supernatant is collected and further processed as described in Subheading 3.1. However, compared to apheresis concentrates, the amount of platelets isolated by this approach might be too low for additional enrichment of phosphorylated, membrane localized peptides. Accordingly, to gain equivalent platelet amounts, at least 1–2 L of donated blood would be required. For the proteomic analysis of resting platelets, it is absolutely mandatory to prevent any contact with calcium ions and glass material. Plastic pipettes and vessels should be used; pH controls should be performed via pH paper strips instead of glass pH electrodes. In addition, sudden vibrations and agitation should be avoided to avert platelet activation.
3. All steps of platelet lysis and plasma membrane purification by aqueous two-phase partitioning should be performed at 4–6°C. The equilibrated saturated polymer phases should be prepared freshly the day before usage and equilibrated at 6°C. 6.3% (w/v) of PEG 3350 and Dextran T500, respectively, in 15 mM Tris–base/H_2SO_4 (pH 7.8) is a well-established working

concentration in our hands, providing reasonable membrane purity with sufficient protein yield (30). Nevertheless, polymer concentration can be adapted to different user needs. For instance, by increasing it to 6.5–6.7% (w/v), contamination of the purified plasma membrane fraction by membranes originating from other subcellular compartments, e.g. mitochondria, can significantly be reduced. By doing so, however, the general protein yield will also decrease distinctly.

4. When working with TiO_2, unwanted copurification of non-phosphorylated acidic peptides can also be reduced upon addition of DHB to the loading buffer. Accordingly, based on the protocols of Larsen et al. (25) or Bodenmiller et al. (27), purified proteolytic peptides are reconstituted in a solution containing 200 mg/mL DHB, 80% (v/v) aqueous acetonitrile, and 2.5% (v/v) aqueous TFA. The rest of the phosphopeptide enrichment procedure is similar to the one described in Subheading 3.4, except for the first two washing steps which are performed with DHB rather than with saturated PA solution. As indicated by the cited study of Bodenmiller et al., however, specificity of phosphopeptide enrichment by DHB-TiO_2 might be significantly lower than the one of PA-TiO_2 (depending on the applied protocol). Moreover, usage of DHB as competition agent may lead to contamination of the ESI ion source (25) and demands respective cleaning. PA turns out to be more compatible with nano-LC-ESI–MS/MS.

5. To address the low frequency of pTyr in MS-based studies, samples can additionally be analyzed by precursor ion scanning (PIS) for the pTyr immonium ion at m/z 216.01 using a Qtrap 4000 mass analyzer.

6. When working with large amounts of proteomic data, automatic data evaluation and in particular assessment of the false discovery rate (FDR) can be facilitated by using so-called target/decoy databases (31). Decoy databases with reversed/randomized sequences of the human Swiss-Prot or the human IPI database can, for instance, be generated with the help of the decoy.pl scripts of Matrix Science (London, UK). MS data are searched against these databases and evaluated via the mass spectrometry-oriented MS-LIMS Project Software (http://genesis.ugent.be/ms_lims/). For that purpose, non-phosphorylated peptides, peptides of less than seven and more than 30 residues, and redundancies have to be removed at first. For a given Mascot score, the FDR is finally calculated by dividing the amount of decoy-peptide hits by the number of target-peptide hits, and thus shall represent the share of false-positive peptide identifications from the forward database.

Acknowledgment

The authors thank the "Ministerium für Innovation, Wissenschaft, Bildung und Forschung" for continuous financial support (031P5800).

References

1. Ruggeri M Z (2002) Platelets in atherothrombosis. *Nature Med* **8**, 1227–1234.
2. Zellner M, Winkler W, Hayden H, et al. (2005) Quantitative validation of different protein precipitation methods in proteome analysis of blood platelets. *Electrophoresis* **26**, 2481–2489.
3. Celotti F, Colciago A, Negri-Cesi P, et al. (2006) Effect of platelet-rich plasma on migration and proliferation of SaOS-2 osteoblasts: role of platelet-derived growth factor and transforming growth factor-β. *Wound Rep Reg* **14**, 195–202.
4. O'Connell S, Impeduglia T, Hessler K, et al. (2008) Autologous platelet-rich fibrin matrix as cell therapy in the healing of chronic lower-extremity ulcers. *Wound Rep Reg* **16**, 749–756.
5. Garcia A, Senis Y A, Antrobus R, et al. (2006) A global proteomics approach identifies novel phosphorylated signaling proteins in GPVI-activated platelets: Involvement of G6f, a novel platelet Grb2-binding membrane adapter. *Proteomics* **6**, 5332–5343.
6. Hernández-Ruiz L, Valverde F, Jimenez-Nuñez M D, et al. (2007) Organellar proteomics of human platelet dense granules reveals that 14-3-3ζ is a granule protein related to atherosclerosis. *J Proteome Res* **6**, 4449–4457.
7. Della Corte A, Maugeri N, Pampuch A, et al. (2008) Application of 2-dimensional difference gel electrophoresis (2D-DIGE) to the study of thrombin-activated human platelet secretome. *Platelets* **19**, 43–50.
8. Zahedi R P, Lewandrowski U, Wiesner J, et al. (2008) Phosphoproteome of resting human platelets. *J Proteome Res* **7**, 526–534.
9. Lewandrowski U, Zahedi R P, Moebius J, et al. (2007) Enhanced N-glycosylation site analysis of sialoglycopeptides by strong cation exchange prefractionation applied to platelet plasma membranes. *Mol Cell Proteomics* **6**, 1933–1941.
10. Persson A, Jergil B (1995) The purification of membranes by affinity partitioning. *FASEB J* **9**, 1304–1310.
11. Moebius J, Zahedi R P, Lewandrowski U, et al. (2005) The human platelet membrane proteome reveals several new potential membrane proteins. *Mol Cell Proteomics* **4**, 1754–1761.
12. Beausoleil S A, Jedrychwoski M, Schwartz D, et al. (2004) Large-scale characterization of HeLa cell nuclear phosphoproteins. *Proc Natl Acad Sci USA* **101**, 12130–12135.
13. Ficarro S B, McCleland M L, Stukenberg P T, et al. (2002) Phosphoproteome analysis by mass spectrometry and its application to Saccharomyces cerevisiae. *Nat Biotechnol* **20**, 301–305.
14. Pinkse M W, Uitto P M, Hilhorst M J, et al. (2004) Selective isolation at the femtomole level of phosphopeptides from proteolytic digests using 2D-nano-LC-ESI-MS/MS and titanium oxide precolumns. *Anal Chem* **76**, 3935–3943.
15. Despopoulos A, Silbernagl S (2003) Color Atlas of Physiology, 5th edn. Thieme, Stuttgart, 88ff.
16. Fujiki Y, Hubbard A L, Fowler S, et al. (1982) Isolation of intracellular membranes by means of sodium carbonate treatment: application to endoplasmic reticulum. *J Cell Biol* **93**, 97–102.
17. Schindler J, Lewandrowski U, Sickmann A, et al. (2006) Proteomic analysis of brain plasma membranes isolated by affinity two-phase partitioning. *Mol Cell Proteomics* **5**, 390–400.
18. Blum H, Beier H, Gross H J (1987) Improved silver staining of plant proteins, RNA and DNA in polyacrylamide gels. *Electrophoresis* **8**, 93–99.
19. Marcus K, Moebius J, Meyer H E (2003) Differential analysis of phosphorylated proteins in resting and thrombin-stimulated human platelets. *Anal Bioanal Chem* **376**, 973–993.
20. Zahedi R P, Begonja A J, Gambaryan S, et al. (2006) Phosphoproteomics of human platelets: A quest for novel activation pathways. *Biochim Biophys Acta* **1764**, 1963–1976.
21. Folkers J P, Gorman C B, Laibinis P E, et al. (1995) Self-assembled monolayers of long chain hydroxamic acids on the native oxides of metals. *Langmuir* **11**, 813–824.

22. Hofer R, Textor M, Spencer N D (2001) Alkyl phosphate monolayers, self assembled from aqueous solution onto metal oxide surfaces. *Langmuir* **17**, 4014–4020.

23. Sano A, Nakamura H (2004) Chemo-affinity of titania for the column-switching HPLC analysis of phosphopeptides. *Anal Sci* **20**, 565–566.

24. Sano A, Nakamura H (2004) Titania as a chemo-affinity support for the column-switching HPLC analysis of phosphopeptides: application to the characterization of phosphorylation sites in proteins by combination with protease digestion and electrospray ionization mass spectrometry. *Anal Sci* **20**, 861–864.

25. Larsen M R, Thingholm T E, Jensen O N, et al. (2005) Highly selective enrichment of phosphorylated peptides from peptide mixtures using titanium dioxide microcolumns. *Mol Cell Proteomics* **4**, 873–886.

26. Sugiyama N, Masuda T, Shinoda K, et al. (2007) Phosphopeptide enrichment by aliphatic hydroxy acid-modified metal oxide chromatography for nano-LC-MS/MS in proteomics applications. *Mol Cell Proteomics* **6**, 1103–1109.

27. Bodenmiller B, Mueller L N, Mueller M, et al. (2007) Reproducible isolation of distinct, overlapping segments of the phosphoproteome. *Nat Methods* **4**, 231–237.

28. Perkins D N, Pappin D J, Creasy D M, et al. (1999) Probability-based protein identification by searching sequence databases using mass spectrometry data. *Electrophoresis* **20**, 3551–3567.

29. Kersey P J, Duarte J, Williams A, et al. (2004) The International Protein Index: An integrated database for proteomics experiments. *Proteomics* **4**, 1985–1988.

30. Lewandrowski U, Wortelkamp S, Lohrig K, et al. (2009) Platelet membrane proteomics: a novel repository for functional research. *Blood* **114**, 10–19.

31. Elias J E, Gygi S P (2007) Target-decoy search strategy for increased confidence in large-scale protein identifications by mass spectrometry. *Nat Methods* **4**, 207–214.

Part V

Bioinformatic Analysis of Blood Proteins and Peptides

Chapter 20

Statistical Design and Analysis of Label-free LC-MS Proteomic Experiments: A Case Study of Coronary Artery Disease

Timothy Clough, Siegmund Braun, Vladimir Fokin, Ilka Ott, Susanne Ragg, Gunther Schadow, and Olga Vitek

Abstract

This chapter presents a case study, which applies statistical design and analysis to an LC–MS-based investigation of subjects with coronary artery disease. First, we discuss the principles of statistical experimental design, and the specification of an Analysis of Variance (ANOVA) model that describes the major sources of variation in the data. Second, we discuss procedures for detecting differentially abundant proteins, estimating protein abundance in individual samples, testing predefined groups of proteins for enrichment in differential abundance, and calculating sample size for a future experiment. The discussion is accompanied by examples of computer code implemented in the open-source statistical software R, which can be followed for an independent implementation of a similar investigation.

Key words: Quantitative proteomics, Statistical design of experiments, Analysis of Variance, Protein quantification, LC–MS, Replication, Randomization, Blocking, Gene set enrichment, Sample size, Coronary artery disease.

1. Introduction

The goal of quantitative proteomics is to monitor the abundance of proteins in biological samples under diverse conditions (1, 2). Liquid chromatography coupled with mass spectrometry (LC–MS) (1) is particularly attractive for discovery-oriented research, as it enables non-targeted, large-scale relative protein quantification. In an LC–MS workflow, proteins in a sample are digested enzymatically into peptides, and subjected to chromatographic separation, ionization, and mass analysis. The intensity of

Richard J. Simpson and David W. Greening (eds.), *Serum/Plasma Proteomics: Methods and Protocols*,
Methods in Molecular Biology, vol. 728, DOI 10.1007/978-1-61779-068-3_20, © Springer Science+Business Media, LLC 2011

the resulting LC–MS features is used for relative quantification of peptides and proteins. Tandem mass spectra are typically acquired as part of the same experiment, and are used to determine the sequence identity of peptides and proteins underlying a subset of the LC–MS features (3). A variety of software tools have been proposed to extract and quantify LC–MS features from the acquired spectra, annotate the features with sequence identity, and align the features across runs (4). Overall, the combined experimental and computational workflows have reached maturity, and are now capable of accurately identifying and quantifying thousands of peptides simultaneously.

Despite the progress in analytical capabilities, LC–MS measurements exhibit a substantial amount of biological and technical variation. For example, in clinical investigations that compare protein abundance between subjects with and without disease, LC–MS measurements reflect the natural variation of protein abundance between subjects, as well as the technical variation introduced during sample collection and handling, and during spectral acquisition. Statistical reasoning is the only approach for deriving objective and reproducible conclusions in the presence of such variation. It should be applied not only when deriving conclusions from the acquired spectra, but also when designing a new experiment.

Statistical experimental design is a protocol that defines all steps of the experiment. At the design stage, statistical reasoning has two goals. First, it ensures that our conclusions avoid *bias*, i.e., systematic errors in interpretation of the data. Second, it ensures that the experiment is *efficient*, i.e., it minimizes the random variation and uncertainty in the experimental data, for a given amount of cost. In the example of a clinical investigation, an experimental design (1) translates the goal of the study into statistical hypotheses, such as comparing the mean of protein abundance between populations of subjects with a disease and controls, (2) selects from the underlying populations the type and number of samples that are appropriate for testing these hypotheses, and (3) allocates the samples to batches of sample processing and spectral acquisition.

Statistical inference derives conclusions from the acquired spectra in the presence of stochastic variation and uncertainty. At the inference stage, statistical reasoning is necessary to characterize the structure of the variation, and to distinguish systematic patterns in the data from the patterns that are likely to occur by random chance. For example, in a clinical LC–MS-based investigation, inference steps may include (1) transformation of feature abundances to the appropriate scale (e.g., log) and normalization of feature abundances across runs, (2) specification of a probabilistic model that describes the major sources of variation, (3) detection of proteins with differential abundance across disease groups, (4) estimation of protein abundance in samples from individual subjects, and (5) testing predefined groups of proteins (such as pathways) for enrichment in differentially abundant proteins across disease groups.

This chapter presents a case study which applies statistical reasoning to the design and analysis of an LC–MS-based investigation of subjects with coronary artery disease. We discuss the motivation underlying various statistical analysis steps and provide examples of computer code implemented in the open-source statistical software R (5) and R-based framework Bioconductor (6). Our purpose is not to present a comprehensive tutorial on statistical methods or a full description of clinically relevant conclusions. Rather, we provide examples of the key statistical analysis steps that can be followed for an independent implementation of a similar investigation (see Note 1).

2. Materials

The case study originated from an investigation of subjects with coronary artery disease at the German Heart Center and at the First Medical Clinic of the Technical University of Munich, Germany. The goal of the study is to compare protein profiles of subjects in four disease groups, namely two groups with acute coronary syndrome [*ST segment elevation myocardial infarction* (*STEMI*) and *non-ST segment myocardial infarction* (*NSTEMI*)] and two groups with angina (*unstable angina* and *stable angina*), against control individuals without coronary artery disease. In years 2005 and 2006, the researchers collected blood samples from 3290 consecutive subjects at a single time point prior to coronary angiography (7). Blood was drawn from a catheter by a licensed physician into EDTA blood collection tubes. Within 30 min of withdrawal, plasma was obtained by centrifugation for 10 min at 2,000 $\times g$ at room temperature and samples were frozen at −80 °C. Sample handling was done following the standard operating procedures by designated laboratory technicians at the Institute for Laboratory Medicine. In addition to the plasma samples, many clinical variables such as age, gender, history of disease, and history of medication, were recorded for each subject as standardized data elements. The researchers aimed at selecting a representative subset of subjects from this larger population, and at comparing their protein profiles as quantified by multiple LC–MS features.

Once a subset of subjects was selected for the study, plasma samples were subjected to albumin depletion, tryptic digest, and mass analysis using liquid chromatography separation coupled with Thermo–Finnigan linear ion-trap mass spectrometry. The procedures were performed at Monarch Life Sciences, Indianapolis, IN.

Depletion was based on an affinity resin column using the Montage Albumin Deplete Kit (Millipore) (8) and was performed by two technicians in two batches of samples.

LC–MS features were quantified, aligned, and annotated with peptide and protein identities as described in (9). When a feature in a run could not be detected, the signal processing workflow reported background intensity signals, and therefore the dataset has no missing values (see Note 2).

In many cases, identified peptides could be mapped to sequences of multiple proteins, in particular to multiple protein isoforms. To reduce the number of peptides with ambiguous mappings to protein isoforms, each peptide was mapped instead to the Entrez gene ID underlying each protein sequence. Only peptides with unambiguous Entrez IDs were kept. Among those, only Entrez IDs of proteins known to be present in plasma in sufficient concentrations were retained for the subsequent analysis. Overall, the procedure identified 77 unique Entrez IDs ranging in concentration from 10^6 to 10^{10} ng/L, with up to 169 peptides per id. For Entrez IDs with over 74 peptides, 74 peptides were randomly selected for further analysis. In the following, we use the terms "Entrez gene ID" and "protein" interchangeably to simplify the presentation.

In addition to the LC–MS measurements, plasma concentrations of 11 of the 77 proteins from the same subjects were measured independently in a clinical setting, with BN Pro Spec Nephelometer (Siemens Healthcare Diagnostics). Agreement of LC–MS quantifications with nephelometry measurements of the 11 proteins was used to evaluate the performance of the analytical procedures.

3. Statistical Methods

3.1. Experimental Design

3.1.1. Translating the Goal of the Study into Statistical Hypotheses

The study of coronary artery disease has three goals. The *first goal* is to compare, separately for each identified protein, the mean abundance in a disease group to the mean abundance in controls. Such differentially abundant proteins may be implicated in the pathophysiological processes of the disease. The *second goal* is to estimate the abundance of the proteins in each sample, on a relative scale that is comparable across subjects. These estimates will be subsequently used as input to unsupervised clustering or supervised classification of the subject profiles. The *third goal* is to examine predefined groups of proteins (for example, proteins that share functional annotations, or belong to the same pathway) for enrichment in differential abundance. Such groups are viewed as "activated" by the processes of the disease, and help derive additional clinical hypotheses.

Although only a subset of the subjects was selected for the study, the researchers would ultimately like to extend the conclusions to the entire population of subjects treated at the German Heart Center and at the First Medical Clinic of the Technical University of Munich at that time. It is therefore important to ensure that the selected subjects can be viewed as representative of this population. Since protein concentration is known to depend on age and gender (10), subject selection should avoid age- and gender-related biases. To avoid confounding effects, subjects who simultaneously suffer from other diseases such as diabetes should be excluded from consideration. Even under these restrictions, subjects with coronary artery disease are heterogeneous, and a variety of factors affect protein abundance. Substantial subject-to-subject variation is therefore anticipated.

Based on the results of this case study, the researchers will conduct subsequent LC–MS investigations using a larger selection of subjects and a more specifically targeted population (for example, by including only a subset of disease groups, or by restricting subjects' clinical characteristics). We use the insights from the present case study to help design such follow-up experiments.

3.1.2. Selection of Subjects for the Study

The three fundamental principles of statistical experimental design are replication, randomization, and blocking (11).

3.1.2.1. Replication

Replication, in this context, requires that the study includes multiple subjects from each disease group (i.e., biological replicates). Replication serves two purposes. First, increasing the number of replicates improves the reliability of estimates such as average protein abundance in a disease group. Second, replication is needed to quantify the natural variation of protein abundance across subjects, and to assess whether patterns in the data, such as differences in protein abundances between groups, are likely to occur by random chance.

3.1.2.2. Randomization

Randomization requires that the study randomly selects subjects from the underlying population. Randomization also serves two purposes. First, it distributes the undesirable and potentially unknown sources of biases roughly equally across subjects in each group. Second, randomization allows us to view the selected subjects as random instances from the population, which is necessary for the application of most statistical analysis techniques.

3.1.2.3. Blocking

Blocking imposes restrictions upon randomization, in order to reduce the bias and variance from known sources of variation. In this study, age and gender are known to influence the abundance of several well-studied proteins. Therefore, blocking amounts to first partitioning the entire population of 3,290 subjects by age, gender, and disease group as shown in Table 1, and then randomly

Table 1
Number of serum samples from subjects with coronary artery disease and controls, available for each combination of age group, gender and disease group

		Disease group				
		Control	Stable angina	Unstable angina	NSTEMI	STEMI
Stratification	≥58 y.o; Female	354	300	49	39	29
	≥58 y.o; Male	701	843	143	86	54
	<58 y.o; Female	80	56	5	5	8
	<58 y.o; Male	264	190	34	23	27

Table 2
Number of serum samples selected for the proteomic experiment. Each disease group has the same number of subjects for each combination of age group and gender

		Disease group				
		Control	Stable angina	Unstable angina	NSTEMI	STEMI
Stratification	≥58 y.o; Female	3	3	3	3	3
	≥58 y.o; Male	3	3	3	3	3
	<58 y.o; Female	2	2	2	2	2
	<58 y.o; Male	2	2	2	2	2

selecting an equal number of subjects across disease groups, within each combination of age and gender. For example, the *Unstable angina* and *NSTEMI* groups in Table 1 have only five younger female subjects, therefore it is only possible to select at most five younger female subjects from each disease group. A larger number of subjects can be selected from all other combinations of age and gender. In this case study, we randomly select three subjects from each stratum that involves older subjects, and two subjects from each stratum that involves younger subjects, as shown in Table 2.

In the present case study, this protocol recruits ten subjects from each disease group. The subjects are randomly selected from the underlying population while using a restricted randomization to ensure an equal number of subjects within each combination of age and gender across disease groups. The choice of the specific number of biological replicates in this and similar studies is discussed in Subheading 3.3 (see Note 3).

3.1.3. Allocation of Experimental Resources

The second application of the principles of replication, randomization, and blocking occurs when allocating the selected samples for preprocessing and spectral acquisition.

3.1.3.1. Replication

In this context, replication involves collecting multiple runs of the same biological sample (i.e., technical replicates). Technical replication helps improve the accuracy of measurements of protein abundance in the sample, and allows us to quantify the extent of run-to-run variation. However, replication of biological samples is the most important type of replication, and it is often acceptable to forgo technical replication (11). Consequently, this study only acquires one technical replicate per subject.

3.1.3.2. Randomization

Randomization in this context implies that all samples are processed in a randomly defined order. This is necessary to avoid biases due to possible, and potentially unknown, experimental artifacts. Randomization distributes the sources of bias roughly equally across samples in each disease group, and prevents misleading conclusions.

3.1.3.3. Blocking

Blocking consists of assigning an equal number of samples from each disease group within experimental units (e.g., trays, plates, or days of spectral acquisition) known to introduce additional variation, in order to equally distribute these unit-specific artifacts. In this case study of coronary artery disease, pilot experiments with the same equipment did not identify major systematic sources of variation. Therefore, blocking was not used, and the order of the samples was completely randomized for both preprocessing and spectral acquisition (see Note 4).

Overall, this protocol produced a dataset which contains an equal number of subjects, and an equal number of replicates, in each disease group. This property of the experimental design is called *balanced*. The balance is desirable because it optimizes the precision of our conclusions as compared to an unbalanced case, and estimates of many quantities of interest simplify to sample averages.

3.2. Statistical Analysis of the Acquired Spectra

3.2.1. Data Transformation and Normalization

3.2.1.1. Data Transformation

It is generally believed that changes in protein abundance are multiplicative in nature. Therefore, when the goal is to compare protein abundances across disease groups, we are interested in the *fold change*, i.e., the ratio of abundances between groups. However, working with the ratios directly is rarely beneficial, for two reasons. First, the ratios do not constitute a natural framework for handling replicated observations in a group, multiple LC–MS features mapped to a protein, and experiments with more than two groups. Second, distributions of feature abundances within a disease group are typically highly skewed. This violates the assumptions of symmetric and (nearly-)Normal distributions, which are assumed by many statistical models.

A logarithm transformation applied to the intensities of LC–MS features helps alleviate these drawbacks. On the logarithmic scale, the quantity of interest is the difference of mean protein abundances between groups, i.e., the *log-fold change*. Many statistical models, such as the one in Subheading 3.2.2, can be used to characterize patterns in the data in terms of differences of means. Moreover, distributions of feature intensities on the log scale are typically more symmetric, which facilitates model-based inference. In this case study we therefore use log-transformed intensities of LC–MS features.

3.2.1.2. Normalization

A variety of factors beyond the control of an experimentalist induce systematic differences of feature intensities between runs. A normalization of the log-transformed intensities is necessary to account for this and to distinguish the true biological changes from experimental artifacts. A popular and easy to implement family of normalization methods assumes that most features in the experiment do not change in abundance across runs. In particular, quantile normalization (12) assumes that the quantiles of the distributions of the normalized intensities are equal across runs. Although quantile normalization was introduced in the context of gene expression microarrays, it has been successfully applied to LC–MS experiments, for example, in (13). The underlying assumption is judged appropriate for the study of coronary artery disease, and we therefore apply quantile normalization to this dataset. When the assumption underlying quantile normalization is judged inappropriate, a different normalization procedure (e.g., involving spiked controls), should be used.

3.2.2. Specification of Probabilistic Model that Describes Major Sources of Variation

3.2.2.1. Model Specification

Figure 1 shows a representative profile of one protein in the coronary artery disease dataset. The protein is quantified by five features; its changes in abundance between groups are relatively small. Due to biological and technical variation, some features produce a somewhat contradictory evidence of differential abundance. A statistical model is necessary to formally characterize the sources of variation of all measurements that pertain to this protein profile, and to distinguish the systematic patterns from noise. Multiple such models exist, and the choice depends on the assumptions that we are willing to make regarding the data. In this case study, we use the fixed effects Analysis of Variance (ANOVA) model advocated in (14). Additional technical details on the model are found in (15, 16).

The fixed effects ANOVA model is shown in Fig. 2. The model simultaneously considers all LC–MS features mapped to a protein, and views them as replicate measurements of protein abundance. The intensity of each peak is decomposed into the contribution of disease group G_i (expressing the systematic difference in protein abundance between groups) and the contribution

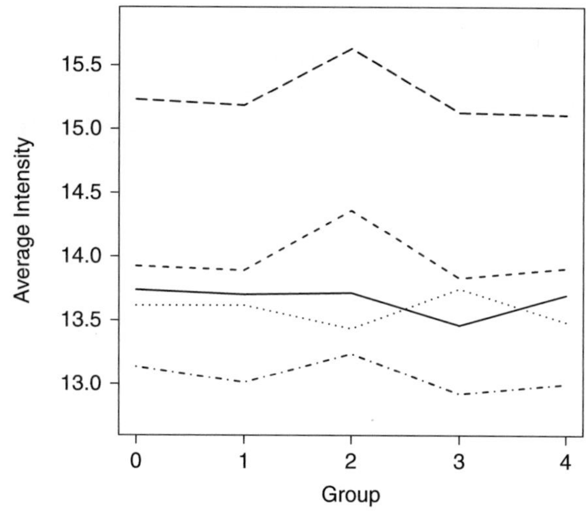

Fig. 1. The quantitative profile of protein 3045 in the coronary artery disease dataset. *X*-axis: disease group. *Y*-axis: log-intensity of a feature, averaged over all samples in the disease group. Each line represents a distinct feature.

Observed feature intensity	=	Overall mean abundance	+	Systematic deviation due to: disease group	+	peptide feature	+	group × feature interaction	+	biological replicate	+	Random measurement error
$yijk$	=	μ	+	G_i	+	F_j	+	$(G{\times}F)_{ij}$	+	$S(G)_{k(i)}$	+	$\epsilon_{ijk} \sim N\left(0, \sigma^2_{Error}\right)$

Fig. 2. Fixed effects ANOVA. $i=1,\ldots,g$ is the index of a disease group, $j=1,\ldots,f$ the index of a feature, and $k=1,\ldots,n$ the index of a subject in a group. In this case study, $g=5$, $n=10$, and f is the number of features in the particular protein. $G_1=F_1=(G\times F)_{i1}=(G\times F)_{1j}=S(G)_{1(i)}=0$ for $i=1,\ldots,g$ and $j=1,\ldots,f$ are identifiability constraints that are required for estimation of the model parameters. σ^2_{Error} is the variance of the measurement error. All random deviations are independent.

of the LC–MS feature F_j that produced this value (expressing the fact that some peptides have a systematically higher ionization efficiency than the others). The model also contains a statistical interaction term $(G\times F)_{ij}$, which describes potential deviations of individual LC–MS features from the average profile due to both biological interferences and experimental artifacts (see Note 5).

The model also specifies the deviation of each subject (i.e., biological replicate, or subject) $S(G)_{k(i)}$ from the overall group mean, which expresses the fact that some individuals have a higher natural abundance of the protein than others. Notation $S(G)_{k(i)}$ is read as "subject within a group," indicating that in comparative experiments, such as this case study, each individual belongs to only one group. A different model, and a different notation, is necessary in experiments with repeated measurements where groups corresponds to time points, and multiple samples from the same individual are collected across time (see Note 6). The last term in the model ε_{ijk} describes the remaining deviations of peak intensity and is viewed as non-systematic replicates of a Normally distributed measurement error with constant variance σ^2_{Error}.

The "fixed effects" in the model name refers to the fact that the individuals selected for the study are considered fixed, and the model limits the scope of our conclusions to these specific individuals. This is appropriate for an initial screening experiment such as this case study, and has been shown to both increase the sensitivity and the specificity of testing, and improve the accuracy of subject quantification (14), which we describe in Subheading 3.2.3 and Subheading 3.2.4. A different "mixed effects" ANOVA model will be necessary for a validation experiment where individuals in the study are viewed as random instances from a larger population of subjects, and one would like to extend the scope of the conclusions to the entire population (see Note 7).

3.2.2.2. Model Fit

The terms in Fig. 2 are estimated from the data separately for each protein, using the standard least squares procedure (15, 16) available in the statistical software R. Once the model is fit, we verify the appropriateness of the assumptions regarding ε_{ijk} using residual plots and Normal quantile–quantile plots.

3.2.3. Model-Based Testing for Differential Abundance

3.2.3.1. Specifying Hypotheses of Interest

We use the model to compare the mean abundance of a protein across pairs of disease groups, on average over all subjects and features. For example, if we denote μ_0 the mean abundance of the protein among controls, and μ_1 the mean protein abundance in the first disease group, the log-fold change is defined as $L=\mu_1-\mu_0$. We test the null hypothesis $H_0 : L=\mu_1-\mu_0=0$ against the alternative $H_a : L=\mu_1-\mu_0\neq 0$.

3.2.3.2. Testing

In the notation of Fig. 2, the log-fold change is estimated from the model as

$$\hat{L} = (\widehat{G}_1 - \widehat{G}_0) + \frac{1}{f}\left(\sum_{j=1}^{f}(\widehat{G\times F})_{1j} - \sum_{j=1}^{f}(\widehat{G\times F})_{0j}\right) + \frac{1}{n}\left(\sum_{k=1}^{n}\widehat{S(G)}_{k(1)} - \sum_{k=1}^{n}\widehat{S(G)}_{k(0)}\right) \quad (1)$$

where "^" indicates that these are model-based quantities estimated from the data. Since this experiment has a balanced design, the model-based estimates of $\hat{L}$ simplify to the difference of sample averages $\hat{L} = \bar{y}_{1..} - \bar{y}_{0..}$. However, in situations where a balanced experiment cannot be achieved (e.g., when some runs or feature intensities are lost), the model-based estimates will differ from the averages, and have been shown to be closer to the true values (14).

The fitted model is also used to estimate the standard error of $\hat{L}$ (i.e., the estimate of its variation). In our special case of a balanced design,

$$\mathrm{SE}\{\hat{L}\} = \sqrt{\frac{2}{nf}\hat{\sigma}^2_{\mathrm{Error}}} \qquad (2)$$

where $\hat{\sigma}^2_{\mathrm{Error}}$ is the estimate of residual variance based on $g(f-1)(n-1)$ degrees of freedom (i.e., data points available after fitting the model). Using these estimates, the null hypothesis is tested with a t-test, which compares the test statistic $t_s = \hat{L}/(\mathrm{SE}\{\hat{L}\})$ to the $Student_{g(f-1)(n-1)}$ distribution, and derives the corresponding p-value. If the p-value of a protein is below a pre-specified cutoff, we reject the null hypotheses. In other words, we state that the observed difference of protein abundance between groups is larger than what would be expected by random chance. In this case study, we perform four sets of t-tests, each set comparing the abundance of the identified proteins in one of the four disease groups to the controls.

3.2.3.3. Multiple Comparisons

Procedures of statistical inference allow us to select a p-value cutoff in a way to control the expected error rate in our conclusions. For example, when testing the null hypothesis for a single protein, the p-value is typically compared to a pre-specified significance level α, where α is the probability of rejecting the null hypothesis given that the null hypothesis is true. Unfortunately, this procedure is only designed for a single test, and does not control the overall error rate in an experiment such as this case study, where we perform multiple tests for all the identified proteins. Therefore, we should select a different, and more conservative, p-value cutoff that controls the experiment-wide error rate. A popular choice is to control the *false discovery rate* (*FDR*), i.e., the expected proportion of false rejections in the reported list of differentially abundant proteins.

Table 3 summarizes the possible outcomes of such a procedure in an experiment with m identified proteins, m_0 of which do not differ in abundance between two groups. R, S, T, U, and V are random variables that depend on the observed data, but only R is actually observed. In the notation of Table 3, the FDR is defined as

$$q = E\left[\frac{V}{\max(R,1)}\right], \qquad (3)$$

where E denotes the expected value of the quantity in brackets, i.e., the "average" value over a large number of repetitions of this experiment.

In this case study, we use the procedure by Benjamini and Hochberg (17), which controls the FDR at a desired level q.

Table 3
Outcomes of testing m null hypotheses $H_0 : \mu_H = \mu_D$ simultaneously for m identified proteins

	# of proteins with no detected difference	# of proteins with detected difference	Total
# true non-diff. proteins	U	V	m_0
# true diff. proteins	T	S	$m_1 = m - m_0$
Total	$m - R$	R	m

m and m_0 are fixed, and $R, S, T, U,$ and V are random. Only m and R are observed

First, for each comparison of a disease group to controls, we order the p-values of the m proteins from the largest $p_{(m)}$ (i.e., the least significant) to the smallest $p_{(1)}$ (i.e., the most significant). Next, we vary j from m to 1, and define the adjusted p-value $\tilde{p}_j$ as in (18)

$$\tilde{p}_j = \min_{k=j,\ldots,m} \left\{ \min\left(\frac{m}{k} p_{(k)}, 1 \right) \right\}.$$

It can be shown that the list of proteins with adjusted p-values $\tilde{p}_i$ below a predefined cutoff has the FDR of at most the cutoff.

3.2.4. Model-Based Protein Quantification in Individual Samples

In addition to comparing average protein abundances across groups, our goal is to obtain a separate quantification of the protein in each subject, from all peptide features, and on a continuous scale that is comparable across runs. We call this procedure *subject quantification*. Results of subject quantification will be subsequently used as input to unsupervised clustering or supervised classification of the samples. The statistical model in Fig. 2 can also be used to perform this task (see Note 8). The model views subject quantification as estimation of the "true" abundance of the protein in each sample from noisy measurements, and helps to appropriately derive this summary. For example, if we denote $\widehat{\mu}_{1 \cdot k}$ as the protein abundance of the kth subject in the first disease group, its model-based estimation is

$$\widehat{\mu}_{1 \cdot k} = \mu + \widehat{G}_1 + \frac{1}{f}\sum_{j=1}^{f} \widehat{F}_j + \frac{1}{f}\sum_{j=1}^{f} \widehat{(G \times F)}_{1j} + \widehat{S(G)}_{k(1)} \quad (4)$$

As before "^" indicates that the quantities in Eq. 4 are estimated from the data. In the case of a balanced design, the model-based quantities simplify to sample averages $\underline{y}_{1 \cdot k}$. In other experiments where balance cannot be achieved, the model-based estimates will differ from sample averages, and they have been shown to be closer to the true values (14).

We confirm the accuracy of model-based quantification by means of independent protein quantification on the same samples. In this study, 11 proteins were measured independently in all samples using nephelometry. We therefore calculate 11 Pearson coefficients of correlation between the model-based quantifications and the nephelometry-based values.

3.2.5. Gene Set Enrichment Analysis

3.2.5.1. Goal of Enrichment Analysis

In addition to testing and quantifying individual proteins, we are interested in conclusions for pre-specified sets of proteins. Typically, such sets of proteins are specified based on prior biological knowledge or on clinical relevance. They can be determined from open-access annotation databases such as Gene Ontology (19) and KEGG (20), commercial databases such as Ingenuity and GeneGo, or by review of the clinical literature. In this case study, we conducted an extensive literature search that involved all the proteins identified by LC–MS. Based on described pathways or interactions, we established ten sets of proteins which share functional annotations and are of interest to the clinical investigators.

Separately for each protein set, and separately for each comparison of a disease group to controls, we test whether the set is "enriched" in the differentially abundant proteins. In other words, we test whether the set contains differentially abundant proteins more frequently than as expected by random chance. A variety of statistical methods have been developed for this task, and most were first applied in the context of gene expression microarrays (21). Therefore, this family of methods is typically referred to as *gene set enrichment analysis* (*GSEA*). Although the context of this case study is proteomics, the same methods apply, and we use the term GSEA for consistency with the existing literature.

Gene set enrichment analysis is important from the clinical perspective, as it allows us to investigate the hypotheses that are more directly relevant to the mechanism of the disease. It is also attractive from the statistical analysis perspective, because it can be more sensitive in detecting small but consistent changes in abundance within the set, and because gene set testing can reduce the number of tests (and therefore the penalty for multiple comparisons) from the large number of proteins to the smaller number of sets (see Note 9).

3.2.5.2. Testing for Enrichment in Differentially Abundant Proteins

In this case study we use the simplest, and probably the most frequently used, Hypergeometric test for enrichment (which is also sometimes called the exact Fisher test). First, we start by determining a set of differentially abundant proteins, separately for each comparison of a disease group against controls, as described in Subheading 3.2.3. Table 4 shows the outcome of the tests in the context of a particular comparison and a particular protein set. The total number of proteins m here is often referred

Table 4
Outcomes of the gene set enrichment analysis (GSEA) for one protein set

	# of proteins with no detected difference	# of proteins with detected difference	Total
# of proteins in the set	$s-K$	K	s
# of proteins not in the set	$(m-s)-(R-K)$	$R-K$	$m-s$
Total	$m-R$	R	m

m is the total number of proteins (also called the "universe"), and s is the total number of proteins in the pre-specified set. The Hypergeometric test is conditional on the number of differentially abundant proteins R. It tests the null hypothesis that the number of differentially abundant proteins in the set K is as expected by random chance, against the alternative hypothesis that K is larger than as expected by random chance

to as the "universe." The Hypergeometric test is conditional on the number of differentially abundant proteins R. We test the null hypothesis that the number K of differentially abundant proteins in the set is as expected by random chance, against the alternative hypothesis that K is larger than as expected by random chance. The p-value of this test is defined as the probability of obtaining K or more differentially abundant products from the set by random chance. It is calculated, in the notation of Table 4, as

$$p = 1 - \sum_{i=0}^{K-1} \frac{\binom{s}{i}\binom{m-s}{R-i}}{\binom{m}{R}} \quad (5)$$

We reject the null hypothesis if the p-value is below a predefined cutoff.

3.2.5.3. Words of Caution

The Hypergeometric test should be applied with caution. First, when testing multiple sets of proteins, the p-values of individual sets in Eq. 5 should be adjusted for multiple comparisons. Since protein sets can overlap, the resulting tests are not independent, and therefore in the case study we use the conservative Bonferroni adjustment for multiple comparisons (18). With this approach, the adjusted p-values $\tilde{p}$ are obtained by multiplying the p-value in Eq. 5 by the total number of protein sets. We then compare the adjusted p-value to the predefined cutoff $\alpha = 0.1$.

Second, the Hypergeometric test is known to be sensitive to the choice of the "universe." If we fix the number of differentially abundant proteins R, and the number of proteins in the protein set s, but increase the size of the universe m by adding proteins that are unrelated to the disease, this will systematically decrease the p-values in Eq. 5. Therefore, inclusion of unrelated proteins in the "universe" can result in false optimism in our conclusions. In the case of LC–MS experiments, the size of the "universe" can be inadvertently inflated, for example, by adding proteins with low confidence of identifications, or by adding noise features. To avoid this, we used a conservative procedure, and limited the "universe" to the gene products identified with at least two unambiguously mapped peptides, and to identified proteins that are known to be present in plasma in sufficient concentrations (see Note 10).

3.3. Planning a Follow-Up Experiment: Sample Size Calculation

At the conclusion of this study, the researchers will conduct follow-up investigations using the same general LC–MS-based workflow. Insights from the present case study can be used to best design such follow-up experiments. In particular, we use summary quantities from the current dataset to calculate the minimal sample size which is required for a future detection of a change in protein abundance between two disease groups (11) (see Note 11).

3.3.1. Sample Size for Comparing Group Means of a Single Protein

To calculate the sample size, we must first specify:

1. $L = |\mu_H - \mu_D|$, the minimal absolute log-fold change in protein abundance that we would like to detect.
2. α, the desired significance level of the test, defined as the probability of rejecting the null hypothesis given that the null hypothesis is indeed true.
3. $1-\beta$, the desired power, defined as the probability of rejecting the null hypothesis given that the alternative hypothesis is true.
4. σ^2, a representative value of the measurement error variance in the future experiment. We use the results of this case study, and set σ^2 to the median estimated error variance $\hat{\sigma}^2_{\text{Error}}$ among all identified proteins.
5. f, a representative number of features mapped to a protein in the future experiment. We use the results of this case study, and set f to the median number of features mapped to all identified Entrez IDs.

The following formula (22) gives an approximate relationship between L and its standard error $\text{SE}\{L\}$

$$\text{SE}\{L\} \leq \frac{L}{z_{1-\beta} + z_{1-\alpha/2}}, \tag{6}$$

where $z_{1-\beta}$ and $z_{1-\alpha/2}$ are respectively the $100(1-\beta)$th and the $100(1-\alpha/2)$th percentiles of the standard Normal distribution. If the planned experiment contains no technical replicates, then we can substitute the expression of the standard error in Eq. 2 into Eq. 6, and solve for the number of biological replicates per disease group n

$$n \geq \frac{2}{f}\left(\frac{z_{1-\beta} + z_{1-\alpha/2}}{L/\sigma}\right)^2. \qquad (7)$$

3.3.2. Sample Size for Comparing Group Means of Multiple Proteins

In proteomic experiments we are rarely interested in changes in abundance of a single protein. Instead, we would like to compare the abundance of all proteins which are simultaneously detected as part of the experiment. The calculation of the sample size, much like the calculation of the error rate for testing in Subheading 3.2.3, must be adjusted to control for the False Discovery Rate in the future experiment. To this end, we first substitute the quantity in (2) above with

(2a) q, the desired level of FDR.

We also specify an additional quantity

6. m_0/m_1, where m_0 is the anticipated number of the true null hypotheses, and m_1 the anticipated number of the true alternative hypotheses in the future experiment, as defined in Table 3. We use the results in this case study, and set m_1 to the number of rejected null hypotheses in the present case study, and m_0 to the number of null hypotheses that we failed to reject.

We then use the procedure by Benjamini and Hochberg (17) to calculate the average significance level α_{ave} over all tests

$$\alpha_{\text{ave}} \leq (1-\beta)\cdot q \frac{1}{1+(1-q)\cdot(m_0/m_1)}, \qquad (8)$$

such that α_{ave} controls the FDR at level q. Finally, we substitute α_{ave} into Eq. 7 to calculate the minimal number of replicates n per disease group.

4. Implementation and Results of the Statistical Analysis Steps

In the following, we provide examples of key programming steps for storage, manipulation, and statistical analysis of data from this experiment using the open-source statistical software R (5) and R-based project Bioconductor (6) (see Note 1).

4.1. Experimental Design

4.1.1. Sample Selection

The initial information regarding all subjects available for the study is stored in a data structure shown in Table 5. The data structure is of the type `data.frame`, and we will call it `population`. Our goal is to randomly select ten subjects from each disease group, such that the number of individuals from each age group and gender is the same in all disease groups. To this end, we add to `population` a new column which combines age group and gender

Table 5
A data structure containing information on all serum samples from subjects with coronary artery disease and controls available at the German Heart Center and at the First Medical Clinic of the Technical University of Munich

Subject id	Age group	Gender	Disease group
1	≥58 y.o	Male	Unstable angina
2	≥58 y.o	Male	STEMI
3	<58 y.o.	Female	NSTEMI
⋮	⋮	⋮	⋮
3288	≥58 y.o	Male	Stable angina
3289	<58 y.o.	Male	Stable angina
3290	<58 y.o.	Male	Control

```
population$AGE.GENDER <- paste(population$AGE,
population$GENDER, sep = "; ")
```

We then create a data structure `populationCounts` shown in Table 1, which contains counts of available samples for all strata of age group and gender

```
populationCounts <- table(population$AGE.
GENDER, population$GROUP)
```

and determine the minimal number of samples per stratum and per disease group

```
minPopulationCount <- apply(populationCounts,
1, min)
```

We can now randomly select samples for the study. For example, in the first stratum (female subjects who are 58 years or older, i.e., the first line in Table 1), we select three subjects from each disease group

```
firstStratum <- population[population$AGE.
GENDER == "≥58 y.o.; Female",]

selectedSubjectIdsInFirstStratum <- unlist(

by(firstStratum$PATIENT, firstStratum$GROUP,
function(x) sample(x, size = 3)))
```

The resulting counts of selected samples in each stratum and each disease group are shown in Table 2.

4.1.2. Allocation of Experimental Resources

If `selectedSubjectIds` contains a vector of subject ids selected for the experiment, then a randomized order of sample processing and spectral acquisition can be obtained using

```
orderOfAcquisition <- sample(
selectedSubjectIds, size = length(selected
SubjectIds), replace = FALSE)
```

4.2. Statistical Analysis of the Acquired Spectra

4.2.1. Data Transformation and Normalization

A typical output of signal processing software for LC–MS experiments is a table of raw feature intensities. The table is typically presented in a "wide" format, i.e., a format where each row represents an aligned LC–MS feature and each column represents an LC–MS run. In this case study, the dataset contains 1,643 features and 50 runs, and therefore the table has 1,643 rows and 50 columns. We denote the R data structure containing this table as `raw.data`; the data structure is of type `matrix`. Logarithm transformation and quantile normalization of this matrix can be obtained by

```
normalizedWide.data <- normalize.quantiles
(log2(raw.data), copy = TRUE)
```

4.2.2. Model Specification

To fit the fixed effects ANOVA model in Subheading 3.2.2, we restructure the normalized data into the "long" format. In this format, each row represents a feature intensity within a single run, and columns contain information on both features and runs. We denote the R implementation of such a data structure as `normalizedLong.data`, of type `data.frame`. In this case study, `normalizedLong.data` has 1,643 ×50 = 82,150 rows. The first and last three rows of the data structure are shown in Table 6. `NORM_LOG_INT` is the log-transformed and

Table 6
Data structure containing normalized log-transformed intensities

ENTREZ	FEATURE	PATIENT	GROUP	NORM_LOG_INT
1	19	1	0	16.91
1	19	2	0	16.65
1	19	3	0	16.31
⋮	⋮	⋮	⋮	⋮
116844	3579	8	4	13.86
116844	3579	9	4	14.86
116844	3579	10	4	14.26

Each row represents a feature intensity within a single run, and columns contain information on both features and runs normalized intensity, of type `numeric`. The remaining columns are of type `factor`.

The formatted data can be used to fit the model in Fig. 2, separately for each protein, using standard ANOVA implementations available in R. For example, the model fit for protein 8542 is obtained using

```
fit.8542 <- lm(NORM_LOG_INT ~ GROUP + FEATURE +
GROUP:FEATURE + GROUP/PATIENT,

normalizedLong.data[normalizedLong.
data$ENTREZ == 8542,])
```

The summary of the model fit outputs 60 estimated parameters; the names of each are shown in the "Parameter" column of Table 7.

The assumption of Normality of the error terms is visually verified using a Normal quantile–quantile plot, which can be obtained using

```
qqnorm(y = fit$residuals, datax = FALSE, ylab =
"Residuals")
```

If points on the plot form approximately a straight line, then the assumption is appropriate. The assumption of constant variance of the error terms is visually verified by plotting, from the fitted model, the residual versus the predicted value for each feature intensity. The plot can be obtained using

```
plot(fit$fitted.values, fit$residuals, xlab =
"Predicted Values", ylab = "Residuals")
```

If the residuals are distributed roughly equally above and below the horizontal line, and have a similar spread across the entire range predicted values, then the assumption of constant variance is appropriate.

The summary also reports the standard errors of the estimated parameters, and the *p*-values of *t*-tests that compare each estimate to 0. However, these estimates and tests are not of direct relevance to the goal of the case study. Instead, we are interested in model-based summaries, such as comparisons of the estimated group means, that we discuss next.

4.2.3. Model-Based Testing for Differential Abundance

Suppose, for example, that we test $H_0 : \mu_1 - \mu_0 = 0$ versus $H_a : \mu_1 - \mu_0 \neq 0$, where (in notation of Table 1) μ_1 represents the mean of protein abundance of subjects in the *Stable angina* group, and μ_0 represents the mean protein abundance among controls. We test the hypothesis for protein 8542, which has three features. As discussed in Subheading 3.2.3, the difference between group means can be estimated from the data as a linear combination of the estimated parameters. To obtain this quantity, one must

Table 7
Coefficients of model parameters for estimating $L = \mu_1 - \mu_0$ and for quantifying protein abundance in subject 2 of group 1, for a protein with three features, and ten subjects per disease group

Parameter	Coefficient for Testing	Coefficient for Quantification	Parameter	Coefficient for Testing	Coefficient for Quantification
(Intercept)	0	1	GROUP0:PATIENT5	−1/10	0
GROUP1	1	1	GROUP1:PATIENT5	1/10	0
GROUP2	0	0	GROUP2:PATIENT5	0	0
GROUP3	0	0	GROUP3:PATIENT5	0	0
GROUP4	0	0	GROUP4:PATIENT5	0	0
FEATURE4227	0	1/3	GROUP0:PATIENT6	−1/10	0
FEATURE6374	0	1/3	GROUP1:PATIENT6	1/10	0
GROUP1:FEATURE4227	1/3	1/3	GROUP2:PATIENT6	0	0
GROUP2:FEATURE4227	0	0	GROUP3:PATIENT6	0	0
GROUP3:FEATURE4227	0	0	GROUP4:PATIENT6	0	0
GROUP4:FEATURE4227	0	0	GROUP0:PATIENT7	−1/10	0
GROUP1:FEATURE6374	1/3	1/3	GROUP1:PATIENT7	1/10	0
GROUP2:FEATURE6374	0	0	GROUP2:PATIENT7	0	0
GROUP3:FEATURE6374	0	0	GROUP3:PATIENT7	0	0
GROUP4:FEATURE6374	0	0	GROUP4:PATIENT7	0	0
GROUP0:PATIENT2	−1/10	0	GROUP0:PATIENT8	−1/10	0
GROUP1:PATIENT2	1/10	1	GROUP1:PATIENT8	1/10	0

GROUP2:PATIENT2	0	0	GROUP2:PATIENT8	0	0
GROUP3:PATIENT2	0	0	GROUP3:PATIENT8	0	0
GROUP4:PATIENT2	0	0	GROUP4:PATIENT8	0	0
GROUP0:PATIENT3	–1/10	0	GROUP0:PATIENT9	–1/10	0
GROUP1:PATIENT3	1/10	0	GROUP1:PATIENT9	1/10	0
GROUP2:PATIENT3	0	0	GROUP2:PATIENT9	0	0
GROUP3:PATIENT3	0	0	GROUP3:PATIENT9	0	0
GROUP4:PATIENT3	0	0	GROUP4:PATIENT9	0	0
GROUP0:PATIENT4	–1/10	0	GROUP0:PATIENT10	–1/10	0
GROUP1:PATIENT4	1/10	0	GROUP1:PATIENT10	1/10	0
GROUP2:PATIENT4	0	0	GROUP2:PATIENT10	0	0
GROUP3:PATIENT4	0	0	GROUP3:PATIENT10	0	0
GROUP4:PATIENT4	0	0	GROUP4:PATIENT10	0	0

explicitly specify the coefficients of the linear combination. According to Eq. 1, the coefficients of all the model-based parameters associated with group 1 are positive, and the coefficients of the parameters associated with group 0 are negative. The absolute value of the coefficients of the two disease groups is 1. Since this protein has three features, the absolute value of all the coefficients of the interaction terms associated with these disease groups is 1/3. Similarly, since there are ten subjects per group, the absolute value of the coefficients of subjects from these disease groups is 1/10. The remaining coefficients are equal to 0. The full set of coefficients is shown in Table 7; it is applicable to all the identified proteins.

We denote a vector of the coefficients in Table 7 as coefficients. The numeric estimate of the linear combination, as well as its standard error and corresponding *p*-value, are obtained using

```
library(gmodels)

pairwiseComp <- estimable( fit.8542, coefficients)
```

When tests are performed for multiple proteins in parallel, the *p*-values must be adjusted for multiple comparisons as discussed in Subheading 3.2.3. If p is a data structure of type `data.frame`, such that its rows are genes and its four columns are *p*-values comparing each of the four disease groups to the controls, then adjustment for multiple comparisons, separately for each type of test, can be obtained as

```
p.BH <- apply(p, 2, function(x) p.adjust(x,
method = "BH"))
```

We reject the null hypothesis, and claim differential abundance, when an adjusted *p*-value is below an FDR cutoff of 0.05. In total, 59 proteins were found to be differentially abundant in at least one of the four comparisons in this case study.

4.2.4. Model-Based Protein Quantification in Individual Samples

Similarly to testing, we use the model to derive subject-level protein quantifications as discussed in Subheading 3.2.4. For example, to estimate the abundance of protein 8542 for the second subject in the first disease group (*Stable angina*), we create a linear combination of the estimated parameters. According to Eq. 4, the coefficients associated with the intercept, with the corresponding disease group, and with the corresponding subject, are all equal to 1. Since the protein has three features, the coefficients associated with the interaction term are 1/3. The remaining coefficients are 0. The full set of coefficients is shown in Table 7, and it is applicable to all the identified proteins. We denote a vector of the coefficients in Table 7 as `subjectQuantCoef`. The numeric estimate of the linear combination is obtained using

```
subjectQuant <- estimable( fit.8542, subjectQuantCoef )$Estimate
```

We further consider the 11 proteins for which independent nephelometry measurements were made for each subject. For each protein, we calculate the Pearson correlation coefficient between the subject-level abundances and the nephelometry measurements. The median coefficient of correlation over the 11 proteins is 0.667.

4.2.5. Gene Set Enrichment Analysis

We test the predefined groups of proteins for enrichment in differential abundance between each of the disease groups and controls. For example, we consider a predefined group of 11 proteins sharing the annotation *Complement*, and the comparison between the first disease group (*Stable angina*) and the group of controls. In this case, in the notation of Table 4, $K=4$, $s=11$, $m-s=66$, and $R=36$. We calculate the p-value of the Hypergeometric test using

```
p <- phyper(K, s, m-s, R, lower.tail = FALSE)
```

The test yields a p-value of 0.660, which indicates an insufficient evidence that the *Complement* group is enriched in proteins that are differentially abundant between *Stable angina* and the controls. Overall, two out of the ten predefined protein groups were found enriched in proteins that are differentially abundant in at least one disease group as compared to controls.

4.3. Planning a Follow-Up Experiment: Sample Size Calculation

To calculate the sample size necessary for a follow-up experiment, we specify

1. $L=|\mu_H-\mu_D|$, the minimal absolute log-fold change in protein abundance that we would like to detect. We set $L=0.3$, which on the original scale corresponds to the fold change of $2^{0.3}=1.23$.
2. q, the desired level of FDR. We set $q=0.05$.
3. $1-\beta$, the desired power. We set $(1-\beta)=0.8$.
4. σ^2, a representative value of the measurement error variance in the future experiment. We use the results of this case study, and set $\sigma^2=0.09794$, i.e., the median error variance of all proteins.
5. f, a representative number of features mapped to a protein in the future experiment. We use the results of this case study, and set $f=14$, the median number of features mapped to all identified Entrez IDs.
6. m_0/m_1, where m_0 is the anticipated number of the true null hypotheses, and m_1 the anticipated number of the true alternative hypotheses in the future experiment. In the present case study, the proportion of null hypotheses that we fail to

reject, on average over the four comparisons of each disease group to controls, is $m_0/(m_0+m_1)=0.55$. Solving for m_0/m_1 yields $m_0/m_1=1.222$.

Substituting these quantities into Eq. 8 results in the average significance level over all tests $\alpha_{ave}=0.0185$, which is required to control the FDR at $q=0.05$. α_{ave} is smaller than the customary $\alpha=0.05$ cutoff, confirming that adjustments for multiple comparisons typically result in more conservative tests. Finally, substituting these values into Eq. 7 gives the required sample size for an experiment with two groups.

To illustrate the results, we vary the log-fold change L from 0.1 to 0.5, which on the original scale corresponds to fold changes from 1.1 to 1.4, and consider three experimental scenarios. The first scenario is an experiment where the majority of proteins do not change in abundance between disease groups, i.e., $m_0/(m_0+m_1)=0.90$. This translates into $m_0/m_1=9$ and, from Eq. 8, into $\alpha_{ave}=0.004$. The second scenario corresponds to a more moderate proportions of unchanging proteins, i.e., $m_0/(m_0+m_1)=0.75$. This translates into $m_0/m_1=3$ and $\alpha_{ave}=0.01$. The final scenario corresponds to the present case study, where nearly half of the identified proteins change in abundance, i.e., $m_0/(m_0+m_1)=0.55$, which translates to $m_0/m_1=1.222$ and $\alpha_{ave}=0.0185$.

Figure 3 shows the required number of individuals per disease group to achieve the power of $(1-\beta)_{ave}=0.8$ in these three scenarios. Detection of smaller fold changes and experiments with a

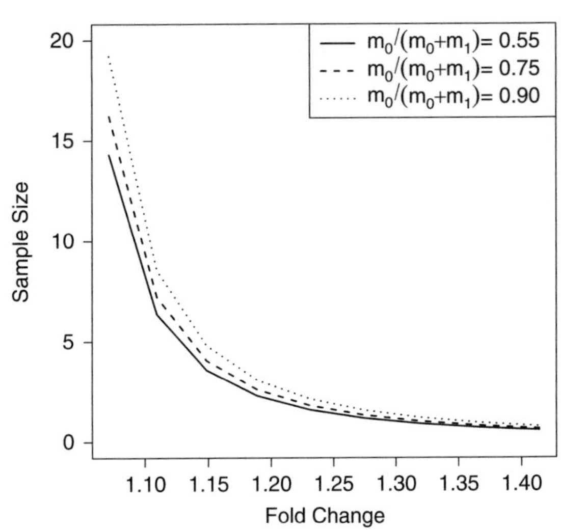

Fig. 3. Estimated sample size required to detect changes in a two-group comparison. *X*-axis: fold change between the two groups. *Y*-axis: number of individuals per group. The curves are obtained by fixing power $(1-\beta)_{ave}=0.8$, and FDR $q=0.05$.

smaller proportion of differentially abundant proteins require a larger sample size. Similarly, by varying the FDR level q, one can show that a more conservative FDR requires a larger sample size.

5. Notes

1. These steps cannot be applied blindly and in a "push button" mode to all investigations and to all types of mass spectrometry-based workflows. We recommend consulting a statistician whenever possible in order to adapt these steps to the goals and technical details of the specific investigation.
2. The data will be made available at http://www.iupui.edu/~compdiag/research/index.php following publication of clinical findings.
3. The principles of replication, randomization, and blocking used in this case study are general, and hold for all investigations. However, a change in the scientific question, e.g., analysis of time course measurements on the same individual, or the use of multiple combined treatments, typically affects the details of recruitment of individuals in the study.
4. Additional sources of experimental variation, or the use of a different quantitative workflow (e.g., using chemical or metabolic labeling), will require different allocation of samples in space and time for spectral acquisition (11).
5. If deviations of features from the overall abundance profiles are assumed to be small, the interaction term can be removed (23).
6. The ANOVA framework can accommodate a variety of additional experimental designs beyond the case-control. Changes in the experimental design such as repeated measurements, the introduction of technical replication, or the presence of other important sources of variation will require changes or additions to the model. The models can be extended for use with other quantitative workflows, e.g., global labeling and targeted selected reaction monitoring (SRM) workflows.
7. See ref. (14) for details on the mixed effects model and its implications for inference in the context of this case study, and ref. (24) for an application of a mixed effects model in a two-dimensional difference gel electrophoresis time-course study of plasma samples.
8. All models within the ANOVA framework can be used to derive comparisons between conditions of interest or to quantify the abundance of proteins in individual samples. However, the linear combinations of model parameters that lead to

these quantities are model-specific, and should be obtained anew for every change in model specification.

9. From the statistical analysis perspective, the predefined groups of proteins can remain fixed across all investigations. However, changes in the biological or clinical context of the experiment may require the specification of different groups with a problem-specific focus. In all cases, it is advantageous to use the smallest number of predefined groups, in order to reduce the penalty for multiple comparisons.

10. A variety of statistical GSEA methods have been proposed to test for enrichment of the groups in differentially abundant proteins, and some of these methods alleviate the drawbacks of the Hypergeometric test (25). Many methods have an open-source implementation in R.

11. Estimates of sample size strongly depend on the experimental design, and also on the probability model that will be used to analyze the acquired data. The estimates are also sensitive to the extent of residual variation, which is specific to the protocol of sample selection, handling, and spectral acquisition. It is important to take these aspects into consideration when calculating the required sample size.

References

1. Aebersold R and Mann M (2003). Mass spectrometry-based proteomics. *Nature*, **422**, 198–207.
2. Gstaiger M and Aebersold R (2009). Applying mass spectrometry-based proteomics to genetics, genomics and network biology. *Nature Reviews Genetics*, **10**, 617–27.
3. Nesvizhskii AI, Vitek O, and Aebersold R (2007). Analysis and validation of proteomic data generated by tandem mass spectrometry. *Nature Methods*, **4**, 787–797.
4. Mueller LN, Brusniak M, Mani DR, et al. (2008). An assessment of software solutions for the analysis of mass spectrometry based quantitative proteomics data. *Journal of Proteome Research*, 7, 51–61.
5. R Development Core Team (2009). *R: A Language and Environment for Statistical Computing*. R Foundation for Statistical Computing, Vienna, Austria.
6. Gentleman R, Carey V, Huber W, et al., eds. (2005). *Bioinformatics and Computational Biology Solutions Using R and Bioconductor*. Springer Verlag.
7. Ragg S, Fokin V, Podgorski K, et al. (2007). Proteomic profiling of plasma samples in coronary artery disease. *Circulation*, **116**, II.575.
8. Higgs RE, Knierman MD, Gelfanova V, et al. (2005). Comprehensive label-free method for the relative quantification of proteins from biological samples. *Journal of Proteome Research*, **4**, 1442–1450.
9. Higgs RE, Knierman MD, Gelfanova V, et al. (2008). Label-free LC-MS method for the identification of biomarkers. *Methods in Molecular Biology*, **428**, 209–230.
10. Kemper M and Levinson SS (1997). Serum Proteins in Clinical Medicine, Vol. 1, Laboratory Section, 1st ed. Robert Richie, ed., Olga Navolotskaia, asst. ed. Foundation for Blood Research, PO Box 190, Scarborough ME 04070-0190. *Clin Chem*, **43**(3), 550a-551.
11. Oberg AL and Vitek O (2009). Statistical design of quantitative mass spectrometry-based proteomic experiments. *Journal of Proteome Research*, **8**, 2144–2156.
12. Bolstad BM, Irizarry RA, Astrand M, et al. (2003). A comparison of normalization methods for high density oligonucleotide array data based on variance and bias. *Bioinformatics*, **19**, 185–193.
13. Patil ST, Higgs RE, Brandt JE, et al. (2007). Identifying pharmacodynamic protein markers of centrally active drugs in humans: A pilot

study in a novel clinical model. *Journal of Proteome Research*, **6**, 955–66.

14. Clough T, Key M, Ott I, et al. (2009). Protein quantification in label-free LC-MS experiments. *Journal of Proteome Research*, **8**, 5275–5284.
15. Kutner M, Nachtsheim C, Neter J, et al. (2004). *Applied Linear Statistical Models.* McGraw-Hill/Irwin, New York, 5th edition.
16. Montgomery DC (2000). *Design and Analysis of Experiments.* John Wiley and Sons, New York, 5th edition.
17. Benjamini Y and Hochberg Y (1995). Controlling the false discovery rate: a practical and powerful approach to multiple testing. *Journal of the Royal Statistical Society, Series B*, **57**, 289–300.
18. Dudoit S, Shaffer JP, and Boldrick JC (2003). Multiple hypothesis testing in microarray experiments. *Statistical Science*, **18**, 71–103.
19. Ashburner M, Ball CA, Blake JA, et al. (2000). Gene Ontology: Tool for the unification of biology. *Nature Genetics*, **25**, 25–29.
20. Kanehisa M and Goto S (2000). KEGG: kyoto encyclopedia of genes and genomes. *Nucleic Acids Research*, **28**, 27–30.
21. Ackermann M and Strimmer K (2009). A general modular framework for gene set enrichment analysis. *BMC Bioinformatics*, **10**, 47.
22. Rao PV (1998). *Statistical Research Methods in the Life Sciences.* Brooks/Cole Publishing Company, Pacific Grove, CA.
23. Karpievitch Y, Stanley J, Taverner T, et al. (2009). A statistical framework for protein quantitation in bottom-up MS-based proteomics. *Bioinformatics*, **25**, 2028–2034.
24. Corzett TH, Fodor IK, Choi MW, et al. (2010). Statistical analysis of variation in the human plasma proteome. *Journal of Biomedicine and Biotechnology*, 2010. doi:10.1155/2010/258494.
25. Zuber V and Strimmer K (2009). Gene ranking and biomarker discovery under correlation. *Bioinformatics*, **25**, 2700.

Chapter 21

Data Management in Mass Spectrometry-Based Proteomics

Lennart Martens

Abstract

Mass spectrometry-based proteomics has made the transition from the analysis of a single or only a few proteins per experiment to a high-throughput platform for discovery/validation that can analyze several hundreds to thousands of proteins in a single experiment. This increase in analytical capability hinged on four main components: (1) innovative methodologies, (2) improved instruments, (3) ready availability of comprehensive protein sequence databases, and (4) development of sophisticated software algorithms to match fragmentation mass spectra against these databases. But as the throughput of the approach increased, so did the necessity to manage and automate the data processing workflow. Indeed, modern instruments generate tens of thousands of fragmentation spectra per hour, providing a substantial bioinformatics challenge. This chapter provides insight into the specifics of this challenge, by looking at a typical workflow, the data types and user roles involved, and a broad overview of available software solutions. Finally, the increasingly important link between a local data management system and the global, centralized dissemination of proteomics data is discussed.

Key words: Proteomics, Mass spectrometry, Database, Data management

1. Introduction

Mass spectrometry-based proteomics has grown considerably in power over the past 10 years. The advent of high-throughput analytical methodologies (1), the development of improved instrumentation (2), the availability of extensive protein sequence databases (3, 4), and the creation of sophisticated search algorithms to link fragmentation (MS/MS) spectra to peptides (5) have allowed the field to evolve from analyzing a single or only a few proteins per experiment into a robust platform that can deliver information about hundreds or even thousands of proteins in a sample (6). With this increased ability to generate data, the need for automated data processing and management has become ever more pressing for modern proteomics laboratories. Indeed, with

Richard J. Simpson and David W. Greening (eds.), *Serum/Plasma Proteomics: Methods and Protocols*,
Methods in Molecular Biology, vol. 728, DOI 10.1007/978-1-61779-068-3_21, © Springer Science+Business Media, LLC 2011

Table 1
A list of all online resources mentioned in the main text

Resource	URL
OLS website	http://www.ebi.ac.uk/ols
mzML format	http://www.psidev.info/index.php?q=node/257
mzIdentML format	http://www.psidev.info/index.php?q=node/40
OMSSA Parser	http://omssa-parser.googlecode.com
XTandem Parser	http://xtandem-parser.googlecode.com
MascotDatfile	http://mascotdatfile.googlecode.com
Peptizer	http://peptizer.googlecode.com
PRIDE	http://www.ebi.ac.uk/pride
PeptideAtlas	http://www.peptideatlas.org
GPMDB	http://gpmdb.thegpm.org
Peptidome	http://www.ncbi.nlm.nih.gov/peptidome
PRIDE Converter	http://pride-converter.googlecode.com
OLS Dialog	http://ols-dialog.googlecode.com/

instruments routinely delivering tens of thousands of MS/MS spectra per 1 h LC run, basic data management solutions such as spreadsheets have reached the limits of their applicability. We will here discuss the issue of proteomics data management by several topics: starting from a description of a typical data processing workflow, the various data types encountered are explained. The next section will detail organizational aspects such as user roles and project and workflow management, followed by a broad outline of the solutions that exist today. In the final section, the link between the local data management system and centralized, publicly available proteomics data repositories is explored.

For quick reference, all online resources mentioned in this text can be found in Table 1, and a compiled list of existing data management systems (both free and commercial) can be found in Table 2.

2. The Workflow of Data Processing in Mass Spectrometry-Based Proteomics

In a typical proteomics experiment, data processing takes up the tail of the overall experimental workflow. Sample processing and preparation remain wet-lab efforts, with the mass spectrometer

Table 2
List of free and commercial data management solutions for proteomics

	Name	URL
Free systems	ms_lims	http://ms-lims.googlecode.com
	MASPECTRAS	https://maspectras.genome.tugraz.at/maspectras
	Proteios	http://www.proteios.org
	CPAS	https://proteomics.fhcrc.org/CPAS
	SBEAMS	http://www.sbeams.org/Proteomics
	LIPAGE	http://proteomeback.tmig.or.jp/2D
	PRIME	http://www.prime-sdms.org
Commercial systems	MASCOT Integra	http://www.matrixscience.com/integra.html
	Protecs	http://www.decodon.com/Solutions/Protecs
	ProteinScape	http://www.protagen.com
	Biotracker	http://www.ocimumbio.com
	LabWare LIMS	http://www.labware.com
	Nautilus LIMS	http://www.thermo.com/informatics
	ProteinCenter	http://www.proxeon.com
	Proteus	http://www.genologics.com
	PASS	http://www.i-a-inc.com
	Sapphire	http://www.labvantage.com
	Sorcerer2	http://www.sagenresearch.com
	WinLIMS.NET	http://www.qsius.com

as the capping stone at the end of a typical wet-lab workflow. The mass spectrometer, however, marks an important shift in the overall experimental workflow, as the output of the instrument is numerical data, which signals the transition from in vitro to in silico work. Regardless of this sharply defined boundary between wet-lab and subsequent data processing work, a data management solution must keep track of the sample processing workflow as well, and this incursion of the bioinformatics data management into the in vitro world often provides specific challenges, which are further discussed in the sections below.

Once the mass spectra acquired by the mass spectrometer have been obtained, data processing most often starts off with one or more preprocessing steps, in which the numerical spectral data are subjected to signal processing algorithms, which dramatically reduces the data size. Three typical preprocessing steps are noise filtering, peak picking (often involving peak centroiding), and charge deconvolution (7). In the next step, the data are analyzed by specialized software that attempts to identify proteins or peptides from the spectra, effectively transforming numerical, spectral data into amino acid sequences. Most modern proteomics methodologies identify peptides at this stage rather than proteins (1), as they base identity on fragmentation spectra of peptides rather than peptide mass fingerprints of proteins. Peptide identification algorithms are not perfect however, and they will report a number of false positives among the identified peptides. As a result, it is now a routine requirement for manuscript authors to provide at least an estimate of the number of false positives that are likely to be found in the dataset (the false discovery rate, or FDR). Additionally, these algorithms will have missed reliable identifications as well, resulting in false-negative results. While there are no guidelines on how to estimate the prevalence of these missed identifications, a thorough data analysis pipeline will attempt to investigate these false negatives as well. The peptides identified are only part of the result, as experimentalists are typically more interested in proteins or genes than peptides. In order to convert peptide information into protein (or gene) information, an additional step is necessary to map the identified peptides back to proteins; the so-called protein inference step (8). Many search algorithms transparently perform this step in the background, and hide the details from the user, even though this step is in fact the most ambiguous and error-prone processing step in the entire workflow (9).

In many cases, the qualitative analysis, aimed at identifying the components in a mixture, is neither the end point nor the true goal of the experiment. Indeed, many experiments are aimed at establishing (relative) quantitative information about the proteins in the sample (10). Several distinct methods can be employed to attain this goal experimentally (11), and various corresponding types of data processing need to be performed after mass spectrometric sample analysis (12). Since these experiments are often aimed at discovering significant differences between samples (e.g., patients versus controls), their a priori design is often not trivial, and neither is data analysis (13). Furthermore, the additional processing required by quantitative workflows can be particularly error prone at different stages of the data analysis, including the initial signal processing applied to the raw detector output (14). Finally, the quantitative data typically need to be linked to inferred proteins, and the relevant quantitative information needs to be stored in the data management system as well.

Obviously, throughout the workflow, various different types of data are encountered, which are listed and discussed in the next section.

3. Data Types Encountered in the Data Processing Workflow

3.1. Experimental Metadata

As mentioned in the previous section, the transition from wet lab, in vitro experiments to in silico data in digital form, is made by the mass spectrometer, so it seems logical to start a discussion of data types at this stage. This would however be a mistake, since such a narrow focus on the digital information alone would remove the data from its context. Indeed, it is vital that the acquired data is given an appropriate placeholder (such as an "experiment"), which can contain relevant metadata such as sample and protocol information. Since such information cannot be automatically extracted (most of the work is done by hand), the first data type encountered is thus text-based user input. This data type is also one of the more complicated ones to work with, since it is often stored as so-called free text, taking the form of a simple text box in which users can type any information they want. This approach has certain advantages, such as the ease with which it can be programmed, and its ability to store any and all descriptive information with ease. Unfortunately, there are also important problems associated with this approach. First and foremost, the unstructured nature of free text fields relegates them to a passive store of information at best, meaning that upon retrieval, the data can be read and understood by a human, but that automated querying or interpretation of the data is practically impossible to achieve. A further complication consists of the many synonyms or abbreviations that users may use when referring to the same underlying concept. A typical example is sample organism of origin; some users might enter "human," others "*Homo sapiens*" (typically using varying capitalization schemes), and every now and then, typographic errors such as "humn" might crop up as well. Due to the open nature of free text entry, synonym usage and propensity for typographical error are allowed by the format, and remain uncorrected.

In order to address the above issues, and to enable both structure and consistency in the annotation of data, controlled vocabularies (CV) and ontologies are typically used (15). To annotate the organism of origin for instance, the NEWT taxonomy CV (16) could be used instead of free text, providing a value of 9606 for human. Interestingly, this almost trivial substitution of a concept name by a concept identifier brings several key advantages. First of all, the abstracted identifier can be linked to many different names, allowing 9606 to represent both "human" as well as "*Homo sapiens*," among others. Second, the use of a single, unique identifier per concept allows automated validation of the number used, limiting the amount of typographical errors that can occur. In practice, however, few people will remember the numbers for the concepts (or "terms") they wish to annotate their experiments with. User interfaces have therefore been developed to allow users to use free text descriptions to quickly locate the appropriate CV terms (17, 18). Tools such as the OLS website and OLS Dialog rely expressly on the presence of synonyms for each term, and search the user input against all possible synonyms for each term, seamlessly locating 9606 regardless of whether the user input was "human" or "*Homo sapiens.*" The same mechanism can be used in data retrieval from the data management system, where a user can find all human-derived samples by querying using any of the known synonyms of 9606. The metadata has thus been structured, and has become automatically queryable as well.

The benefits of using CV terms do not end there, however. An often overlooked, but extremely powerful feature of CV terms is the presence of links between the terms. These can take various forms, but commonly used ones include "part of," "is a," and "develops from." These directed relationships allow the CV to communicate implicitly that 9606 (*Homo sapiens*) in NEWT "is a" 40674 (Mammalia), for instance. Through relationships between terms, any term used to annotate data will convey much more context than the mere term itself. For instance, annotating a proteomics dataset as BRENDA Tissue Ontology term BTO:0000232 (cerebellum), instantly makes clear that the dataset is derived from BTO:0000142 (brain) as well. It is therefore quite easy to develop a data retrieval option for users in which a query for brain data will display the cerebellum dataset as well, a feature that can already be found in the PRIDE database (19), for instance.

Metadata is thus captured as text, either in free text (appropriate for user-friendly naming of experiments for instance), or using CV terms. While the latter require a bit more infrastructure to function in a user-friendly way, they provide the required context in a structured, queryable, and highly efficient way.

3.2. Mass Spectra

Mass spectral data is essentially numeric in nature, and is typically binary encoded in a proprietary vendor-specific (and often even instrument-specific) format. This data type is called raw data (20), but may not be as pristine as its name suggests. Indeed, depending on the vendor, the instrument type, and the acquisition settings, the "raw" data may already be a processed signal. Irrespective of whether any type of processing took place, most users will not use these proprietary files for downstream processing. Instead, the raw files are typically heavily processed to yield text-based peak lists (20). As pointed out before, often-used processing steps include smoothing, noise filtering, deisotoping, and charge deconvolution (7). Peak lists can take a variety of forms, but can usually be interconverted with relative ease. Certain formats, such as pkl and dta, provide only minimalist information: precursor m/z, charge (if known) and a list of fragment ion peaks as m/z, and intensity pairs. More extensive formats, such as the Mascot Generic File (mgf) format provide optional placeholders for limited metadata as well, useful in recording additional information, such as retention time. The optional nature of these additional annotations does, however, imply that many such files may not contain this type of information, a distinct drawback of the format from a data consumer point of view. In all of these peak list formats, detailed information on the signal processing applied is also inherently lost, thus largely decoupling the peak lists from the actual binary data they were derived from. As a result, several proposals have been made to provide a fully descriptive and detailed format for mass spectrometry data, including MS2 (21), mzXML (22), and mzData (23). Of these formats, mzXML and mzData were the most popular, especially mzXML, as it provided the basic file format for the Trans Proteomic Pipeline (TPP) set of open source software tools. Few instruments output any of these standards directly however, necessitating the use of conversion software, typically not written by the vendors themselves. As the formats started to show their age, it was therefore decided to combine efforts, and to build a single, robust XML-based format for mass spectrometry data. The result is the mzML format (24), and while third-party convertors are still required for legacy instruments, most vendors have implemented direct mzML export from their instrument software packages. Additionally, software support for reading mzML files is widely available, and a complete table of compatible software can be found online (http://www.psidev.info/index.php?q=node/257). It is well worth noting that mzML relies heavily on CV for metadata information, while the actual spectral data is binary encoded to ensure a lossless representation of the peak information. An important advantage of mzML is that it can be validated both structurally (through standard XML validation) and semantically. The latter implies that not only can a minimal amount of annotation be ensured, but that the semantic accuracy of the annotations can be validated as well (e.g., the semantic validation will flag the annotation of an instrument ion source as a "time-of-flight," which is an analyzer, as incorrect) (25).

3.3. Peptide and Protein Identification Data

Output formats for peptide and protein identification data have followed a similar evolution as seen in the mass spectra. Each search engine typically comes with its own output format, many of which are text-based (often formatted as MIME or XML) and therefore readable. The differences in formatting also include differences in the level of available annotation, however. Information on the fragment ions actually matched by the algorithm during identification, for instance, can be found in only a few algorithms. Despite the availability of open source libraries to read the various output formats (e.g., (26, 27)), this fragmentation of information is far from ideal. As with mass spectral data, common formats have therefore been proposed, including SQT (21), pepXML (28), and protXML (29). The latter two proved the most popular as they are part of the TPP data formats. Yet similar to the situation for mass spectra, these various formats captured very little metadata, and could not cope with increasingly sophisticated, multipass searches in a cohesive way. As a result, a combined effort with broad support from the community was set up to develop the mzIdentML format as a unified standard for capturing search engine output data (http://www.psidev.info/index.php?q=node/40). mzIdentML is an XML format based on the FuGE model (30), and provides similar benefits as the aforementioned mzML for mass spectral data, allowing it to be self-contained, and enabling structural and semantic validation through the use of CV. In fact, the mzML and mzIdentML-CV are merged into a single entity, since there is substantial overlap and reuse between them. A list of software that implements the mzIdentML standard can be found online (http://www.psidev.info/index.php?q=node/408).

3.4. Quantitative Data

The standardization of quantitative proteomics data currently remains a work in progress. With many of the methodologies only just reaching maturity, and new methodologies undoubtedly set to appear in the next years, quantitative analyses present perhaps the fastest evolving area of proteomics. As a result, no common formats have yet been proposed, and existing algorithms all adhere to custom-built output formats, which are often text based. Despite the ongoing evolution of this aspect of proteomics, the same group that developed mzIdentML is starting an initiative to develop mzQuantML, a unified format for quantitative proteomics data.

4. Organizational Aspects of Data Management

Data management automatically comes with two important corollaries: data storage and data retrieval. In both cases, access privileges come into play. It would be undesirable, for instance, to have a colleague delete or overwrite your data, or to have a collaborator look at data that is private to your group or to another, unrelated collaborator. In order to ensure that data access is granted to the correct users, most data management systems will implement a privileges system, in which four types of access are usually distinguished: insertion of data, updates to existing data, retrieval of data, and deletion of data. The privileges system typically employs a group-based architecture. In this setup, rights can be granted directly to individuals (usually special users, such as the project owner or database administrator) or to customizable groups of individuals, allowing for considerable flexibility. Some data management systems also enforce privacy of data through encryption, which helps safeguard the information in case of hardware theft or hacked database computers. This practice can be especially useful if sensitive, clinical patient data is stored in a data management infrastructure.

Apart from their data access privileges, users also tend to differ in their expertise. A mass spectrometrist obviously has a different specialization than a bioinformatician. Many data management systems will take the different roles taken up by users in to account, providing specific modules for specific roles. The ms_lims (31) system, for instance, provides a uniform interface for data storage for mass spectrometrist, regardless of instrument type. The role of "mass spectrometrist" in ms_lims is thus linked to a single interface.

A third meta-aspect of a data management system is that it can be used for workflow management and prioritization. Data typically move through several steps [sample preparation, acquisition on mass spectrometer, spectral data storage, search(es) of mass spectra to identify peptides and proteins, storage of search results, and postidentification processing (e.g., validation (32))]; at each of these steps, the system can inform relevant users of the progression in state of the project. Simultaneously, a supervisor or project lead can choose to prioritize particular projects in specific phases, for instance when multiple samples are ready for mass spectrometric analysis, and the instruments are all occupied.

The final important issue with all data management system is data backup. Ideally, backups should enable the data management system to be restored after any kind of damage (be it a crashed hard drive or a fire in the facility). It is therefore important to ensure that a reasonable and reliable backup process is in place, that the backups are stored safely (preferably off-site, since this minimizes the risk of total data loss in case of fire), and that recovery is tried and tested (there is nothing worse than finding out that your backup turns out to be useless after disaster strikes). Typical backup solutions rely on magnetic tape drives, but these can be expensive and are rather difficult to set up properly. With the advent of cheap cloud computing and storage, however, quite a few firms now offer services that backup databases into the cloud, ensuring that they are safe in the event of local calamity. These solutions can be very competitive on price, and provide a high level of safety as well.

5. Existing Solutions for Managing Proteomics Data

Various solutions exist today to handle proteomics data. Quite a few of these are developed by research labs or informatics groups working closely with research labs, and these tend to be free and lightweight, at the price of being rather tailored to a particular setting or being relatively feature-poor. On the other end of the spectrum are the full-blown LIMS systems that find their origin in operation-wide data tracking for chemical or pharmaceutical corporations. Needless to say, these systems are extremely heavyweight, are packed with features, security, and auditing measures, and come at a hefty price. Remarkably, these latter packages often provide a poor fit to research environments, as these tend to evolve quickly whereas the corporate LIMS systems are understandably geared toward standard operating procedures. An interesting middle ground is occupied by smaller, field-specific companies that provide relatively affordable systems that strike a good balance between feature set, price, and user friendliness.

If a free solution is chosen, an important aspect is the availability of support. While the available support can never match that provided by a commercial company, certain systems will have been developed by groups that more readily deal with user requests and problems than others. Contacting the group(s) responsible is therefore always a good first step in assessing a prospective system.

We here do not provide a detailed review of available systems, as the following reviews have already considered and evaluated several of these, and should therefore be of interest (33, 34). Table 2 provides a compiled list of data management solutions based on these review articles.

6. Public Dissemination of Locally Managed Data

A final important aspect of local data management solutions is their ability to interface with public, online repositories of proteomics data. Indeed, journals are increasingly demanding that proteomics data in support of submitted manuscripts is made publicly available through such repositories. A variety of repositories exists (see refs. 35–37 for reviews) today, but perhaps the most prominent ones are the PRIDE repository at the EMBL – European Bioinformatics Institute in Cambridge, UK (38), and Peptidome at the NCBI in Maryland, USA (39). While some existing data management systems support direct output to PRIDE (40, 41), overall export functionality remains scarce. To this end, the PRIDE Converter application was developed to automatically process and annotate proteomics data from spectrum files and search engine output files (42), as well as from the ms_lims system. Peptidome, on the other hand, can process certain output files directly, and also accepts specifically formatted spreadsheets. These latter conversion solutions remain suboptimal however, since good data management systems capture a considerable amount of highly relevant data and metadata on a project, and thus constitute the best possible point of origin for a submission to a public repository. It is therefore hoped that implementation of direct export functionality will soon be implemented in existing or to-be-developed systems, as this will surely benefit the community enormously over time.

Acknowledgement

LM thanks Joël Vandekerckhove for his support.

References

1. Gevaert K., Van Damme P., Ghesquière B. et al. (2007) A la carte proteomics with an emphasis on gel-free techniques. *Proteomics* 7, 2698–2718.
2. Domon B. and Aebersold R. (2006) Mass spectrometry and protein analysis *Science* **312**, 212–217.
3. Hubbard T., Aken B., Ayling S. et al. (2009) Ensembl 2009 *Nucleic acids research* **37**, D690–7.
4. (2009) The Universal Protein Resource (UniProt) 2009 *Nucleic acids research* **37**, D169–74.
5. Sadygov R.G., Cociorva D. and Yates J.R. (2004) Large-scale database searching using tandem mass spectra: looking up the answer in the back of the book *Nat Methods* **1**, 195–202.
6. Aebersold R. and Mann M. (2003) Mass spectrometry-based proteomics *Nature* **422**, 198–207.
7. Gentzel M., Köcher T., Ponnusamy S. and Wilm M. (2003) Preprocessing of tandem mass spectrometric data to support automatic protein identification *Proteomics* **3**, 1597–1610.
8. Nesvizhskii A.I. and Aebersold R. (2005) Interpretation of shotgun proteomic data: the protein inference problem *Mol Cell Proteomics* **4**, 1419–1440.
9. Martens L. and Hermjakob H. (2007) Proteomics data validation: why all must provide data *Mol Biosyst* **3**, 518–522.
10. Panchaud A., Affolter M., Moreillon P. and Kussmann M. (2008) Experimental and computational approaches to quantitative proteomics: status quo and outlook *J Proteomics* **71**, 19–33.
11. Fenselau C. (2007) A review of quantitative methods for proteomic studies. *J Chromatogr B Analyt Technol Biomed Life Sci* **855**, 14–20.
12. Mueller L.N., Brusniak M., Mani D.R. and Aebersold R. (2008) An assessment of software solutions for the analysis of mass spectrometry based quantitative proteomics data *J Proteome Res* **7**, 51–61.
13. Karp N.A. and Lilley K.S. (2007) Design and analysis issues in quantitative proteomics studies *Proteomics* **7** **Suppl 1**, 42–50.
14. Vaudel M., Sickmann A. and Martens L. (2010) Peptide and Protein Quantification: a Map of the Minefield. *Proteomics*, in press.
15. Martens L., Palazzi L.M. and Hermjakob H. (2008) Data standards and controlled vocabularies for proteomics *Methods Mol Biol* **484**, 279–286.
16. Phan I.Q.H., Pilbout S.F., Fleischmann W. and Bairoch A. (2003) NEWT, a new taxonomy portal *Nucleic Acids Res* **31**, 3822–3823.
17. Côté R.G., Jones P., Apweiler R. and Hermjakob H. (2006) The Ontology Lookup Service, a lightweight cross-platform tool for controlled vocabulary queries. *BMC Bioinformatics* **7**, 97.
18. Barsnes H., Cote R.G., Eidhammer I. and Martens L. (2010) OLS Dialog: An open-source front end to the Ontology Lookup Service *BMC Bioinformatics* **11**, 34.
19. Vizcaíno J.A., Côté R., Reisinger F., Barsnes H., Foster J.M., Rameseder J. et al. (2010) The Proteomics Identifications database: 2010 update *Nucleic Acids Res* **38**, D736–42.
20. Martens L., Nesvizhskii A.I., Hermjakob H., Adamski M., Omenn G.S., Vandekerckhove J. et al. (2005) Do we want our data raw? Including binary mass spectrometry data in public proteomics data repositories. *Proteomics* **5**, 3501–3505.
21. McDonald W.H., Tabb D.L., Sadygov R.G., MacCoss M.J., Venable J., Graumann J. et al. (2004) MS1, MS2, and SQT-three unified, compact, and easily parsed file formats for the storage of shotgun proteomic spectra and identifications *Rapid Commun Mass Spectrom* **18**, 2162–2168.
22. Pedrioli P.G.A., Eng J.K., Hubley R., Vogelzang M., Deutsch E.W., Raught B. et al. (2004) A common open representation of mass spectrometry data and its application to proteomics research *Nat Biotechnol* **22**, 1459–1466.
23. Orchard S., Jones A.R., Stephan C. and Binz P. (2007) The HUPO pre-congress Proteomics Standards Initiative workshop. HUPO 5th annual World Congress. Long Beach, CA, USA 28 October-1 November 2006 *Proteomics* **7**, 1006–1008.

24. Deutsch E. (2008) mzML: a single, unifying data format for mass spectrometer output *Proteomics* **8**, 2776–2777.
25. Montecchi-Palazzi L., Kerrien S., Reisinger F., Aranda B., Jones A.R., Martens L. et al. (2009) The PSI semantic validator: a framework to check MIAPE compliance of proteomics data *Proteomics* **9**, 5112–5119.
26. Helsens K., Martens L., Vandekerckhove J. and Gevaert K. (2007) MascotDatfile: an open-source library to fully parse and analyse MASCOT MS/MS search results. *Proteomics* **7**, 364–366.
27. Barsnes H., Huber S., Sickmann A., Eidhammer I. and Martens L. (2009) OMSSA Parser: an open-source library to parse and extract data from OMSSA MS/MS search results *Proteomics* **9**, 3772–3774.
28. Keller A., Nesvizhskii A.I., Kolker E. and Aebersold R. (2002) Empirical statistical model to estimate the accuracy of peptide identifications made by MS/MS and database search *Anal Chem* **74**, 5383–5392.
29. Nesvizhskii A.I., Keller A., Kolker E. and Aebersold R. (2003) A statistical model for identifying proteins by tandem mass spectrometry *Anal Chem* **75**, 4646–4658.
30. Jones A.R., Miller M., Aebersold R., Apweiler R., Ball C.A., Brazma A. et al. (2007) The Functional Genomics Experiment model (FuGE): an extensible framework for standards in functional genomics. *Nat Biotechnol* **25**, 1127–1133.
31. Helsens K., Colaert N., Barsnes H., Muth T., Flikka K., Staes A. et al. (2010) ms_lims, a simple yet powerful open source LIMS for mass spectrometry-driven proteomics *Proteomics*, in press.
32. Helsens K., Timmerman E., Vandekerckhove J., Gevaert K. and Martens L. (2008) Peptizer, a tool for assessing false positive Peptide identifications and manually validating selected results *Mol Cell Proteomics* **7**, 2364–2372.
33. Piggee C. (2008) LIMS and the art of MS proteomics *Anal Chem* **80**, 4801–4806.
34. Stephan C., Kohl M., Turewicz M., Podwojski K., Meyer H.E. and Eisenacher M. (2010) Using laboratory information management systems (LIMS) as central part of a proteomics data workflow *Proteomics*, in press.
35. Mead J.A., Shadforth I.P. and Bessant C. (2007) Public proteomic MS repositories and pipelines: available tools and biological applications *Proteomics* **7**, 2769–2786.
36. Mead J.A., Bianco L. and Bessant C. (2009) Recent developments in public proteomic MS repositories and pipelines *Proteomics* **9**, 861–881.
37. Riffle M. and Eng J.K. (2009) Proteomics data repositories *Proteomics* **9**, 4653–4663.
38. Martens L., Hermjakob H., Jones P., Adamski M., Taylor C., States D. et al. (2005) PRIDE: the proteomics identifications database *Proteomics* **5**, 3537–3545.
39. Slotta D.J., Barrett T. and Edgar R. (2009) NCBI Peptidome: a new public repository for mass spectrometry peptide identifications *Nat Biotechnol* **27**, 600–601.
40. Hartler J., Thallinger G.G., Stocker G., Sturn A., Burkard T.R., Körner E. et al. (2007) MASPECTRAS: a platform for management and analysis of proteomics LC-MS/MS data *BMC Bioinformatics* **8**, 197.
41. Häkkinen J., Vincic G., Månsson O., Wårell K. and Levander F. (2009) The proteios software environment: an extensible multiuser platform for management and analysis of proteomics data *J Proteome Res* **8**, 3037–3043.
42. Barsnes H., Vizcaíno J.A., Eidhammer I. and Martens L. (2009) PRIDE Converter: making proteomics data-sharing easy. *Nature Biotechnology* **27**, 598-599.

Chapter 22

Bioinformatics Challenges in the Proteomic Analysis of Human Plasma

Joseph M. Foster and Lennart Martens

Abstract

Mass spectrometry has become the method of choice for studying proteins in complex mixtures in a qualitative and quantitative fashion. The application of mass spectrometry-based proteomics analyses on plasma has correspondingly been established as an important method for disease-associated biomarker discovery and validation. Yet despite being a readily available human sample, plasma poses several important challenges to the proteomics researcher. With a focus on bioinformatics aspects, this chapter will discuss the problems involved in analyzing plasma proteomics data, along with the scope of solutions available through specialised tools and sophisticated analysis methods.

Key words: Mass spectrometry, Proteomics, Plasma, Data analysis, Bioinformatics

1. Introduction

Human blood plasma is a yellow liquid that acts as a solvent for blood cells, and is composed of 90% water and a highly complex array of solvated proteins, minerals, sugars, and other molecules. Blood plasma is also the vehicle for excretory transport and endocrine signalling, coming into contact with all bodily tissues throughout its journey of the body while acting as a buffer to maintain osmotic pressure (1). It is this widespread contact that leads to the absorption of products of tissue leakage, apoptosis, and cell-surface reactions which increases both the complexity of the plasma and its relevance for biomarker discovery.

Despite its enormous complexity, studying the plasma proteome offers many benefits. Perhaps most importantly, plasma is one of the most routinely collected samples in a clinical setting as it is very easy to obtain compared to other bodily fluids such as synovial fluid or cerebrospinal fluid, making it available in relatively

Richard J. Simpson and David W. Greening (eds.), *Serum/Plasma Proteomics: Methods and Protocols*,
Methods in Molecular Biology, vol. 728, DOI 10.1007/978-1-61779-068-3_22, © Springer Science+Business Media, LLC 2011

large quantities (2). In addition, established standard processing protocols are well documented and have been widely adopted, which is not always the case for more readily available fluids with clinical relevance such as urine, where the challenges consist of the removal of contaminating cells, prevention of microbial growth, and low concentrations (2).

The many benefits of plasma as a sample have led to a strong focus on plasma proteome analysis in biomarker discovery studies, together with efforts aimed at the full characterisation of plasma and the development of efficient sample and data processing protocols (3–5). One of the better-known studies is the Human Proteome Organisation's (HUPO) Plasma Proteome Project 1 (2003–2005), identifying 15,710 out of a potential 56,530 protein sequences from the human International Protein Index (IPI version 2.21) (6). According to the PRoteomics IDEntifications Database (PRIDE) (7), the total number of identified human proteins in plasma has risen to 27,351 out of 86,379 sequences (based on human IPI version 3.65) and is likely to rise further as additional results from the PPP2 collaboration (8) are deposited (http://www.ebi.ac.uk/pride).

Repeat high-throughput proteomics experiments show a decreasing return of novel protein identifications with increasing iterations (9, 10), which illustrates that replicate analyses may not be the most cost-effective way to analyse the plasma proteome in a single laboratory.

Indeed, repeating an experiment ad infinitum is not an option for an individual research group trying to answer specific questions. A further complication lies in the biases that can be introduced by employing a single standardised sample processing protocol. Distinct experimental protocols more consistently sample a specific subset of proteins than others (11). Additionally, very crudely processed samples without sophisticated sample preparation steps, such as depletion of abundant proteins, clearly yield unique identifications (11). This paradox can most likely be explained by the occurrence of transitive unspecific depletion (12). Because of the protocol-based nuances in protein detection, it is no surprise that combining data obtained from a variety of sample depletion and separation methods yields a substantially higher coverage of the proteins in plasma than can be obtained in any individual analysis. It is, therefore, extremely useful to archive any and all plasma analysis results in centralised databases such as PRIDE (7) or the Human Plasma Peptide Atlas (13). These repositories will over time be able to provide a comprehensive view on the plasma proteome that builds upon and transcends the individual contributing analyses. This chapter describes the problems encountered in plasma proteomics with an emphasis on the bioinformatics challenges they present. It will also provide solutions for these problems, and their scope. By clearly understanding the technical challenges and limitations encountered when analysing plasma, the reader will be able to design experimental and data analysis protocols that optimally suit the goals of the study.

2. Problems and Solutions

2.1. Peptide–Protein Mapping

Following acquisition of data on a mass spectrometer, fragmentation spectra are processed by identification software to find peptides they most likely represent. While providing a detailed background of how this process works is outside the scope of this chapter (see ref. 14 for a full review of this topic), a brief explanation is given below. Typically, theoretical spectra are generated for all peptides derived from an in silico digest of each entry in a sequence database such as IPI; experimental spectra are subsequently correlated against theoretical spectra in order to assign best matches. A less commonly used method employs direct de novo sequencing (15) and is, therefore, independent of any sequence database. A spectrum is considered as identified if a peptide match can be obtained that fulfils certain acceptance criteria. It is noteworthy that the majority of spectra (70–90% of all recorded spectra) typically remain unidentified, having not met the stringent acceptance criteria set by the software or the user. For the spectra where a reliable match could be found, the identified peptides are associated with protein entries in a sequence database, a process that differs between software and that obviously relies heavily upon the sequence search database provided by the user. The way in which this protein inference step is performed has significant effects on final results (15) (Fig. 1).

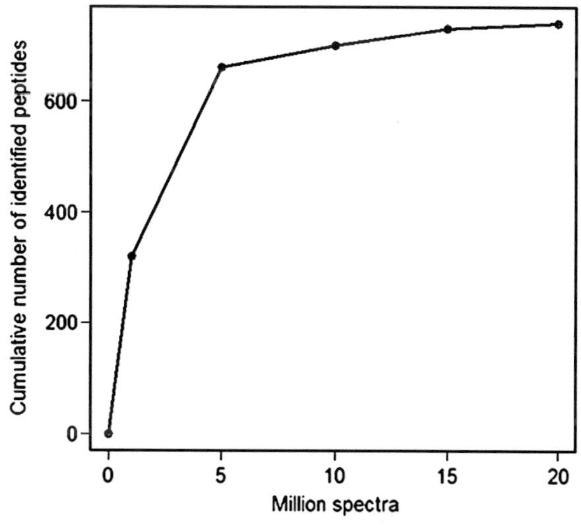

Fig. 1. Repeat high-throughput proteomics experiments show a decreasing return of novel protein identifications with increasing iterations. Adapted from Ref. (9). Copyright (2003) National Academy of Sciences, U.S.A.

Three main strategies are in common use for generating lists of identified proteins from peptides (16). The first one relies on the principle of maximum parsimony; the smallest explanatory set of proteins that explain the identified peptides is reported. In the case where a peptide can be assigned to more than one protein, the protein with the highest coverage is typically reported (see Fig. 2 "Minimal set"). This strategy is, for instance, employed by the ProteinProphet software (17). The "Maximal set" (see Fig. 2 "Maximal set"), the second strategy, encompasses all potential explanations of a peptide and makes no judgement as to the presence of these proteins in the sample. This strategy has been employed in the HUPO Liver Proteome Project (18). The final protein identification strategy (see Fig. 2 "Minimal set, maximal annotation") relies on information about a protein's annotation. Scoring proteins by the amount of available experimental evidence (usually cDNA or direct Edman sequencing) and using carefully curated data sources such as UniProtKB/SwissProt (19) provides a minimal set that preferentially represents a high-confidence peptide to protein mapping, as opposed to reporting proteins that are theoretically translated from the genome with no experimental evidence of transcription or translation. The HUPO PPP (3) provides an example of this data processing method. It is important to note that none of these options are strictly speaking superior; each one has its role depending on the study in hand.

The sequence database against which peptides are mapped also plays a critical role in the composition of the protein list

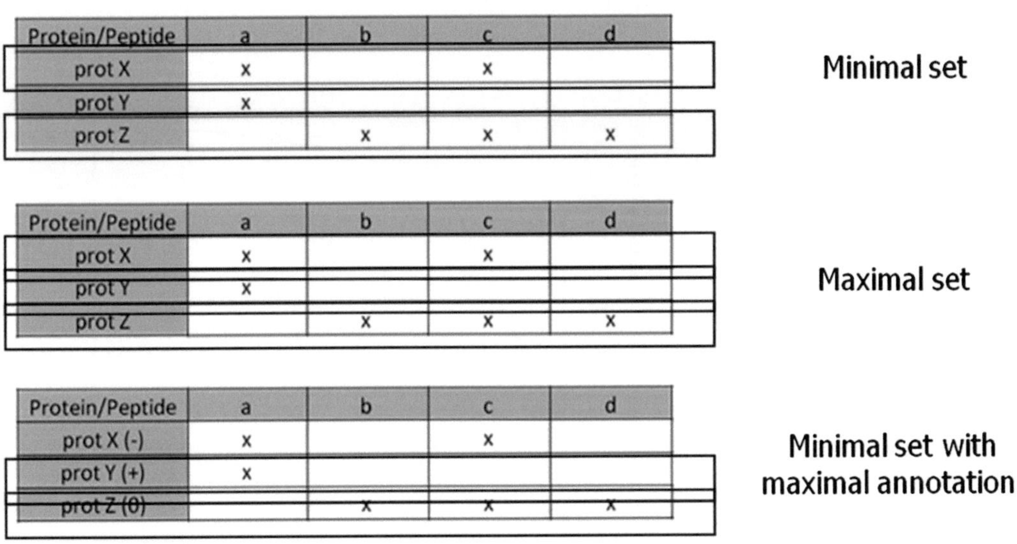

Fig. 2. Adapted from Ref. (16). Reproduced by permission of The Royal Society of Chemistry, three strategies for reporting identified proteins. See text for details.

reported and, therefore, represents a prime target for improved mapping. For single-lab projects investigating blood plasma where the focus is not on discovering novel proteins, new sequence search databases can be generated that contain only curated plasma proteins, thus reducing mapping complexity significantly by representing a narrow-focus version of the "Minimal set, maximal annotation" strategy described earlier. Since protein inference is based entirely on identified peptides, gains can also obviously be made at the level of peptide identification. Recent studies of many different peptide identification algorithms have highlighted how different these can be in rejecting or accepting an identification (20, 21). Correspondingly, approaches have been designed that unify the many identification algorithms and then combine their predictions of identified proteins in order to generate a consensus list for subsequent filtering (22, 23). The final word on this topic is still out, and may remain out as the problem of protein inference from shotgun data is essentially an ill-posed problem, for which a single unambiguous solution is not available. It is, therefore, important to understand the various ways in which peptides can be used to infer protein lists, and to interpret the results accordingly. Indeed, when comparing protein lists obtained using different inference algorithms, in different databases, and/or at different times, it is typically better to simply remap all peptides to proteins using a common algorithm, against a specific sequence database (24).

2.2. Proteolytic Degradation

Many secreted proteins that end up in plasma have undergone a process of maturation, in which signal peptides or propeptides have been cleaved off for targeting or activation purposes. The corresponding cleavage sites tend to be well-defined and when known, these can be found as annotations on the protein sequences in well-curated databases such as UniProtKB/SwissProt (19). Furthermore, it has already been established that plasma collects unwanted material and debris from cells and tissues throughout the body, and it is therefore no surprise that it also contains enzymes to degrade and to recycle these waste biomolecules. In the case of proteins, the recycling process is performed by proteases and peptidases. It is mainly the activity of non-specific aminopeptidases and carboxypeptidases (collectively called exoproteases) that sequentially degrade the termini of peptides; this generates a vastly more complex sample (25) than is typically expected by researchers and peptide identification algorithms alike. The resulting degraded protein sequences are termed "ragged" sequences. Using the peptide "YSFVATAER" as an example, the possible products of amino-terminal

(N-terminal) and carboxy-terminal (C-terminal) proteolytic degradation are as follows:

N-terminal	C-terminal
YSFVATAER	YSFVATAER
SFVATAER	YSFVATAE
FVATAER	YSFVATA
VATAER	YSFVAT
ATAER	YSFVA
TAER	YSFV
AER	YSF
ER	YS
R	Y

Such sequential degradation happens continuously in human plasma and can easily continue during sample recovery, preparation, and storage unless very stringent criteria are met (25). Depending on the time taken from recovery to analysis in the mass spectrometer, these degradation processes can lead to very different views of the same plasma proteome (25). Several recommendations have been made to counteract ex vitro proteolytic degradation in order to reduce sample complexity, such as the addition of anticoagulants such as EDTA and storage with protease inhibitors (4). While these approaches reduce ex vivo proteolytic degradation, the effect is still visible (primarily due to the in vivo proteolytic degradation that occurs continuously in humans) and so informatics techniques were developed to accommodate this issue and enable the identification of previously unidentifiable spectra.

As mentioned previously, search engines take a reference sequence database against which they identify spectra. Such a reference database (either a peptide database or a protein database which is subsequently digested in silico) acts as the search space for all potential peptides in the sample. Importantly, this database and the processing offered on it by the search engine define what peptides can be identified. In the case of proteolytic degradation, a distinction can be made between the known maturation cleavages and the unspecific exoprotease cleavages. While the former sites can be found in highly annotated databases such as UniProtKB/SwissProt (19), this information is typically ignored by search engines that blindly digest the protein sequences using the specified in vitro protease. The actual sequences of the mature N- or C-terminal peptides of the protein in question are, therefore, not considered by search engines, and can never be identified.

In the latter case of unspecific exoproteases, no solid predictions can be made at all for the resulting ragged peptides, and essentially all possible forms of the degradation products need to be considered. To encompass the entire scope of the protein degradation problem, the Database on Demand (26) tool (DoD; http://www.ebi.ac.uk/pride/dod) provides a universal solution by exploiting the common functionality of all search engines in being able to take a pre-created protein/peptide database in a FASTA format. Based on the stand-alone DBToolkit (27) application, DoD offers an intuitive web interface that allows the creation of customised peptide/protein sequence databases with a wide range of processing capabilities and source databases, including UniProtKB/Swiss-Prot (19), UniProtKB/TrEMBL (19), and IPI (28). Users can combine these source databases in any combination and then generate a workflow to process them sequentially into a single sequence database in FASTA format that represents the most accurate search space for the identification in the context of the sample. For plasma, a standard sequence database generation protocol could be as follows:

- Selecting a source database.
- Filtering of the source database for human sequences only.
- Applying a second filter to perform in silico maturation of the protein sequences (effectively removing annotated signalling or pre- and propeptides from the sequences in the database).
- Enzymatic digestion with the correct in vitro protease (typically trypsin) of all proteins, allowing for up to two missed cleavages.
- "Ragging of the database", in which amino acids are sequentially removed from the digested proteins to model (unspecific) proteolytic degradation (the tool offers the choice of N-terminal or C-terminal ragging).
- Filtering peptides for acceptable mass limits detectable by the mass analyzer (600–5,000 Da are the empirically derived default values that will work well for most scenarios).

Using such a processed database has been shown to increase the number of identified peptides by as much as 50% (29), highlighting the importance of generating a suitable search space for peptide identification to avoid the pitfalls of proteolytic degradation and other posttranslational processing events. Additionally, identifying degradation products can be a great aid to quantify peptides and proteins accurately, as a single protein can be represented by many forms of the same degraded peptide (e.g. the peptide YSFVATAER might be found in the plasma sample in the following combination of degradation products: 10% YSFVATAER, 7% SFATAER, 22% FATAER, and 61% TAER). Combining the intensities or spectral counts (depending on the quantification methodology employed) of the various degraded forms will thus result in a more realistic quantification of the peptide, and thus of the protein.

2.3. Modifications

Posttranslational modification (PTM) of amino acids represents one of the primary stages of regulation of protein activity and modulation of cellular function. These modifications are often transient, highly variable, and quite numerous, with well-known modifications such as phosphorylation already established as important in the regulation of clinically relevant targets such as the tumour suppressor protein p53 (30). Modifications are becoming increasingly relevant as public tools are made available to catalogue them, and the positions on proteins where they have been identified in projects such Unimod (31) and PSI-MOD (32) provide accessible locations and data formats for viewing protein modification data in an attempt to standardise future software development. Apart from the biologically relevant PTMs, typical sample processing procedures either intentionally or accidentally introduce further in vitro or artefactual modifications that can either be controlled, such as the alkylation of cysteines with iodoacetamide, or uncontrolled and variable, such as spontaneous methionine oxidation or pyroglutamate formation.

Unimod, which captures both in vivo PTMs and in vitro artefactual modifications, contains 608 distinct modifications to date, with at least one modification present for each natural amino acid. Taking the simple case of in vitro methionine oxidation and in vivo tyrosine phosphorylation with the tryptic peptide "MSNYESFTMDYDTMSWTCGNQDYK", there are six individual sites for modification (three methionine residues and three tyrosine residues), resulting in a potential 49 different combinations of modification states for the peptide. If threonine and serine phosphorylation would also be taken into account, the number of possible modifications skyrockets to 2,401. Traditional search engines, which largely rely on matching theoretical spectra against the experimentally obtained spectra, need to calculate all possible modification combinations for each tryptic peptide derived from the database, quickly resulting in an insurmountable computational overhead when considering several variable modifications. Furthermore, certain PTMs such as glycosylation that are commonly found in plasma (33) can have a variety of masses, depending on the actual composition of the sugar tree. While three amino acids can be combined into six possible sequences, three hexose monosaccharides can theoretically generate 1,056 possible structures. PTMs thus pose a significant bioinformatics challenge to the investigation of plasma using mass spectrometry. Several specialised algorithms have been developed to tackle the problem of identifying PTMs, however, and they can be generalised into two approaches: restrictive, and unrestrictive or blind algorithms.

Restrictive PTM searching is dependent on a predefined set of modifications: the user specifies a set of mass delta values that are caused by expected modifications. Early approaches involved the generation of virtual databases of all theoretical modified peptides for a select few modifications (34). However, as mentioned above, the problem became obvious as when more modifications were taken into account, the approach became computationally unfeasible. A refinement of this is carried out by performing the database search in two distinct steps. A first step identifies only unmodified peptides, or peptides carrying very few specified modifications, while the second step exhaustively compares the unidentified spectra against only the protein sequences identified in the first step. This second pass search is either done automatically on a user-defined set of modifications (35) or separately on an entire catalogue of modifications (36). Since the search database is reduced to protein sequences that were identified in the primary search in order to reduce computing time, no new proteins can be identified in this approach. The main benefits are the detection of modifications on the already identified proteins, and the ability to possibly increase protein sequence coverage by identifying additional (modified) peptides.

An unrestrictive or blind PTM search presents two important improvements upon the more traditional approaches: Firstly, no prior knowledge of the modifications present in the sample is required, and secondly, it can even be independent of a database of known modifications, providing the advantage that completely new modifications can be discovered. Various approaches exist, but they typically all rely on comparing experimental spectra against similar experimental spectra and then identifying the unmatched mass deltas for similar spectra with different precursor masses. In a process similar to sequence alignment, spectral alignment compares two spectra and scores their edit distance to judge the best explanation for the difference (37). Scoring of the most likely explanation of a sequence is often accompanied by a PTM frequency matrix – similar to a PAM matrix in sequence alignment that gives preference to common modifications as opposed to less reliable but lower distance explanations (38). This process can either be performed on an identified spectrum that is aligned to an unidentified spectrum in an effort to find novel modifications of a particular sequence (39, 40), or in an all-against-all approach without prior peptide identification (41). In the latter approach, overrepresented mass shifts are highlighted which can subsequently be traced back to a modification database or can be identified as new modifications (41).

Between the two main peptide identification methods, database searching and de novo sequencing, lies a less frequently used technique which encompasses the essence of both. High-quality regions of fragmentation spectra are de novo sequenced to produce high-confidence short sequence tags for a spectrum (42). Peptides that match this sequence tag are compiled into a new filtered database for searching; this huge reduction in search space represents a 1,500-fold increase in efficiency. The corresponding reduction in required computation provides time for an exhaustive multi-PTM search in the flanking regions of the tag for each spectrum (the tag itself is typically considered to be correct and immutable) (43, 44). While unrestrictive and quick, this method relies on accurate de novo sequencing of short sequence stretches, which can be challenging and often results in ambiguity. Additionally, the sequence tag should be relatively uncommon in the database, especially if multiple ambiguous tags can be extracted from a single spectrum. Yet, the ability to anchor a core stretch of sequence to a shortlist of candidate peptides does allow exhaustive variation to be applied to the unsequenced parts of the peptide, and discrimination between possible options can finally be performed by matching the theoretical fragmentation spectra of the candidate (modified) peptides to the experimental spectrum.

2.4. Dynamic Range

It is fairly well understood in the field that detection of peptides derived from low-abundance proteins by mass spectrometry is thwarted mainly by peptides of high-abundance proteins that can competitively interfere with their ionisation and detection. The dynamic range of protein concentration is of particular importance in plasma, where serum albumin is in the region of 45 mg/mL, while interleukins have been detected in the 1 pg/mL region, providing a dynamic range that spans ten orders of magnitude (Fig. 3).

It is important to note that it is not the very low concentrations of proteins as such that restrict their measurement, as modern FTICR-MS instruments, for instance, can comfortably detect molecules in the attomolar (10^{-18}) concentration range. The real challenge in plasma proteomics lies in detecting low-abundance peptides against the overwhelming background of high-abundance peptides. The problem becomes all too clear when analysis of public proteomics data in general reveals that 50% of all identified spectra are matched against peptides derived from only 88 genes. The dynamic range problem not only affects the detectability of low-abundance peptides, but also confounds the relative quantification of identified peptides and proteins. Indeed, as the $\log_2$ ratio of expected peptide concentration to measured peptide concentration deviates from zero (which is equivalent to equal concentration in the log scale), both the accuracy of the measurement and the precision (standard deviation) of the measurement deteriorate (45). As a result, meaningful comparison of the relative abundance of two peptides or proteins can only be carried out reliably when they are separated by fewer than four orders of magnitude.

While bioinformatics solutions to address the problem of dynamic range remain elusive, there have been some radical changes

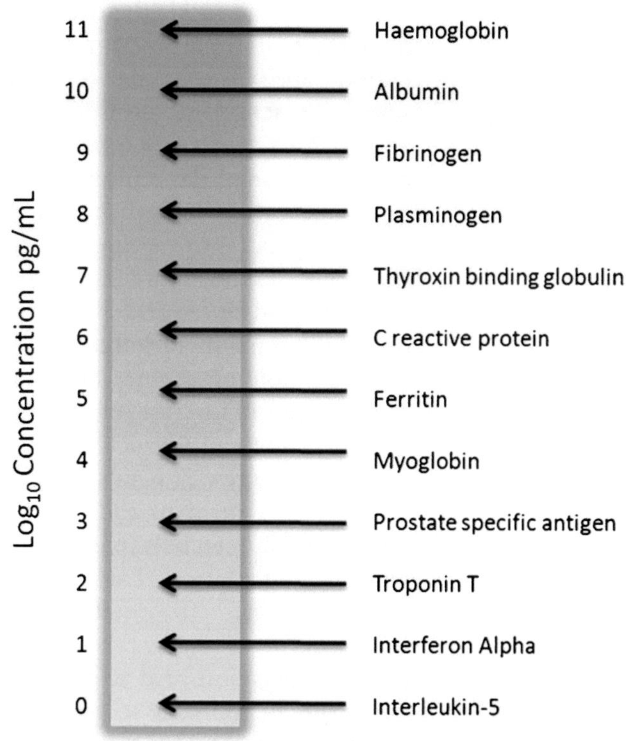

Fig. 3. Protein abundance spans 12 orders of magnitude in blood plasma, with classical plasma proteins forming the high abundance region while products of tissue leakage cluster in the mid-to-low concentration region.

in sample preparation and instrumentation. During the HUPO PPP, major efforts were made to identify the most successful sample preparations steps, leading to the highest number of unique protein identifications covering the widest possible dynamic range. Results showed that early unspecific solvent precipitation protocols for depleting serum albumin could be vastly improved upon with affinity-based methods, particularly immunoaffinity depletions using specific antibodies to remove the proteins of high abundance (12). An important caveat here lies in the fact that regardless of the sophistication of the depletion method, the next most abundant proteins are always in line to mask the lowest abundance proteins, especially in plasma where a large amount of proteins of average abundance remain four to five orders of magnitude more abundant than the low-abundance proteins. While depletion approaches can increase the overall chances of detecting proteins in plasma, the performance of these technologies with regard to low-abundance proteins remains limited (46). Existing data also show that no single depletion technique gives the entire view of a complex proteome, and that analyses without a prior depletion technique yield unique identification of interesting proteins that would otherwise be subjected to unwanted co-depletion with the proteins selected against.

One way to tackle the dynamic range problem is to step away from the traditional shotgun approach for whole proteome elucidation, and instead analyse plasma using a technique called selected reaction monitoring (SRM). SRM employs a targeted approach to detect and quantify proteins, and is especially applicable to those of low abundance. Typically performed on a triple quadrupole mass spectrometer, peptides undergo a two-stage selection process to capture and detect a peptide with high specificity. The selections are on precursor ion mass/charge (*m/z*) in the first quadrupole and after fragmentation in the second quadrupole, on fragment ion *m/z* in the third quadrupole. It differs from a whole-proteome shotgun approach in the requirement that proteins of interest (biomarkers, modified proteins, isoforms, etc.) are known in advance. These proteins of interest can then be specifically targeted for analysis by selecting peptides that uniquely identify these proteins and that have properties that make them amenable for detection (e.g. containing charge-carrying residues, while having relatively few hydrophobic stretches) (47) (Fig. 4).

SRM comes with specific bioinformatics challenges, involving the selection of suitable precursor and corresponding product ions (termed transition sets) to identify a protein of interest uniquely while also being likely to be detected. It is well established that different peptides have different properties that govern their detectability in a mass spectrometer (48), and selection of candidate peptides for SRM can be performed using two main approaches. The first of these relies on the use of empirical evidence extracted from public repositories to inform which peptides have been previously detected, and which transitions were the most prominent for these precursors. The second approach relies on a purely theoretical selection of candidates from a sequence database, and is based selection of peptides with a favourable detectability. There are several services available for the selection of transition sets (49, 50) that involve quality assessment of spectra to determine the most likely combination of peptides and fragment ions, returning a ranked list of suitable candidates to give the best chance of detection and quantification of a protein of interest.

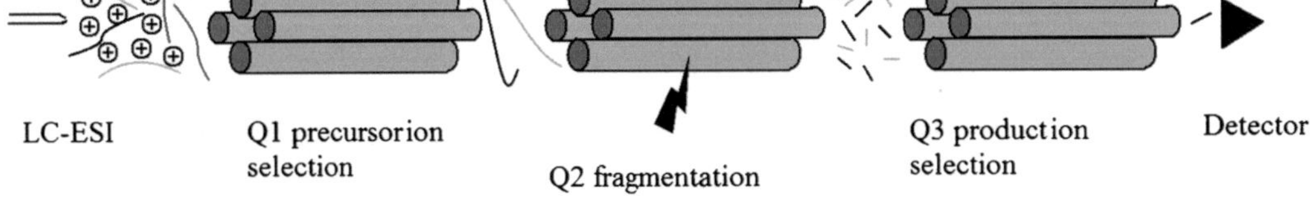

Fig. 4. The principal elements of a Selected Reaction Monitoring experiment. A digested protein sample is ionised by electrospray ionisation to produce precursor peptides; a targeted precursor peptide *m/z* of interest is selected for in quadrupole 1 (Q1); precursor peptides with the selected *m/z* are fragmented in quadrupole 2 (Q2) to produce product ions; product ions enter quadrupole 3 (Q3) where they are selected for based on a particular targeted product ion *m/z* and are subsequently detected.

While the SRM approach itself is not novel, having been utilised successfully for many years in the small molecules field (51), the application of SRM to proteomics samples is still in its infancy. It is, therefore, likely that additional software tools and instrumental protocols will emerge as time progresses that utilise this technology to drill ever deeper into the plasma proteome (52).

References

1. Chiras, D., D. (2005) Human Biology, 5th Edition
2. Anderson NL. (2003) The Human Plasma Proteome: History, Character, and Diagnostic Prospects. *Mol. Cell. Proteomics* **1**, 845–867.
3. Omenn GS. (2004) The Human Proteome Organization Plasma Proteome Project pilot phase: Reference specimens, technology platform comparisons, and standardised data submissions and analyses. *Proteomics* **4**, 1235–1240.
4. Rai AJ, Gelfand Ca, Haywood BC, et al. (2005) HUPO Plasma Proteome Project specimen collection and handling: towards the standardization of parameters for plasma proteome samples. *Proteomics* **5**, 3262–77.
5. de Roos B, Duthie SJ, Polley AC, et al. (2008) Proteomic methodological recommendations for studies involving human plasma, platelets, and peripheral blood mononuclear cells. *J. Proteome Res.* **7**, 2280–90.
6. Omenn GS, States DJ, Adamski M, et al. (2005) Overview of the HUPO Plasma Proteome Project: results from the pilot phase with 35 collaborating laboratories and multiple analytical groups, generating a core dataset of 3020 proteins and a publicly-available database. *Proteomics* **5**, 3226–45.
7. Martens L, Hermjakob H, Jones P, et al. (2005) PRIDE: the proteomics identifications database. *Proteomics* **5**, 3537–45.
8. Omenn GS, Aebersold R, Paik Y. (2009) 7th HUPO World Congress of Proteomics: launching the second phase of the HUPOPlasma Proteome Project (PPP-2) 16-20 August 2008, Amsterdam, The Netherlands. *Proteomics* **9**, 4–6.
9. Sickmann A, et al. (2003) The proteome of Saccharomyces cerevisiae mitochondria. *PNAS* **100**, 13207–13212.
10. Gevaert K, Van Damme P, Ghesquière B, et al. (2007) A la carte proteomics with an emphasis on gel-free techniques. *Proteomics* **7**, 2698–718.
11. Klie S, Martens L, Vizcaíno JA, et al. (2008) Analyzing Large-Scale Proteomics Projects with Latent Semantic Indexing. *J. Proteome Res.* **7**, 182–191.
12. Stempfer R, Kubicek M, Lang IM, Christa N, Gerner C. (2008) Quantitative assessment of human serum high-abundance protein depletion. *Electrophoresis* **29**, 4316–23.
13. Deutsch EW, Eng JK, Zhang H, et al. (2005) Human Plasma PeptideAtlas. *Proteomics* **5**, 3497–500.
14. Sadygov RG, Cociorva D, Yates JR. (2004) Large-scale database searching using tandem mass spectra: looking up the answer in the back of the book. *Nat. Meth.* **1**, 195–2022.
15. Nesvizhskii AI, Vitek O, Aebersold R. (2007) Analysis and validation of proteomic data generated by tandem mass spectrometry. *Nat. Methods* **4**, 787–797.
16. Martens L, Hermjakob H. (2007) Proteomics data validation: why all must provide data. *Molecular bioSystems* **3**, 518–22.
17. Nesvizhskii AI, Keller A, Kolker E, Aebersold R. (2003) A statistical model for identifying proteins by tandem mass spectrometry. *Anal. Chem.* **75**, 4646–58.
18. He F. Human liver proteome project: plan, progress, and perspectives. (2005) *Mol. Cell. Proteomics* **4**, 1841–8.
19. The UniProt Consortium. The Universal Protein Resource (UniProt) 2009. (2009) *Nucleic acids research* **37**, D169–74.
20. Kapp EA, Schütz F, Connolly LM, et al. (2005) An evaluation, comparison, and accurate benchmarking of several publicly available MS/MS search algorithms: sensitivity and specificity analysis. *Proteomics* **5**, 3475–90.
21. Balgley BM, Laudeman T, Yang L, Song T, Lee CS. (2007) Comparative evaluation of tandem MS search algorithms using a target-decoy search strategy. *Mol. Cell. Proteomics* **6**, 1599–608.
22. http://www.openms.de

23. Stephan C, Reidegeld KA, Hamacher M, et al. (2006) Automated reprocessing pipeline for searching heterogeneous mass spectrometric data of the HUPO Brain Proteome Project pilot phase. *Proteomics* **6**, 5015–29.
24. Martens L, Müller M, Stephan C, et al. (2006) A comparison of the HUPO Brain Proteome Project pilot with other proteomics studies. *Proteomics* **6**, 5076–86.
25. Yi J, Kim C, Gelfand Ca. (2007) Inhibition of Intrinsic Proteolytic Activities Moderates Preanalytical Variability and Instability of Human Plasma. *J. Proteome Res.* **6**, 1768–1781.
26. Reisinger F, Martens L. (2009) Database on Demand - An online tool for the custom generation of FASTA-formatted sequence databases. *Proteomics* **9**, 4421–4424.
27. Martens L. (2005) DBToolkit: processing protein databases for peptide-centric proteomics. *Bioinformatics* **21**, 3584–3585.
28. Kersey PJ, Duarte J, Williams A, et al. (2004) The International Protein Index: an integrated database for proteomics experiments. *Proteomics* **4**, 1985–8.
29. Gevaert K, Goethals M, Martens L, et al. (2003) Exploring proteomes and analyzing protein processing by mass spectrometric identification of sorted N-terminal peptides. *Nature Biotechnology* **21**, 566–569.
30. Meek D. (1998) Multisite Phosphorylation and the Integration of Stress Signals at p53. *Cellular Signalling* **10**, 159–166.
31. Creasy DM, Cottrell JS. (2004) Unimod: Protein modifications for mass spectrometry. *Proteomics* **4**, 1534–1536.
32. Montecchi-Palazzi L, Beavis R, Binz P, et al. (2008) The PSI-MOD community standard for representation of protein modification data. *Nature Biotechnology* **26**, 864–866.
33. Gornik O, Lauc G. (2008) Glycosylation of serum proteins in inflammatory diseases. *Disease markers* **25**, 267–78.
34. Yates JR, Eng JK, McCormack AL, Schieltz D. (1995) Method to correlate tandem mass spectra of modified peptides to amino acid sequences in the protein database. *Anal. Chem.* **67**, 1426–36.
35. Craig R, Beavis RC. (2004) TANDEM: matching proteins with tandem mass spectra. *Bioinformatics* **20**, 1466–7.
36. Creasy DM, Cottrell JS. (2002) Error tolerant searching of uninterpreted tandem mass spectrometry data. *Proteomics* **2**,1426–34.
37. Pevzner PA, Tang CL. (2000) Mutation-Tolerant Protein Identification by Mass Spectrometry. *Journal of Computational Biology* **7**, 777–787.
38. Tsur D, Tanner S, Zandi E, Bafna V, Pevzner PA. (2005) Identification of post-translational modifications by blind search of mass spectra. *Nat. Biotech.* **23**, 1562–7.
39. Tanner S, Pevzner Pa, Bafna V. (2006) Unrestrictive identification of post-translational modifications through peptide mass spectrometry. *Nat. Prot.* **1**, 67–72.
40. Falkner Ja, Falkner JW, Yocum AK, Andrews PC. (2008) A spectral clustering approach to MS/MS identification of post-translational modifications. *J. Proteome Res.* **7**, 4614–22.
41. Savitski MM, Nielsen ML, Zubarev Ra. (2006) ModifiComb, a new proteomic tool for mapping substoichiometric post-translational modifications, finding novel types of modifications, and fingerprinting complex protein mixtures. *Mol. Cell. Proteomics* **5**, 935–48.
42. Mann M, Wilm M. (1994) Error-tolerant identification of peptides in sequence databases by peptide sequence tags. *Anal. Chem.* **66**, 4390–9.
43. Tabb DL, Saraf A, Yates JR. (2003) GutenTag: high-throughput sequence tagging via an empirically derived fragmentation model. *Anal. Chem.* **75**, 6415–21.
44. Tabb DL, Ma Z, Martin DB, Ham AL, Chambers MC. (2008) DirecTag: accurate sequence tags from peptide MS/MS through statistical scoring. *J. Proteome Res.* **7**, 3838–46.
45. Gevaert K, Impens F, Ghesquière B, et al. (2008) Stable isotopic labelling in proteomics. *Proteomics*, **8**, 4873–85.
46. Echan LA, Tang H, Ali-Khan N, Lee K, Speicher DW. (2005) Depletion of multiple high-abundance proteins improves protein profiling capacities of human serum and plasma. *Proteomics* **5**, 3292–303.
47. Mallick P, Schirle M, Chen SS, et al. (2007) Computational prediction of proteotypic peptides for quantitative proteomics. *Nat. Biotech.* **25**, 125–131.
48. Lange V, Picotti P, Domon B, Aebersold R. (2008) Selected reaction monitoring for quantitative proteomics: a tutorial. *Molecular systems biology* **4**, 222.
49. Mead Ja, Bianco L, Ottone V, et al. (2009) MRMaid, the web-based tool for designing multiple reaction monitoring (MRM) transitions. *Mol. Cell. Proteomics* **8**, 696–705.
50. Sherwood Ca, Eastham A, Lee LW, et al. (2009) MaRiMba: A Software Application for Spectral Library-Based MRM Transition List Assembly. *J. Proteome Res.* **8**, 4396–405.

51. Zakett D, Flynn RG, Cooks RG. (1978) Chlorine isotope effects in mass spectrometry by multiple reaction monitoring. *J. Physical Chem.* **82**, 2359–2362.

52. Anderson NL, Anderson NG, Pearson TW, et al. (2009) A human proteome detection and quantitation project. *Mol. Cell. Proteomics* **8**, 883–6.

Chapter 23

Using the Human Plasma PeptideAtlas to Study Human Plasma Proteins

Terry Farrah, Eric W. Deutsch, and Ruedi Aebersold

Abstract

PeptideAtlas is a web-accessible database of LC–MS/MS shotgun proteomics results from hundreds of experiments conducted in diverse laboratories, with all data processed via a uniform analysis pipeline. A total of 91 experiments on human plasma and serum are included. Using the PeptideAtlas web interface, users can browse and search the Human Plasma PeptideAtlas for identified peptides and identified proteins, view spectra, and select proteotypic peptides. Users can easily view supporting information such as chromosomal mapping, estimated abundances, and sequence alignments. Herein, the reader is instructed in the use of the Human Plasma PeptideAtlas through an illustrated exploration of cytokine receptors in plasma.

Key words: Plasma, PeptideAtlas

1. Introduction

Shotgun proteomics using liquid chromatography, tandem mass spectrometry (LC–MS/MS) is currently the most powerful tool available for discovering proteins present in human plasma (1). As the technique develops, more and more proteins can be identified in a single experiment. To compile the most comprehensive list of human plasma proteins, one approach is to collect all proteins identified in various individual proteomics experiments, as was done by the Human Proteome Organization (HUPO (2)) via the Human Plasma Proteome Project (HPPP) in 2003–2005 (3). In HPPP Phase I, protein identifications from 18 laboratories were combined, and all proteins identified by at least two laboratories were used to generate a list of 3,020 proteins (4).

One deficiency of this approach is that each laboratory interprets its data in its own manner, resulting in protein lists that are

Richard J. Simpson and David W. Greening (eds.), *Serum/Plasma Proteomics: Methods and Protocols*,
Methods in Molecular Biology, vol. 728, DOI 10.1007/978-1-61779-068-3_23, © Springer Science+Business Media, LLC 2011

not comparable. HPPP Phase II (5) aims to address this by collecting raw data, rather than protein identifications. The PeptideAtlas project (6), a key participant in this effort, collects raw data from experiments conducted in many different laboratories (including HPPP data) and processes them using a common computational pipeline. To date, PeptideAtlas has collected and interpreted 91 experiments from human plasma, yielding over three million identified spectra and 20,709 distinctly identified peptides, which provide evidence for at least 2,170 different proteins (7). While HPPP Phase I resulted in more protein identifications, the PeptideAtlas protein identifications have a higher confidence, with a false discovery rate (FDR) of 1%.

PeptideAtlas is accessible to the public via a web interface. For human plasma (as well as a number of other species proteomes and subproteomes), it provides a large database of proteins, peptides, and spectra, with supporting data such as probabilities, FDRs, genome mappings, sequence alignments, links to other databases, uniqueness of peptide-protein mappings, observability of peptides, predicted observable peptides, estimated protein abundances, and cross-references to other databases. The PeptideAtlas also provides many useful methods for accessing the data; the user may search by protein or peptide, or may construct a query to retrieve proteins or peptides with certain characteristics.

This chapter provides an introduction to using the Human Plasma PeptideAtlas by guiding the reader through an illustrated example.

2. PeptideAtlas Construction

First, we provide a description of how data are added to PeptideAtlas, illustrated in Fig. 1. PeptideAtlas is organized into various *builds*, each encompassing data from a single proteome or subproteome. Each build begins with raw LC–MS/MS spectra contributed by the community. Data can be deposited into one of several repositories or sent directly to the PeptideAtlas project. We then search these spectra against a sequence database (8), a spectral library (9), or both. Each search assigns a peptide identification and score to each spectrum. Search results are mapped to a comprehensive reference protein database (for human builds, this is a combination of Swiss-Prot (10), Ensembl (11), and IPI (12)), and post-processed using the Trans-Proteomic Pipeline (13), a suite of software tools developed at the Institute for Systems Biology for assigning a probability of being correct to each peptide–spectrum match (PSM), distinct peptide identification, and protein identification. The Trans-Proteomic Pipeline

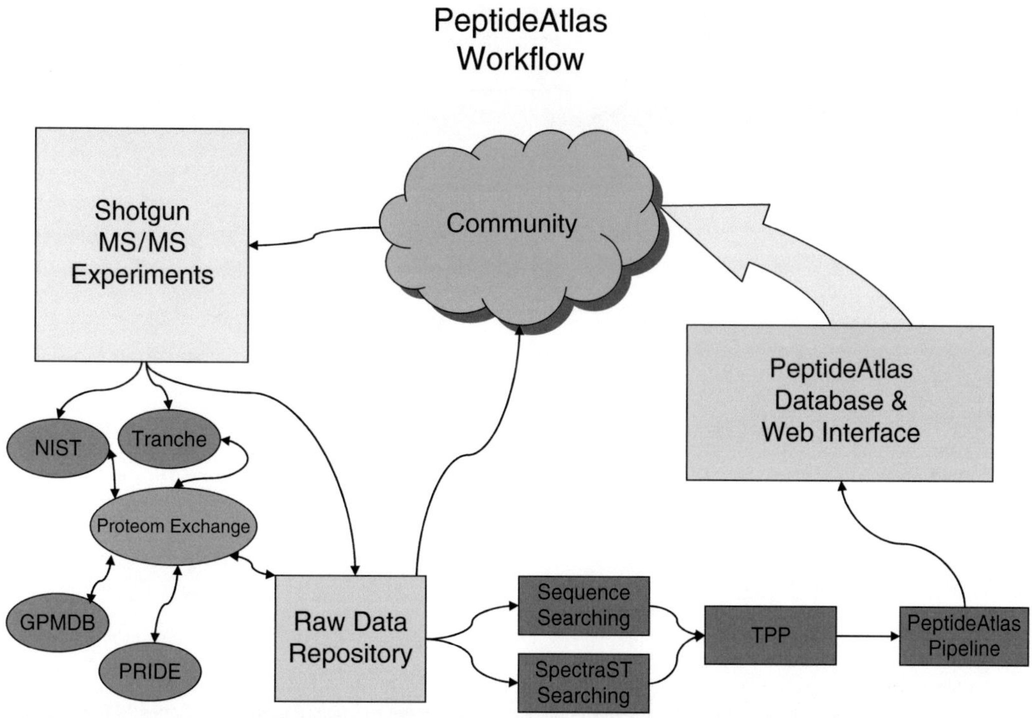

Fig. 1. PeptideAtlas is built from data provided by the community and is itself a community resource. Data can be submitted to one of several repositories, or sent directly to the PeptideAtlas project. The data are then processed via a uniform pipeline that includes searching, validating by the Trans-Proteomic Pipeline, and post-processing. The results, as well as the raw data, are stored in a database and made accessible via a web interface. (Reproduced from (17) with permission from Humana Press.).

includes the tools PeptideProphet (14), InterProphet (15), and ProteinProphet (16). Finally, other software processing tools are applied to store PSM, peptide, and protein identifications in PeptideAtlas, and to generate and store supporting data such as genome mappings and predicted proteotypic peptides. The atlas build is then made available to the community.

PeptideAtlas and its web interface are continually under development, and data are constantly being added. Thus, what the reader sees when using PeptideAtlas may not always exactly match what is described in this tutorial.

3. Using PeptideAtlas

3.1. PeptideAtlas Web Interface

Go to http://www.peptideatlas.org. The front page provides basic search functionality and displays PeptideAtlas news. To access the full functionality of PeptideAtlas, click GO, without typing anything in the search box. You will see the page shown in Fig. 2.

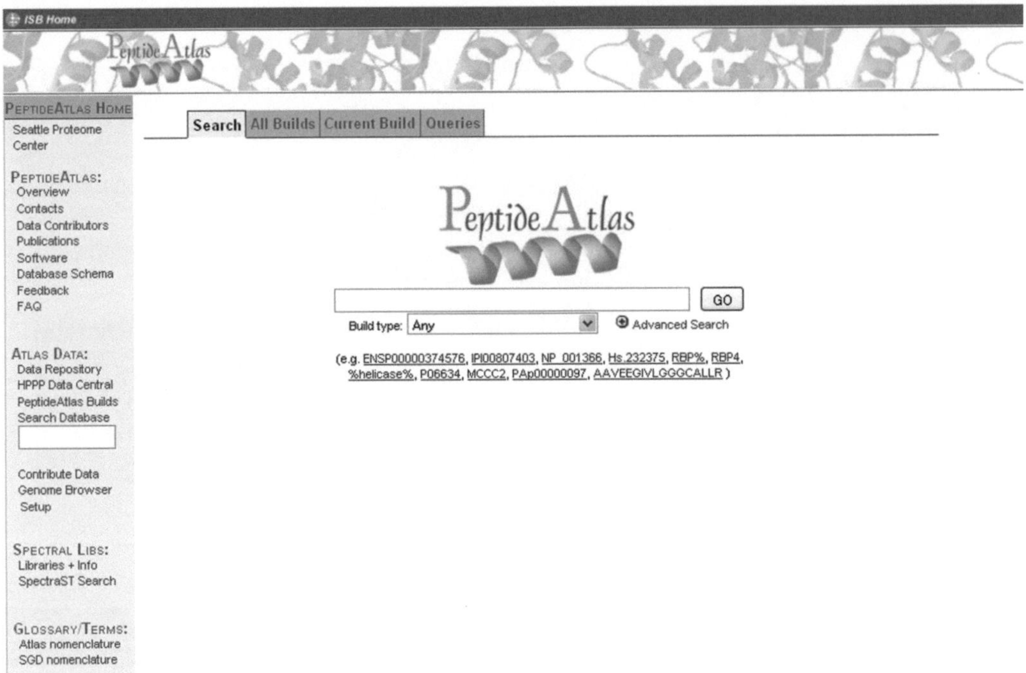

Fig. 2. PeptideAtlas web interface. Near the *top* of the display are four tabs representing four categories of functionality. Search functionality is shown. Users may enter a protein accession, gene name, protein description, peptide accession, or peptide string to access quickly all matching peptides in all of the latest PeptideAtlas builds or in a particular build.

Near the top are four gray tabs representing four categories of functionality:

Search. Keyword search within a single build or across all the most current builds.

All Builds. Allows the user to view all available builds, to select a particular build for functions under the Current Build tab, and to see a peptide's presence across all builds.

Current Build. Allows the user to obtain information on a specific peptide or protein in the currently selected build.

Queries. Allows the user to retrieve information on a set of peptides, proteins, or transitions that satisfy user-specified criteria.

To see the functions that are available for each tab, place the cursor over the tab. Many of these options will be reviewed in detail later in the tutorial.

In the left sidebar are links that will take you to auxiliary pages within PeptideAtlas. Here are descriptions of some key pages:

Overview. Description of how PeptideAtlas is constructed.

Publications. What to cite if you use PeptideAtlas in your research.

Data Repository. Links for downloading much of the raw data used to construct PeptideAtlas.

HPPP Data Central. Background and links for the HUPO HPPP.

PeptideAtlas Builds. Lists and download links for all available PeptideAtlas builds.

Search Database. A conveniently accessible search bar with the same functionality as the Search tab (described below).

Contribute Data. How to contribute your own data to the PeptideAtlas project.

Libraries+Info. Access to spectral libraries created from data in PeptideAtlas builds as well as libraries from National Institute of Standards and Technology (NIST) (18) and links to other spectral library search resources. You can download the libraries available here and use them to perform spectral library searches of your own data.

SpectraST Search. A web server that allows you to perform a spectral library search using the SpectraST software (9).

After exploring these links, click *Search Database* in the left navigation bar to return to the Search functionality.

3.2. Search for a Protein by Keyword

Using the Search functionality, you can enter a protein accession, a peptide sequence, a gene name, or a keyword or phrase, and retrieve links to all matching proteins. You can search one build or all current builds.

To illustrate the use of PeptideAtlas, we will investigate the presence of cytokine receptors in plasma. Cytokines are signaling molecules secreted by specific cells of the immune system and interact with receptors on the surfaces of other cells, thereby initiating various responses. Cytokine receptors perform their primary functions while embedded in the surfaces of cells. However, under some conditions, cytokine receptor extracellular portions are shed and may then perform a variety of secondary functions. More generally, it has been hypothesized that nearly every human protein will appear in plasma under some conditions.

Click the Build type dropdown menu. You will see a list of all current builds, including two human plasma builds. The more encompassing of these is named "Human Plasma." Select Human Plasma, click Tabular Results, click the + (plus) sign next to Advanced Search, and check Protein/Gene Name. Finally, to search for cytokine receptor names that contain the word "interleukin", type interleukin in the search box and click GO.

Note that not all of the proteins listed in the results are actually observed in this atlas, but rather, the list includes all proteins within the database that match your search criteria. The rightmost column, N Peptide obs, lists the number of observations of each protein. To focus on observed proteins, perform a descending sort on the last column by clicking the downward-pointing gray triangle.

3.3. Protein View

The third most frequently observed interleukin receptor protein in the Human Plasma PeptideAtlas is interleukin-6 receptor subunit beta (*IL6ST*). Click the identifier link (P40189-2) for that result. You will be taken to the PeptideAtlas Protein View (Fig. 3). Note that a different primary tab, Current Build, is now highlighted, and that the secondary tab, Protein, is selected.

For each protein in the reference proteome for a given build, a dynamic Protein View page summarizes the information available for that protein. The page is segmented into several collapsible sections that can be easily minimized by clicking the small icon in the orange section header.

The top section provides basic information about the protein, including alternative names (most of which are hyperlinked to external databases), the total number of spectra (observations), and distinct peptides that map to the protein. To learn more about interleukin-6 receptor subunit beta, click the second UniProt link (P40189-2). Here, we see that this protein is also known by several other names, including the commonly used name gp130. For the purpose of this tutorial, we will refer to it as *IL-6R-beta*. The UniProt page also describes the structure and

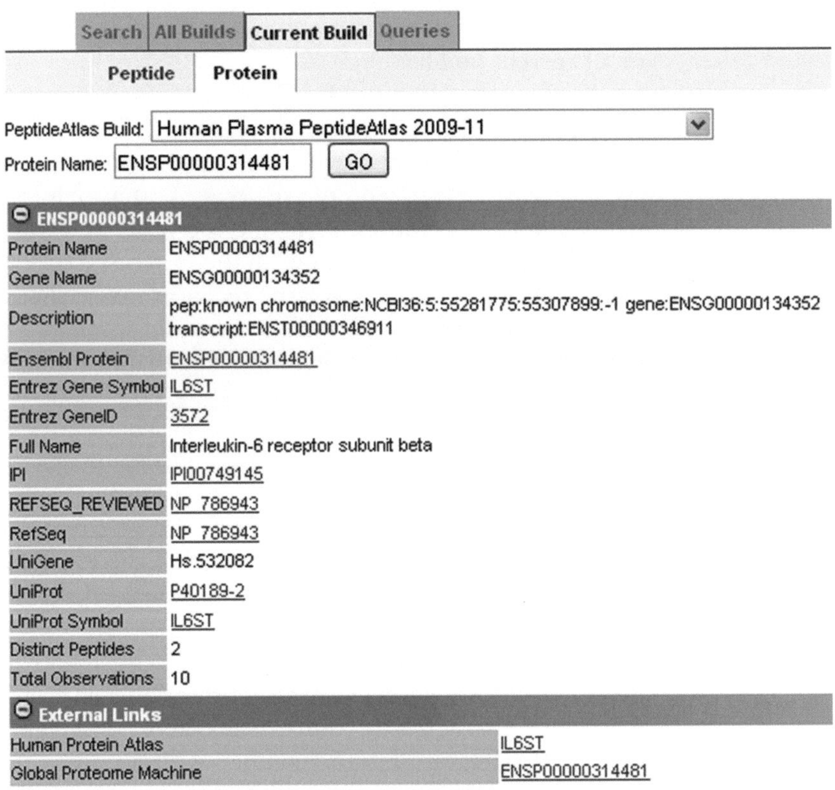

Fig. 3. Protein View, *top section*. For any protein included in a particular PeptideAtlas build, users may retrieve a page such as this displaying alternative names (most of which are clickable hyperlinks to external databases), the full name of the protein, and the number of distinct peptides and spectra (observations) that map to this protein within this build. The Protein View provides additional information further down on the page, as shown in Figs. 4 and 5.

function of this molecule and provides literature references. We see that Isoform 1 is a membrane protein, whereas the smaller Isoform 2 (P40189-2) is secreted. Because the protein we are examining lists P40189-2 as its synonym, we know it is the smaller, secreted isoform. According to this page, IL-6R-beta is a signal-transducing molecule participating in the receptor systems for IL6 and several other cytokines.

Going back to the top section of the Protein View page (Fig. 3), we see that two distinct peptides have been identified as IL-6R-beta from a total of ten observations (spectra). Just below is an External Links section that takes you to other peptide and protein atlases: the Human Protein Atlas, which lists antibodies available for the protein, and the Global Proteome Machine, which is another database that collects and displays MS/MS search results to the community.

The following two sections, Sequence Motifs and Sequence, summarize the peptide coverage of the protein (Fig. 4). A graphical diagram, similar to a genome browser view, summarizes all the peptides that map either uniquely or redundantly to the protein, plus information on segments unlikely to be observed with mass spectrometers, as well as signal peptides and transmembrane domains, where available. The observed peptides are highlighted in red in the actual protein sequence. We see that for IL-6R-beta, the two observed peptides are in the C-terminal half of the chain, covering only 12.7% of the likely observable sequence. It is not surprising that the protein coverage is so low, given that we expect

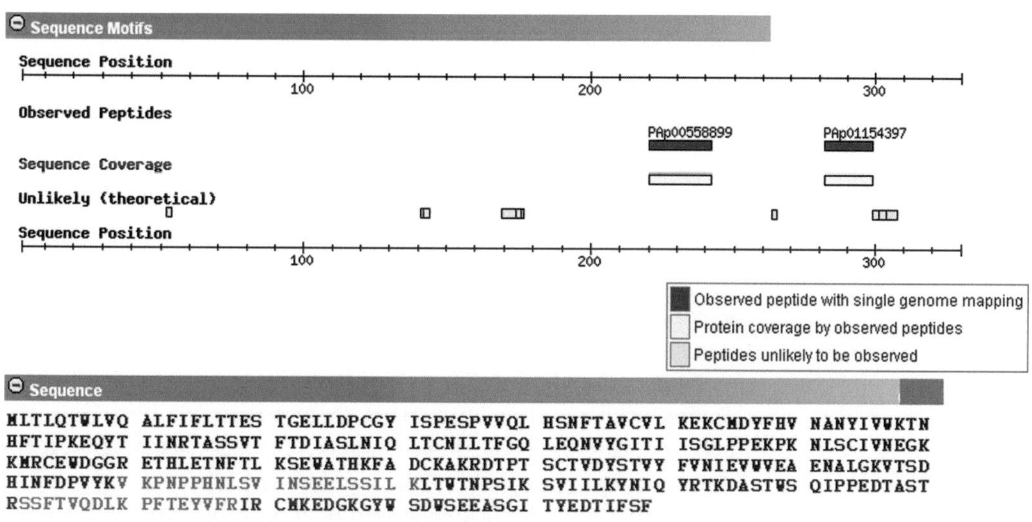

Fig. 4. More details from the Protein View page. The Sequence Motifs section displays a linear representation of the protein, with differently *shaded bars* depicting observed peptides and peptides that are unlikely to ever be observed due to small size.

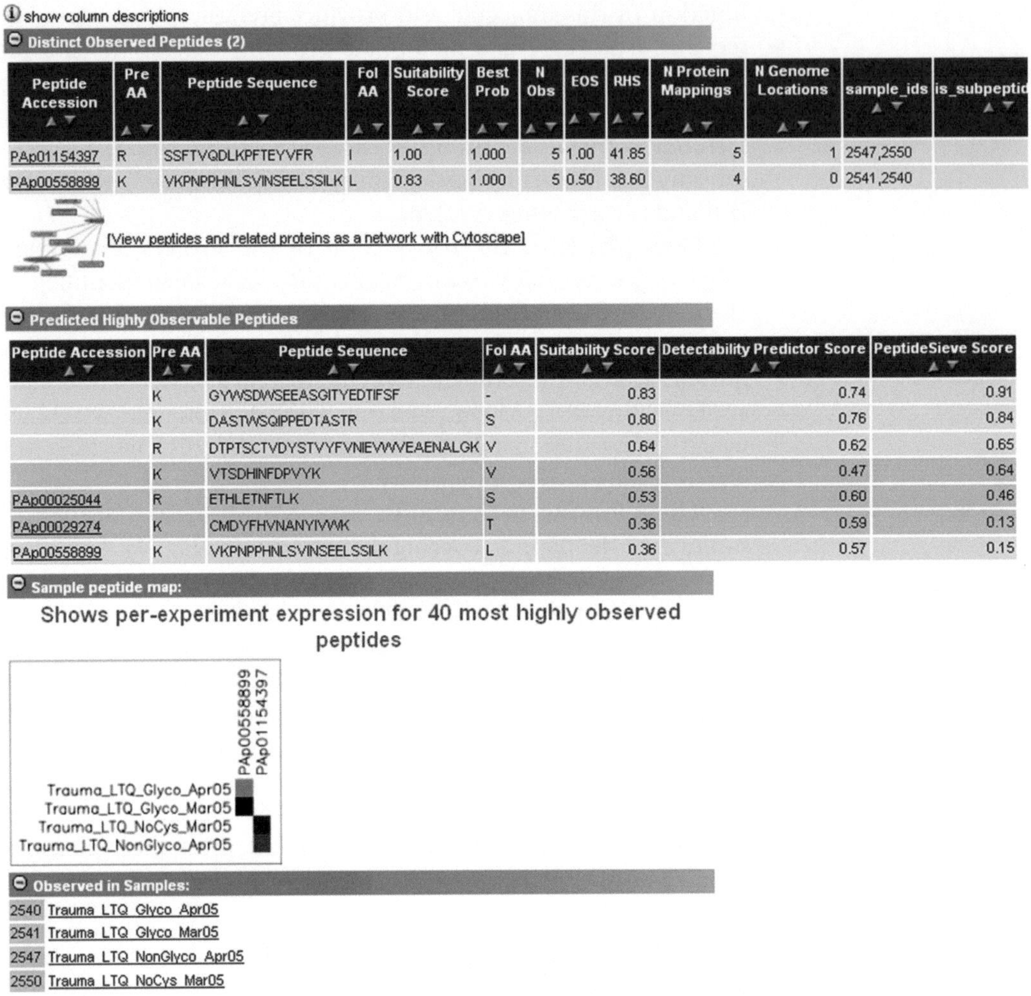

show column descriptions

Distinct Observed Peptides (2)

Peptide Accession	Pre AA	Peptide Sequence	Fol AA	Suitability Score	Best Prob	N Obs	EOS	RHS	N Protein Mappings	N Genome Locations	sample_ids	is_subpeptid
PAp01154397	R	SSFTVQDLKPFTEYVFR	I	1.00	1.000	5	1.00	41.85	5	1	2547,2550	
PAp00558899	K	VKPNPPHNLSVINSEELSSILK	L	0.83	1.000	5	0.50	38.60	4	0	2541,2540	

[View peptides and related proteins as a network with Cytoscape]

Predicted Highly Observable Peptides

Peptide Accession	Pre AA	Peptide Sequence	Fol AA	Suitability Score	Detectability Predictor Score	PeptideSieve Score
	K	GYWSDWSEEASGITYEDTIFSF	-	0.83	0.74	0.91
	K	DASTWSQIPPEDTASTR	S	0.80	0.76	0.84
	R	DTPTSCTVDYSTVYFVNIEVWVEAENALGK	V	0.64	0.62	0.65
	K	VTSDHINFDPVYK	V	0.56	0.47	0.64
PAp00025044	R	ETHLETNFTLK	S	0.53	0.60	0.46
PAp00029274	K	CMDYFHVNANYIVWK	T	0.36	0.59	0.13
PAp00558899	K	VKPNPPHNLSVINSEELSSILK	L	0.36	0.57	0.15

Sample peptide map:

Observed in Samples:

2540	Trauma LTQ Glyco Apr05
2541	Trauma LTQ Glyco Mar05
2547	Trauma LTQ NonGlyco Apr05
2550	Trauma LTQ NoCys Mar05

Fig. 5. The last four sections in the Protein View page. First, observed peptides are listed, with links to further details for each and to a Cytoscape representation of their protein mappings. Second, predicted highly observable peptides are listed, which are of use in the design of targeted proteomics experiments. Third, a graphical map shows which peptides are observed in which samples, with *darker shading* denoting more observations. Finally, links to the samples are provided.

cytokine receptor proteins to be present at very low abundance in plasma, and only under certain conditions.

Next is the Distinct Observed Peptides section, which lists all the observed peptides and maps them to the protein (Fig. 5). Each peptide has a PeptideAtlas accession of the form PApxxxxxxxx; a peptide with a given sequence has the same accession in any PeptideAtlas build. The table displays several attributes of the peptides, including the number of times they were observed in the selected build (N Obs), highest PeptideProphet probability among all observations (Best Prob), theoretically calculated hydrophobicity (RHS), and the samples in which the peptides were observed. The Empirical Observability Score (EOS) and Suitability Score (ESS) metrics are listed as well. The EOS reflects the likelihood that if the protein is detectable in the sample, it will be detected via that peptide. The ESS represents a ranking of how suitable the peptide is as a reference or proteotypic peptide. The score includes information about the total number of observations, the EOS, and the best probability of identification. It also includes penalties if the peptides are not fully tryptic or contain missed cleavages or undesirable residues that might impact the suitability of the peptide for targeting (such as methionine, which is variably oxidized).

Immediately below is a Cytoscape (19) link, allowing users to see which observed peptides are also observed in related proteins. For IL-6R-beta, this network (Fig. 6) shows the two peptides listed in the Protein View page, PAp01154397 and PAp00558899, connected to the protein we are examining (ENSP00000314481) and 10 additional proteins. The network also includes two peptides not seen in the Protein View page, PAp00415855 and PAp01420893, that each also map to some of these 10 additional proteins.

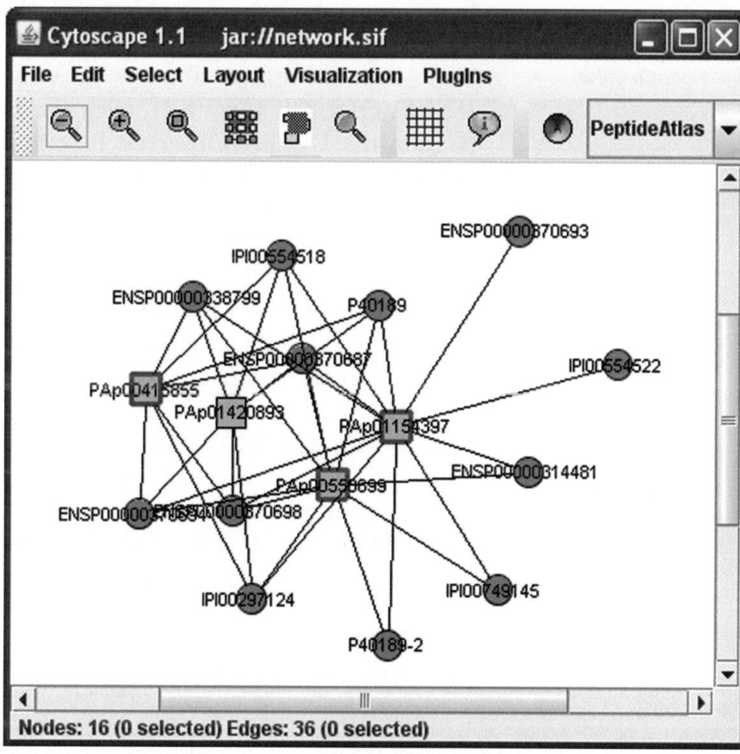

Fig. 6. Cytoscape (19) depiction of peptide–protein mapping. *Squares* near the center denote peptides and *circles* near the periphery denote proteins. Here, one peptide maps to all 12 proteins shown, while the other peptides map to fewer proteins. The user can grab and move the nodes. The Cytoscape graphic is hyperlinked from the Protein View page.

Below the Cytoscape link is the Predicted Highly Observable Peptides section, listing theoretical peptides for the protein. Each protein is digested *in silico* and both the PeptideSieve (20) and DetectabilityPredictor (21) software tools are used to predict which peptides might be easily detectable in an electrospray mass spectrometry platform. For low-abundance or otherwise hard-to-detect proteins, these theoretical predictions are useful. For IL-6R-beta, we see in the left column that three theoretical peptides have peptide accessions, indicating that they are actually observed somewhere in the PeptideAtlas, although not necessarily in this build.

Next, under the heading Sample peptide map is a graphical depiction of the abundance of each peptide in each sample. Darker shades denote more observations.

Finally, the Observed in Samples section provides links to the samples in which the protein was observed. Note that IL-6R-beta was observed only in samples from trauma patients. IL-6R-beta is known to regulate cell growth and differentiation and to play an important role in immune response. Observation of IL-6R-beta in plasma after trauma may lead an expert to hypothesize further about IL-6R-beta's role in the body's response to trauma. To learn more about these samples, click the links.

3.4. Peptide View

To learn more about the peptide PAp01154397 (SSFTVQDLKPFTEYVFR), click the link on the Protein View page in the Distinct Observed Peptides section, which will take you to the Peptide View page (Fig. 7).

For each peptide observed in the data for a given build, this dynamic Peptide View page summarizes the available information. Like the Protein View page, it is segmented into several collapsible sections. The first section provides a number of attributes for the peptide, including predicted hydrophobicity and pI, as well as the number of spectra supporting the identifications. We can see that this peptide is found in other builds also (a total of three, as of this writing); click on the numeral to see more information in a new browser tab using the peptide summary view (Fig. 8).

As of this writing, this peptide is found in the current (2009-11) human plasma build, the previous build (2009-05), and in a special human plasma build containing only non-glycocapture

Fig. 7. Peptide View page. Information on any peptide within a build can be retrieved by either the PeptideAtlas peptide accession or peptide sequence. The top section of the page shows general information about the peptide and its presence in the selected PeptideAtlas build. Below that, each distinct genome mapping is detailed, with the last line listing mappings to sequences from protein databases that do not provide genomic information. Third is a listing of all the different modifications and charge states that were observed, with graphical depictions of the consensus spectra for each. Fourth is a listing of each individual spectrum that was identified to this peptide, with links to graphical depictions of the actual spectra. Finally, links are provided to the samples in which this peptide was observed.

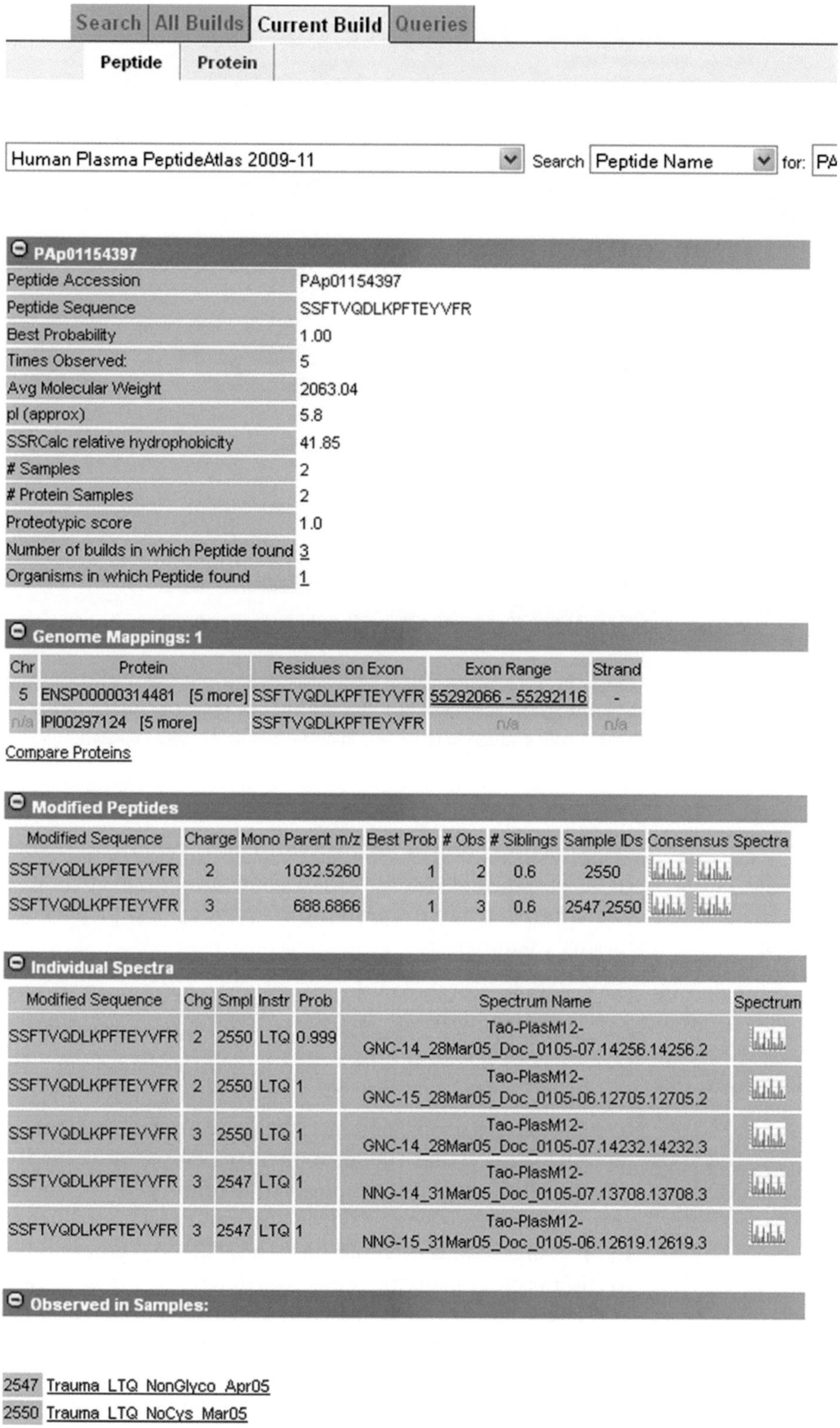
Search
All Builds
Current Build
Queries
Peptide
Protein
Human Plasma PeptideAtlas 2009-11
Search
Peptide Name
for:
PAp01154397
Peptide Accession PAp01154397
Peptide Sequence SSFTVQDLKPFTEYVFR
Best Probability 1.00
Times Observed: 5
Avg Molecular Weight 2063.04
pI (approx) 5.8
SSRCalc relative hydrophobicity 41.85
Samples 2
Protein Samples 2
Proteotypic score 1.0
Number of builds in which Peptide found 3
Organisms in which Peptide found 1
Genome Mappings: 1
Chr Protein Residues on Exon Exon Range Strand
5 ENSP00000314481 [5 more] SSFTVQDLKPFTEYVFR 55292066 - 55292116 -
n/a IPI00297124 [5 more] SSFTVQDLKPFTEYVFR n/a n/a
Compare Proteins
Modified Peptides
Modified Sequence Charge Mono Parent m/z Best Prob # Obs # Siblings Sample IDs Consensus Spectra
SSFTVQDLKPFTEYVFR 2 1032.5260 1 2 0.6 2550
SSFTVQDLKPFTEYVFR 3 688.6866 1 3 0.6 2547,2550
Individual Spectra
Modified Sequence Chg Smpl Instr Prob Spectrum Name Spectrum
SSFTVQDLKPFTEYVFR 2 2550 LTQ 0.999 Tao-PlasM12-GNC-14_28Mar05_Doc_0105-07.14256.14256.2
SSFTVQDLKPFTEYVFR 2 2550 LTQ 1 Tao-PlasM12-GNC-15_28Mar05_Doc_0105-06.12705.12705.2
SSFTVQDLKPFTEYVFR 3 2550 LTQ 1 Tao-PlasM12-GNC-14_28Mar05_Doc_0105-07.14232.14232.3
SSFTVQDLKPFTEYVFR 3 2547 LTQ 1 Tao-PlasM12-NNG-14_31Mar05_Doc_0105-07.13708.13708.3
SSFTVQDLKPFTEYVFR 3 2547 LTQ 1 Tao-PlasM12-NNG-15_31Mar05_Doc_0105-06.12619.12619.3
Observed in Samples:
2547 Trauma_LTQ_NonGlyco_Apr05
2550 Trauma_LTQ_NoCys_Mar05

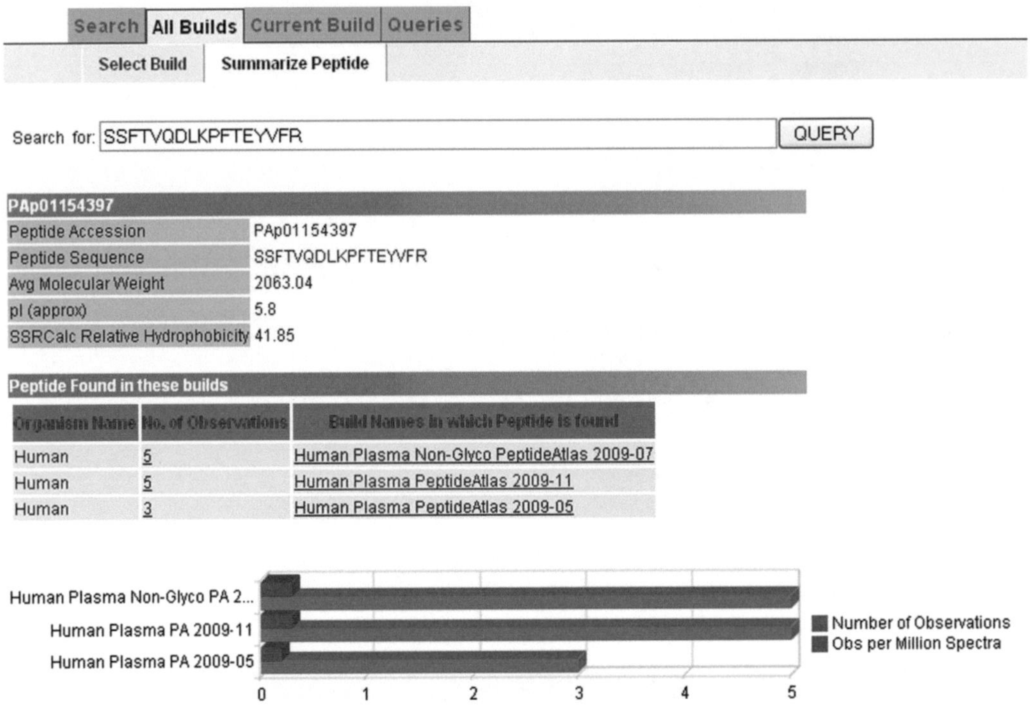

Fig. 8. Peptide summary view. Whereas the Peptide View shows information on a peptide within a specific PeptideAtlas build, the peptide summary view shows all the builds in which this peptide was observed. The *bar graph* shows the abundance of the peptide in each build, both in absolute terms and in observations per million spectra.

experiments (Human Plasma Non-Glyco PeptideAtlas 2009-07). In all builds, it is seen less than once per million spectra. Return to the browser tab displaying the Peptide View page (Fig. 7) for further examination.

The second section, Genome Mappings, displays the peptide-to-protein and chromosomal mapping information. Since the peptide-to-protein mapping can be multiplex and confusing, this section tries to simplify the mapping information. For IL6-RB, there are two rows in the Genome Mappings table, but only the row for the Ensembl sequence database contains mapping information. We see that this protein maps to six different Ensembl entries, all of which correspond to a single genome location on chromosome 5. If you click the chromosomal coordinates in the Exon Range column (55292066–55292116), you will be taken to the Ensembl page for those coordinates in a new browser tab.

The Compare Proteins link displays an alignment of all the proteins to which a peptide maps, highlighting the peptides observed for each isoform or different protein (Fig. 9).

In the sequence alignment for this peptide, we see the same 12 proteins that we saw in the Cytoscape display. The lettered superscripts indicate which sequences are in a set of identical sequences.

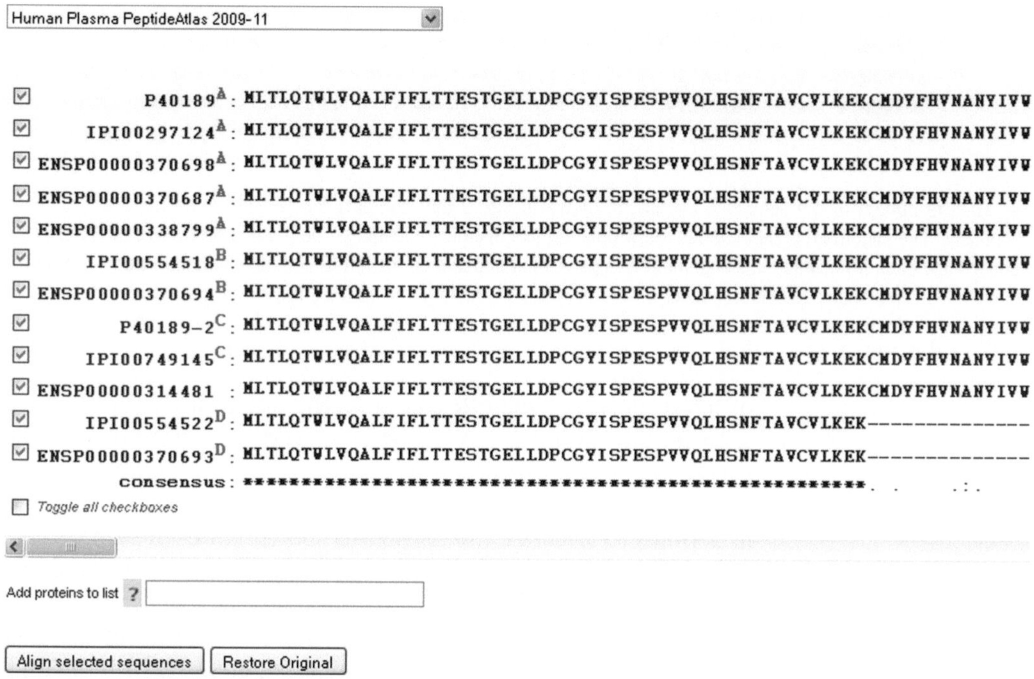

Fig. 9. A sequence alignment of all proteins to which a peptide maps, accessible by clicking Compare Proteins from the Peptide View page (Fig. 7). Lettered superscripts denote sets of identical proteins. For example, all sequences marked "A" are identical. In this illustration, all the sequences are identical in the displayed region (the N-terminal region). The user can scroll to the *left* to see the rest of the alignment (see Fig. 10).

To show just one of each set, de-select all checkboxes except the uppermost for each letter. Also, leave selected the sequence with no superscript. Then click "Align selected sequences" to align only those sequences (Fig. 10).

Scrolling to the right, find the first two highlighted peptides. The green one is the one we are currently investigating. The blue one is the other observed peptide for ENSP00000314481. Note that the bottom sequence does not contain the blue peptide. We see that there are a total of four variants (actual or predicted) that contain both peptides. Two of them, P40189 and P40189-2, are identified with Swiss-Prot accessions and correspond to the two splice variants we saw in the UniProt entry for IL-6R-beta. Swiss-Prot is a highly curated protein subset of UniProt; it includes only proteins with good evidence for existence.

Scrolling further to the right, two peptides are highlighted in blue. Neither of these appears in ENSP00000314481; these are the two additional peptides we saw in the Cytoscape display. Indeed, one of the sequences to which the peptides map is P40189, the longer splice variant described in the UniProt entry.

Going back to the Peptide View (Fig. 7) using your browser's back button, the next section, Modified Peptides, lists all of the

```
HINFDPVYKVKPNPPHNLSVINSEELSSILKLTWTNPSIKSVIILKYNIQYRTKDASTWSQIPPEDTASTRSSFTVQDLKPFTEYVFRIRCMKE
HINFDPVYKVKPNPPHNLSVINSEELSSILKLTWTNPSIKSVIILKYNIQYRTKDASTWSQIPPEDTASTRSSFTVQDLKPFTEYVFRIRCMKE
HINFDPVYKVKPNPPHNLSVINSEELSSILKLTWTNPSIKSVIILKYNIQYRTKDASTWSQIPPEDTASTRSSFTVQDLKPFTEYVFRIRCMKE
HINFDPVYKVKPNPPHNLSVINSEELSSILKLTWTNPSIKSVIILKYNIQYRTKDASTWSQIPPEDTASTRSSFTVQDLKPFTEYVFRIRCMKE
---------LKPNPPHNLSVINSEELSSILKLTWTNPSIKSVIILKYNIQYRTKDASTWSQIPPEDTASTRSSFTVQDLKPFTEYVFRIRCMKE
 .  ..   .:************************************************************************************
```

Fig. 10. The portion of the sequence alignment shown in Fig. 9 that contains the two observed peptides for these sequences. The alignment is restricted to show only the five distinct sequences. The peptide that was described in the Peptide View page from which this alignment was generated appears on the left (Green highlighting in web interface); it is found in all sequences depicted. Another observed peptide is highlighted on the left (Blue in web interface); it is seen in only four of the five sequences.

different observed peptide ions, i.e., the different charge states or mass modifications that were observed. For each peptide ion, the predicted monoisotopic precursor m/z is listed along with the number of observations, samples in which they were seen, and hyperlinks to visualize the consensus spectra for each peptide ion. We see that this peptide is observed in charge states 2 (two observations, one sample) and 3 (three observations, two samples).

The next section, Individual Spectra, lists every spectrum that supports the identification of this peptide, along with individual attributes of the match such as the probability of being correctly identified. View the first spectrum in a new browser tab by clicking its spectrum icon (Fig. 11).

Finally, at the bottom is a listing of the experiments that included the peptide, along with some simple charts depicting the relative number of spectral counts in each experiment.

3.5. Select a Build

You can construct queries to retrieve proteins or peptides fulfilling certain criteria from a specified build. Before doing so, you should select the build by going to All Builds > Select Build (Fig. 12).

By default, only *default builds* are displayed. These are the latest versions of all builds available to the public. Click the small plus-sign icon to see older builds.

As we saw earlier when using Search, two default human plasma builds are available. On this page, their longer, more descriptive names are displayed:

Human Plasma PeptideAtlas 2009-11

Human Plasma Non-glyco PeptideAtlas 2009-07

The first build listed corresponds to the one named Human Plasma in the Search interface, and this is the more comprehensive build. Select it by clicking the radio button to its left. Note that near the top of the page, it now says, “Your current build is: Human Plasma PeptideAtlas 2009-11.”

Before leaving this page, click the hyperlink for this atlas to view its Build Summary Page (Fig. 13).

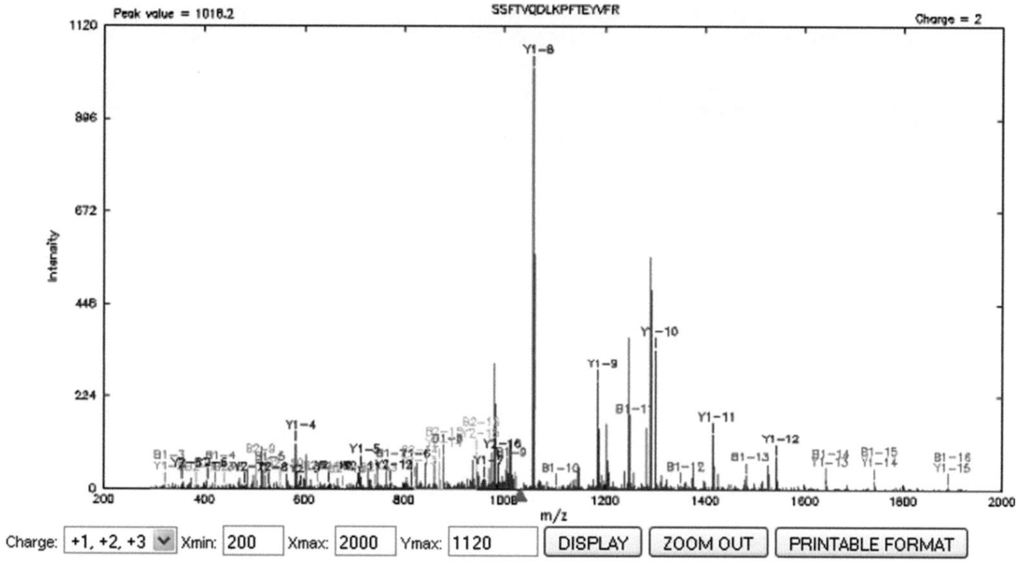

Download spectrum in Format: TSV, Excel
Download peptide data in Format: TSV, Excel

Residue	ion	B	Y	ion	+1
S	B1-1	88.0	--	Y1-17	
S	B1-2	175.1	1977.0	Y1-16	
F	B1-3	322.1	1890.0	Y1-15	
T	B1-4	423.2	1742.9	Y1-14	
V	B1-5	522.2	1641.9	Y1-13	
Q	B1-6	650.3	1542.8	Y1-12	
D	B1-7	765.3	1414.7	Y1-11	
L	B1-8	878.4	1299.7	Y1-10	
K	B1-9	1006.5	1186.6	Y1-9	
P	B1-10	1103.6	1058.5	Y1-8	
F	B1-11	1250.6	961.5	Y1-7	
T	B1-12	1351.7	814.4	Y1-6	
E	B1-13	1480.7	713.4	Y1-5	
Y	B1-14	1643.8	584.3	Y1-4	
V	B1-15	1742.8	421.3	Y1-3	
F	B1-16	1889.9	322.2	Y1-2	
R	B1-17	--	175.1	Y1-1	

Fig. 11. Spectrum view for the spectrum identified as SSFTVQDLKPFTEYVFR, charge 2. A graphical depiction shows peaks identified as Y-ions in *blue* and as B-ions in *red* (*colors* not visible in illustration). Because this identification is for charge 2, it comprises fragment ions in both charge state 1 and charge state 2. Identified peaks for charge state 1 are listed in a chart beneath the spectrum, with *colored text* allowing easy correlation with the peaks. A similar chart for charge state 2 can be viewed by scrolling down.

The top section, Build Overview, gives general information about the build. Most entries are self-explanatory.

For most current builds, the *PSM FDR threshold* describes how the PSMs were selected for inclusion in this build. This build of the Human Plasma PeptideAtlas is unusual in that it is a combination of two other builds, each created with a separate PSM FDR. Thus, this build shows N/A for the PSM FDR threshold. Older builds were created with a PeptideProphet probability threshold, usually of 0.9. This is a very generous threshold that results in more false identifications than that found in the current build of the Human Plasma PeptideAtlas.

Fig. 12. Four primary tabs are displayed on most PeptideAtlas pages, with each allowing access to a set of secondary tabs. The Select Build functionality can be accessed from the primary tab All Builds.

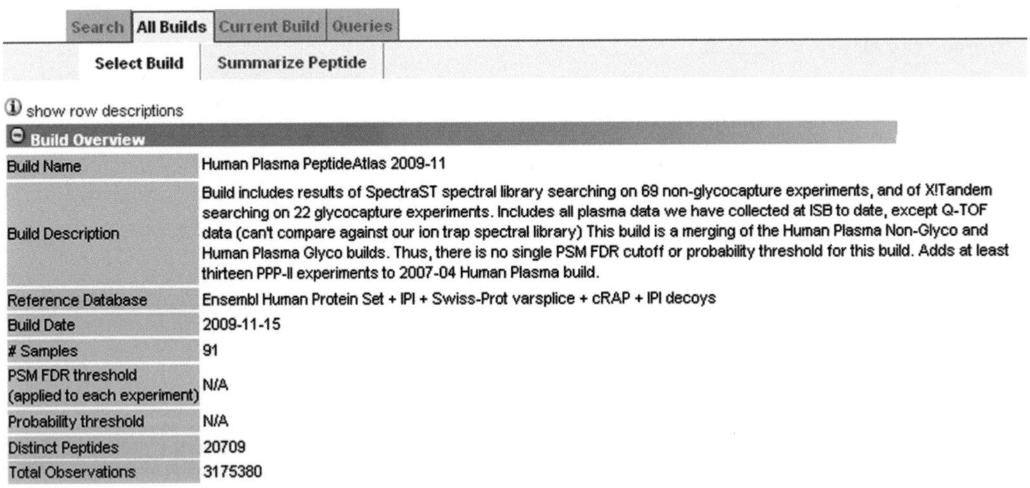

Build Name	Human Plasma PeptideAtlas 2009-11
Build Description	Build includes results of SpectraST spectral library searching on 69 non-glycocapture experiments, and of X!Tandem searching on 22 glycocapture experiments. Includes all plasma data we have collected at ISB to date, except Q-TOF data (can't compare against our ion trap spectral library) This build is a merging of the Human Plasma Non-Glyco and Human Plasma Glyco builds. Thus, there is no single PSM FDR cutoff or probability threshold for this build. Adds at least thirteen PPP-II experiments to 2007-04 Human Plasma build.
Reference Database	Ensembl Human Protein Set + IPI + Swiss-Prot varsplice + cRAP + IPI decoys
Build Date	2009-11-15
# Samples	91
PSM FDR threshold (applied to each experiment)	N/A
Probability threshold	N/A
Distinct Peptides	20709
Total Observations	3175380

Fig. 13. Build Summary Page for the Human Plasma PeptideAtlas, showing information about the build as a whole.

Total observations refers to the total number of PSMs, in other words, the total number of identified spectra.

Further down on the Build Summary page are sections describing the contribution of each sample and a detailed FDR analysis. Click the *show column descriptions* links to learn how to interpret these tables.[1] The Mayu Decoy-Based FDR Analysis

[1] As of this writing, the Sample Contribution table shows data reflecting the number of *multiply observed* peptides, a vestige of the time when the Trans-Proteomic Pipeline was less good at distinguishing between true and false hits; at that time we believed that peptides observed only once were not reliable enough.

table shows, in the 19th column, that this build's protein FDR is 0.00940853, or just under 1%. This is a very high confidence level; of the more than 2,000 *canonical* proteins identified (see Subheading 3.7 below for definition of canonical), we expect only about 20 false identifications.

3.6. Search Proteins

Let's now look up all cytokine receptors in the Human Plasma PeptideAtlas. We can get such a list from Amigo, the Gene Ontology web browser (http://amigo.geneontology.org), by typing *cytokine receptor* in the Amigo search bar, selecting *GO terms*, and clicking Submit Query. Three GO terms will be returned. Click the *286 gene products* link for *cytokine receptor activity* (the exact number may be different). On the resulting page, set two filters: click *protein* under Gene Product Type and *Homo sapiens* under Species. Then click Set Filters. Next, just above the filters section, click *Download all associations information in gene association format.* Your results will be displayed right in the browser. From your browser, save this page in a file with a .tsv extension. Then edit the file (perhaps using Excel) to contain only the UniProt accessions (column 2) and save the file. This yields a list of about 68 accessions, all of which come from the curated Swiss-Prot section of UniProt. Some accessions will be duplicates; in this case, there are about 54 distinct accessions.

Now, go to the Search Proteins function under the Queries tab, upload the file you just saved, and press QUERY (Fig. 14). After a minute or two, you will get a table with one line for each input accession. If an accession does not appear in the reference database for the atlas build, the word UNKNOWN will appear in the *hit* column, and you will need to find another accession to get information on that protein. In this case, because all our input accessions were from Swiss-Prot, and the reference database for this PeptideAtlas build includes Swiss-Prot, UNKNOWN should not appear.

Most of the entries have no observations (n_obs = 0). To bring those that were observed to the top, click the downward-pointing triangle in the n_obs header.

We see that nine distinct cytokine receptors are observed in the Human Plasma PeptideAtlas. The first column, *prot_id*, shows the accession we searched for. The second, *hit*, shows the accession used for that protein in the reference database. The fourth column, *equiv_ids*, shows other accessions from the reference database for that protein sequence. To learn about each protein, click links for each protein under *hits* and the Protein View will be displayed. Or, to learn about all nine cytokines at once, download those nine into a file and look them up using the Browse Proteins function. To do so, go to the bottom of the page. You will find several download formats available (Fig. 15). Click *CSV*, choose

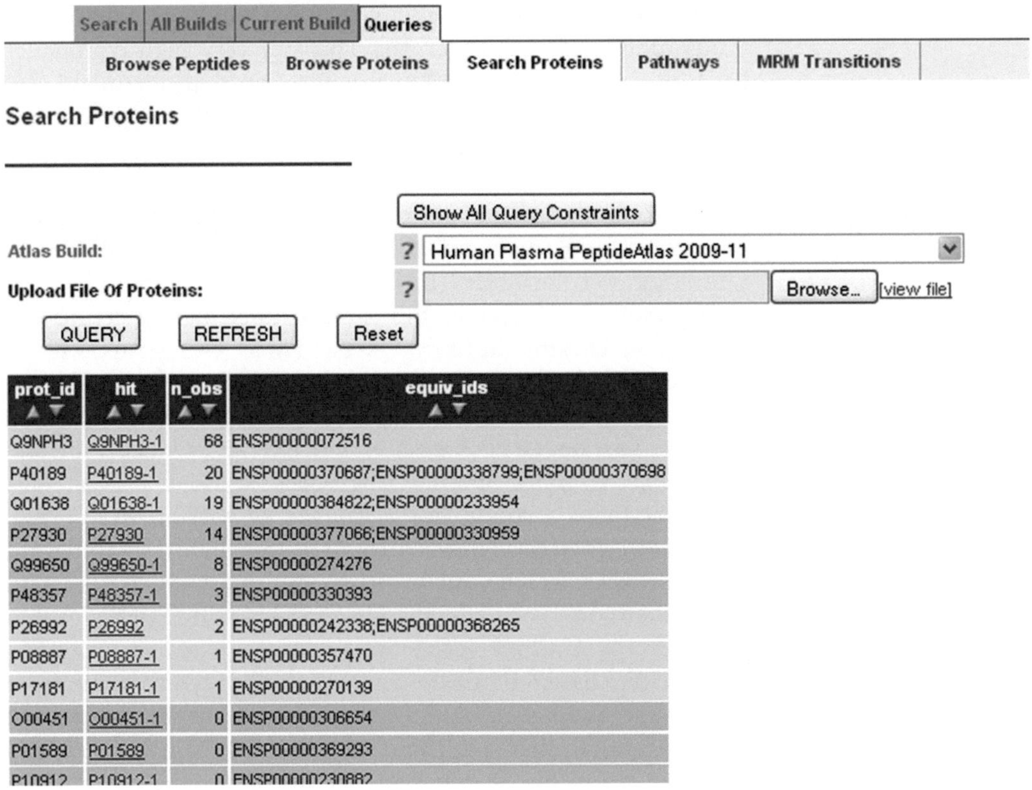

Fig. 14. Search Proteins functionality takes a PeptideAtlas build and a list of protein accessions as input, and returns one line per accession, showing how many times each protein is observed in that build. Accessions for identical sequences in the build's reference database are also shown. If an input accession is not found in the reference database, it is reported as UNKNOWN (not shown).

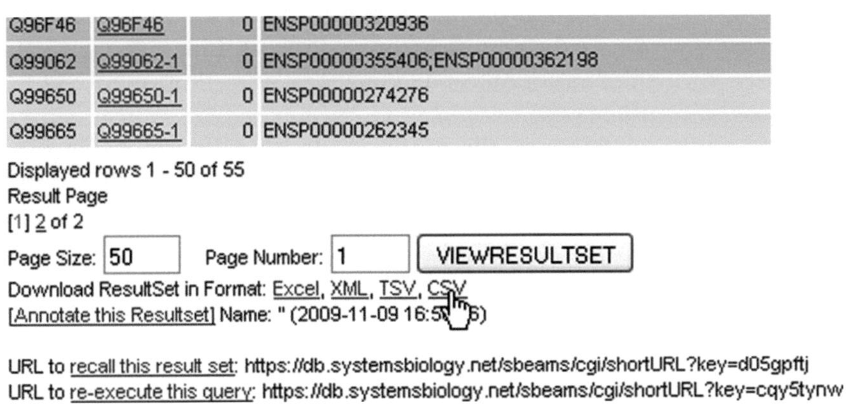

Fig. 15. Results from Browse Peptides, Browse Proteins, and Search Proteins can be downloaded in any of four formats by clicking the links at the bottom of the page.

a filename, and save. Then edit the file so that it contains just the top nine distinct accessions that are in the first column (delete the other rows and columns, including the header row and any duplicates). Be sure to save the final file in a text format (.txt, .csv, or .tsv), rather than in an Excel format.

Now, go to Browse Proteins under the Queries tab (Fig. 16).

3.7. Browse Proteins

This page allows you to view detailed information on a set of proteins. The set can either be uploaded as a list, as for the Search Proteins function, or be specified via constraints, or a combination of both.

Continuing with our example, enter or browse to the file containing the list of nine accessions you just created. Then, remove the default constraints under Presence Level and Redundancy by scrolling to the blank entry at the bottom of each menu and selecting that. Under Display Options, select *Show Estimated Abundances.* Finally, click QUERY.

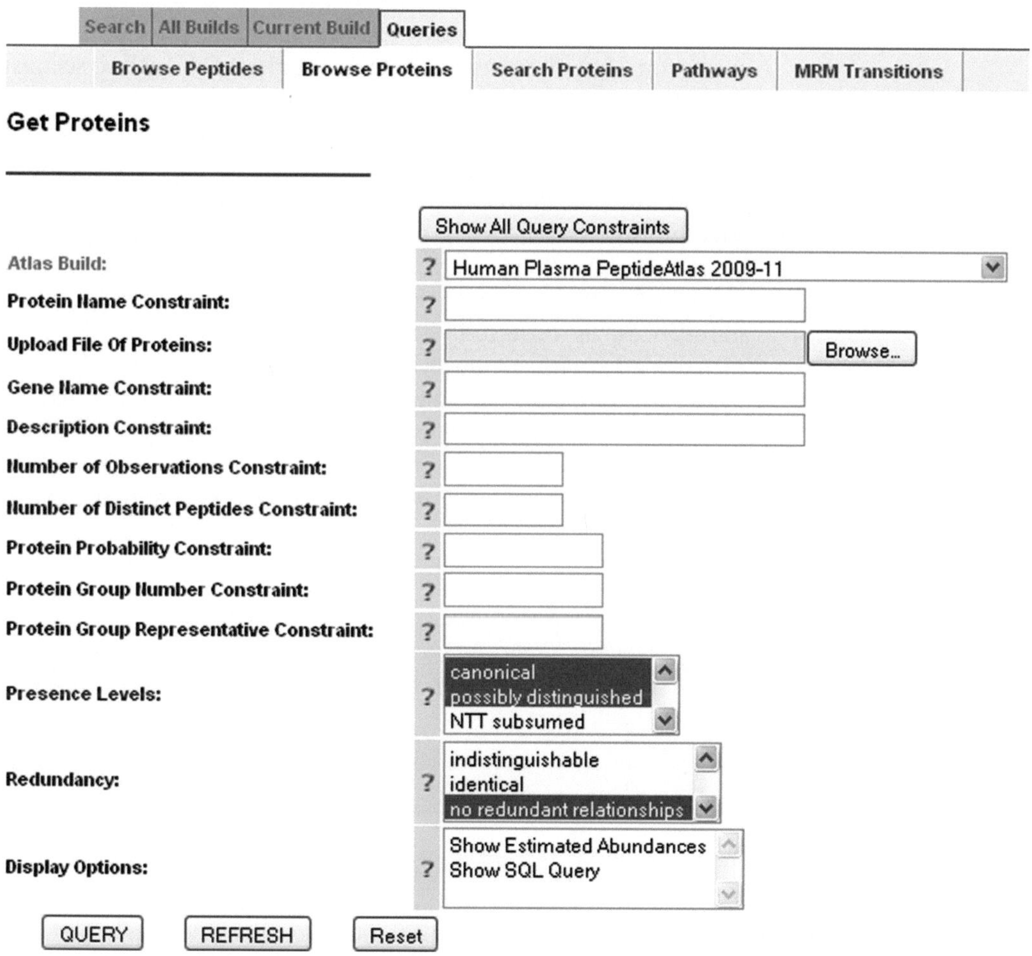

Fig. 16. Browse Proteins functionality. For a particular PeptideAtlas build, a user can retrieve proteins that fulfill a set of constraints. To retrieve all proteins, select no constraints. By default, constraints are specified to show a highly nonredundant list of proteins (shown). This query will return about 2,500 proteins. To retrieve all protein sequences in the reference database that contain observed peptides, deselect *canonical*, *possibly distinguished*, and *no redundant relationships*. This will return about 16,000 protein sequences, many of which are identical to one another. To learn more about any constraint, click its question mark icon.

A table containing about 66 proteins will be returned. This illustrates the redundancy of the reference database; all proteins from the reference database that are either indistinguishable from (same set of observed peptides) or identical to (sequence-identical) one of the nine input proteins are displayed. Only the top 50 are displayed. To see all 66 proteins, type 66 in the Page Size box at the end of the table and select VIEW_RESULT_SET.

Note that nine of the entries are labeled canonical and that they are exactly the ones in our query list. This is not an accident: from among each set of indistinguishable and/or identical database entries, PeptideAtlas selects one to assign a *Presence Level*, preferring Swiss-Prot entries, and further preferring Swiss-Prot entries without splice-variant suffixes. We can apply a Presence Level constraint of *canonical* and request *no redundant relationships* to see only those nine proteins. Select those constraints, plus *Show Estimated Abundances* under Display Options. Then – important – browse to the file again, and press QUERY to see just the nine proteins of interest (Fig. 17).

We can see that the nine cytokine receptor proteins observed in the Human Plasma PeptideAtlas include IL-1 receptor accessory protein, IL-1 receptor-like protein 1, IL-6R-beta, leptin receptor, oncostatin-M-specific receptor subunit beta, IL-1 receptor type II, interferon-alpha/beta receptor alpha chain, IL-6 receptor subunit alpha, and CNTF receptor alpha, and that all are observed at low abundances, as expected. The highest estimated abundance is 210 ng/mL, with an uncertainty of 30×. This means that the estimated abundance must be divided by 30 and multiplied by 30 to obtain a lower and upper bound. Therefore, the abundance for the first protein in the list, IL-1 receptor accessory protein, is estimated to be between 7.3 and 6,900 ng/mL. Abundances in

Biosequence Name	Presence Level	Protein Prophet Prob	Mult Hyp Test Prob	N Obs	N Distinct Peptides	Estimated ng/ml	Uncertainty ng/ml	Redundancy Relationship	Redundant With Respect To	Protein Group	Seq Uniq Prots in Grp	Protein Group Seq Alignmt	Protein Desc
Q9NPH3	canonical	1.000	1.000	23	6	220.000	30x			Q9NPH3	5		Interleukin-1 receptor accessory protein OS=H
Q01638	canonical	1.000	1.000	14	5	210.000	30x			Q01638	4		Interleukin-1 receptor-like 1 OS=Homo sapiens
P40189	canonical	1.000	1.000	12	4	210.000	30x			P40189	5		Interleukin-6 receptor subunit beta OS=Homo s
P48357	canonical	0.999	1.000	3	2	60.000	30x			P48357	6		Leptin receptor OS=Homo sapiens GN=LEPR P
Q99650	canonical	0.999	1.000	8	2	64.000	30x			Q99650	3		Oncostatin-M specific receptor subunit beta O
P27930	canonical	0.964	1.000	10	1	130.000	30x			P27930	1		Interleukin-1 receptor type II OS=Homo sapiens
P17181	canonical	0.964	1.000	1	1	19.000	30x			P17181	4		Interferon-alpha/beta receptor alpha chain OS
P08887	canonical	0.964	1.000	1	1	26.000	30x			P08887	2		Interleukin-6 receptor subunit alpha OS=Homo
P26992	canonical	0.952	1.000	1	1	5.900	10x			P26992	2		Ciliary neurotrophic factor receptor alpha OS=

Displayed rows 1 - 9 of 9

Fig. 17. Results of a Browse Proteins query requesting information on nine cytokine receptors. The accessions for these proteins were specified in an uploaded file. All of the proteins are labeled *canonical*, which means that each is considered to be the most representative protein (or one of the most representative proteins) in its protein group. A link is provided in the 11th column to display all members of any group. Protein probabilities from PeptideProphet (14) and a multiple hypothesis-testing technique (22) are shown. The N Obs column shows the number of identified spectra for each protein. Abundance in ng/mL is estimated using a spectral counting method calibrated to abundances reported in the literature (23). A multiplicative uncertainty factor is provided for each estimated abundance.

the Human Plasma PeptideAtlas are estimated based on the total number of observations (N Obs), calibrated against a set of plasma protein abundances reported in the literature (23).

You may wonder how the abundances of these receptors compare to the abundances of other receptors found in this atlas. To retrieve all higher abundance receptors, enter %receptor% in the Description Constraint, >100 in the Number of Observations Constraint, and *no redundant relationships* in the Redundancy constraint. Also, select the Query Option *Show Estimated Abundances*. Leave all the other constraints unspecified and click QUERY. We see that four proteins are returned. Although all contain the term *receptor* in their descriptions, further investigation shows that only one, Q86VB7 (Scavenger receptor cysteine-rich type 1 protein M130), is actually a receptor, with an estimated abundance of 730 ng/mL (uncertainty 10×). Plasma proteins can be present at over 10^6 ng/mL, so we see that even this receptor has a fairly low abundance. We can conclude that, assuming all receptor proteins contain the word *receptor* in their descriptions, no receptor protein is estimated to have an average abundance in human plasma greater than 730 ng/ml.

Proteins that share many observed peptides are partitioned into *protein groups*. To examine the protein group for IL-6R-beta, our original example, enter its Swiss-Prot accession, P40189, into the Protein Group Representative Constraint, select *Show Estimated Abundances*, leave all other constraints unspecified (be sure to erase the ones you entered for Description and Number of Observations), and click QUERY. You will see all proteins from the reference database that belong to the same protein group as IL-6R-beta (Fig. 18).

Biosequence Name	Presence Level	Protein Prophet Prob	Mult Hyp Test Prob	N Obs	N Distinct Peptides	Redundancy Relationship	Redundant With Respect To	Protein Group	Seq Uniq Prots in Grp	Protein Group Seq Alignmt	
P40189	canonical	1.000	1.000	12	4			P40189	5		Interleukin-6 receptor subunit beta OS=Ho
IPI00554518						identical	ENSP00000370694	P40189	5		IPI:IPI00554518.1|TREMBL:Q5FC04|ENSEM
ENSP00000370694						indistinguishable	P40189	P40189	5		pep:known chromosome:NCBI36:5:55270
ENSP00000338799						identical	P40189	P40189	5		pep:known chromosome:NCBI36:5:55270
ENSP00000370687						identical	P40189	P40189	5		pep:known chromosome:NCBI36:5:55272
ENSP00000370698						identical	P40189	P40189	5		pep:known chromosome:NCBI36:5:55266
IPI00297124						identical	P40189	P40189	5		IPI:IPI00297124.2|SWISS-PROT:P40189-1|T
IPI00554522						identical	ENSP00000370693	P40189	5		IPI:IPI00554522.1|TREMBL:Q5FC05|ENSEM
ENSP00000370693	subsumed	0.000	1.000	4	1		P40189	P40189	5		pep:known chromosome:NCBI36:5:55270
P40189-2	subsumed	0.000	1.000	7	2		P40189	P40189	5		Isoform 2 of Interleukin-6 receptor subunit
ENSP00000314481						indistinguishable	P40189-2	P40189	5		pep:known chromosome:NCBI36:5:55281
IPI00749145						identical	P40189-2	P40189	5		IPI:IPI00749145.1|SWISS-PROT:P40189-2|E

Displayed rows 1 - 12 of 12

Fig. 18. Browse Proteins results for the protein group whose representative is interleukin-6 receptor subunit beta (P40189). Two proteins, labeled *subsumed*, are mapped to by peptide sets that are proper subsets of the peptide sets mapping to P40189. ENSP000000370694, labeled *indistinguishable*, is mapped to by the exactly same peptide set as P40189. Likewise, ENSP00000314481 is indistinguishable from P40189-2. All the other entries in this table are marked *identical*; each is identical in sequence to some other member of this group.

Each protein has a Presence Level or a Redundancy Relationship as follows:

Canonical. The single protein (or, for a few groups, several proteins) that represents the group. Usually canonical proteins have a Swiss-Prot accession. All canonical proteins in an atlas build are well-distinguished from one another; any pair of canonicals will share no more than 80% of their observed peptides. There are 2057 canonical proteins in this build. This number serves as a very conservative count of the number of distinct proteins observed.

Possibly distinguished (none in this group). A protein that contains some observed peptides that distinguish it from the canonical(s), but those distinguishing peptides are fewer than 20% of the total number of observed peptides in the protein. There are 506 possibly distinguished proteins in this build.

Subsumed. A protein for which the observed peptides are a subset of the observed peptides for a canonical or possibly distinguished protein. A subsumed protein is a protein that may have been observed, but there is no peptide to distinguish it from its subsuming protein(s).

NTT-subsumed (none in this group). A protein that has an identical set of peptides to a canonical or possibly distinguished protein, but has fewer tryptic termini. Again, there is no peptide to distinguish it.

Indistinguishable. A protein with observed peptides that are identical to those of a canonical, possibly distinguished, or subsumed protein. Taken together, the canonical, possibly distinguished, subsumed, NTT-subsumed, and indistinguishable proteins of a build comprise an exhaustive list of sequence-unique proteins that may have been observed in the given experiments; for the Human Plasma PeptideAtlas, this list includes over 8,000 proteins.

Identical. A protein from the reference database for the build that is identical in sequence to a protein with one of the above labels.

More detailed explanations of this terminology can be found by clicking *Protein ID terms* near the bottom of the left navigation bar.

It is usually not useful to see the proteins labeled *identical.* To remove them from the display, click *Indistinguishable* in the Redundancy constraint menu, and then click QUERY again. You will then see the five nonidentical proteins in this group. To see the sequence alignment for the group, click the icon under Protein Group Seq Alignmt. This is the same sequence alignment you saw when clicking Compare Proteins from the Peptide View page.

3.8. Design MRM Assays

Multiple reaction monitoring (MRM; also called selected reaction monitoring or SRM) is a proteomics technique that aims to detect the presence of a small number of specific proteins in a sample. For such assays, the mass spectrometer is configured to monitor unique ion signatures, called transitions, of predetermined peptides in order to achieve a detection or confident upper limit for desired peptides to the exclusion of all other peptides. A transition is the combination of precursor m/z value and a product ion m/z value. The design of MRM assays is currently a topic of great interest, and the methods used are under constant development.

PeptideAtlas can be used in conjunction with any number of other tools to select peptides and transitions for such assays efficiently. New features and builds are currently being added to PeptideAtlas to make it even more useful for this task (24). Because the methods used in this field are in such flux, we will only provide a very simplified example here.

Let us design an assay for detecting our example protein, the shorter, secreted isoform of IL-6R-beta. First, we select the peptides we wish to detect. We want to choose *proteotypic* peptides, which are peptides that are easily observable using the available instrument *and* which map uniquely to the protein of interest. If we have the luxury of choosing as many as five peptides, it may make sense to choose the two that are observed, plus the three with the highest suitability scores (refer Fig. 5). We check the N Protein Mappings column for the observed peptides to see if they are unique to our protein. We see that they have four and five mappings, all to splice isoforms of the same gene. So if we design our assay to use these peptides, the assay will be specific for products of this gene, but will not distinguish the shorter isoform from the others. Let us assume that this is acceptable.

We should also check the theoretical peptides to see if they are unique to IL-6R-beta, perhaps using the BLAST tool (http://www.ncbi.nlm.nih.gov/BLAST).

The next task is transition selection. We will probably be executing our MRM assay on a triple-quadrupole instrument. However, the observed peptides were all likely observed on ion trap instruments, because these are the most commonly used instruments for the shotgun experiments contributed to PeptideAtlas. The relative intensities of the fragment ions in triple-quadrupole instruments can be quite different from those in ion traps, but are generally fairly similar (25), so we will ideally synthesize these five peptides, obtain triple quadrupole spectra, and select transitions from those spectra. If this is not possible, we will manually examine the consensus spectra stored in PeptideAtlas for our selected peptides and choose a few higher intensity transitions for each peptide.

A special build of PeptideAtlas, called SRMAtlas, contains only MS/MS spectra from triple-quadrupole instruments. Only a relatively small number of such spectra are currently available in the SRMAtlas; none were available for human plasma at the time of this writing, but should become available very soon. However, the data contained therein provide the best available transitions for the proteins represented in these special builds. SRMAtlas is growing and will eventually contain spectra for most of the proteins in the Human Plasma PeptideAtlas.

4. Summary and Conclusion

PeptideAtlas provides a wealth of information and tools for exploring the human plasma proteome. Much can be learned with this resource that cannot be learned from any single experiment. Here, exploring the world of cytokine receptor proteins, we learned that nine have been observed with high confidence among 91 plasma experiments. We examined one receptor protein, the shorter isoform of interleukin-6 receptor subunit beta, in-depth, and found that within these 91 experiments, it was observed only in samples from trauma patients. Of the two observed peptides that map to this isoform, we learned that one of them appears in all five IL-6R-beta isoforms found in a comprehensive sequence database, while the other appears in only four isoforms. The tool allowed us to examine the spectra underlying these identifications, to confirm their reliability, and/or to select transitions for targeted proteomics assays. We were able to access additional information and data displays such as predicted proteotypic peptides, genome mappings, and sequence alignments easily.

A primary purpose of the PeptideAtlas project is to support the discovery of biomarkers to diagnose and stage human disease, to select effective treatments for an individual, and to assess the progress of treatment. Success in this arena will advance and improve medicine by enabling earlier diagnosis of disease and providing more effective treatment options for patients. Plasma, an easily obtained specimen that contains proteins from many or perhaps all other body tissues, is of tremendous interest for biomarker discovery, and PeptideAtlas facilitates this discovery by integrating data from many diverse experiments and presenting it in a user-friendly and information-rich web resource. Such integrative resources are essential if we are to harness the power of all experimental work done in plasma proteomics and make progress in this hugely challenging task of transforming the practice of medicine.

Acknowledgments

The PeptideAtlas Project has involved a great many contributors. The authors would like to thank the following persons for their contributions to the design and implementation of PeptideAtlas:

Dave Campbell, Nichole King, Luis Mendoza, David Shteynberg, Natalie Tasman, Abhishek Pratap, Pat Moss, Jimmy Eng, Ning Zhang, Frank Desiere, Zhi Sun, and Michael Johnson. The authors would also like to thank Christopher Paulse and Robert West for reviewing this chapter and Julie Bletz for editing.

The authors have been funded in part with Federal funds from the National Heart, Lung, and Blood Institute, and the National Institutes of Health, under contract No. N01-HV-28179, and from PM50 GMO76547/Center for Systems Biology.

References

1. Aebersold, R., and Mann, M. (2003) Mass spectrometry-based proteomics. *Nature 422*, 198–207.
2. Hanash, S., and Celis, J. E. (2002) The Human Proteome Organization: a mission to advance proteome knowledge. *Mol Cell Proteomics 1*, 413–4.
3. Omenn, G. S. (2004) The Human Proteome Organization Plasma Proteome Project pilot phase: reference specimens, technology platform comparisons, and standardized data submissions and analyses. *Proteomics 4*, 1235–40.
4. Omenn, G. S., States, D. J., Adamski, M. et al. (2005) Overview of the HUPO Plasma Proteome Project: results from the pilot phase with 35 collaborating laboratories and multiple analytical groups, generating a core dataset of 3020 proteins and a publicly-available database. *Proteomics 5*, 3226–45.
5. Omenn, G. S., Aebersold, R., and Paik, Y. K. (2009) 7(th) HUPO World Congress of Proteomics: launching the second phase of the HUPOPlasma Proteome Project (PPP-2) 16-20 August 2008, Amsterdam, The Netherlands. *Proteomics 9*, 4–6.
6. Desiere, F., Deutsch, E. W., King, N. L. et al. (2006) The PeptideAtlas project. *Nucleic Acids Res 34*, D655–8.
7. Farrah, T., Deutsch, E. W., Omenn, G. S., et al. A high-confidence human plasma proteome reference with estimated concentrations in PeptideAtlas, submitted.
8. Eng, J., McCormack, A. L., and Yates, J. R. (1994) An Approach to Correlate Tandem Mass Spectral Data of Peptides with Amino Acid Sequences in a Protein Database. *J. Am. Soc. Mass Spectrom. 5*, 976–989.
9. Lam, H., Deutsch, E. W., Eddes, J. S. et al. (2007) Development and validation of a spectral library searching method for peptide identification from MS/MS. *Proteomics 7*, 655–67.
10. Boeckmann, B., Bairoch, A., Apweiler, R. et al. (2003) The SWISS-PROT protein knowledgebase and its supplement TrEMBL in 2003. *Nucleic Acids Res 31*, 365–70.
11. Hubbard, T. J., Aken, B. L., Beal, K. et al. (2007) Ensembl 2007. *Nucleic Acids Res 35*, D610–7.
12. Kersey, P. J., Duarte, J., Williams, A. et al. (2004) The International Protein Index: an integrated database for proteomics experiments. *Proteomics 4*, 1985–8.
13. Keller, A., Eng, J., Zhang, N., Li, X. J., and Aebersold, R. (2005) A uniform proteomics MS/MS analysis platform utilizing open XML file formats. *Mol Syst Biol 1*, 2005 0017.
14. Keller, A., Nesvizhskii, A. I., Kolker, E., and Aebersold, R. (2002) Empirical statistical model to estimate the accuracy of peptide identifications made by MS/MS and database search. *Anal Chem 74*, 5383–5392.
15. Shteynberg, D., Deutsch, E. W., Lam, H., et al. iProphet: Improved statistical validation of peptide identifications in shotgun proteomics, submitted.
16. Nesvizhskii, A. I., Keller, A., Kolker, E., and Aebersold, R. (2003) A statistical model for identifying proteins by tandem mass spectrometry. *Anal Chem 75*, 4646–4658.
17. Deutsch, E. W. (2010) The PeptideAtlas Project. *Methods in Molecular Biology 604*, 319–331.
18. National Institute of Standards and Technology.
19. Shannon, P., Markiel, A., Ozier, O. et al. (2003) Cytoscape: a software environment for integrated models of biomolecular interaction networks. *Genome Res 13*, 2498–504.
20. Mallick, P., Schirle, M., Chen, S. S. et al. (2007) Computational prediction of proteotypic peptides for quantitative proteomics. *Nat Biotechnol 25*, 125–31.

21. Tang, H., Arnold, R. J., Alves, P. et al. (2006) A computational approach toward label-free protein quantification using predicted peptide detectability. *Bioinformatics 22*, e481–8.

22. States, D. J., Omenn, G. S., Blackwell, T. W. et al. (2006) Challenges in deriving high-confidence protein identifications from data gathered by a HUPO plasma proteome collaborative study. *Nat Biotechnol 24*, 333–8.

23. ZLu, P., Vogel, C., Wang, R., et al. (2007) Absolute protein expression profiling estimates the relative contributions of transcriptional and translational regulation. *Nat Biotechnol* **25**, 117–124.

24. Deutsch, E. W., Lam, H., and Aebersold, R. (2008) PeptideAtlas: a resource for target selection for emerging targeted proteomics workflows. *EMBO Rep 9*, 429–34.

25. Sherwood, C. A., Eastham, A., Lee, L. W. et al. (2009) MaRiMba: a software application for spectral library-based MRM transition list assembly. *J Proteome Res 8*, 4396–405.

Appendix A

Standard Operating Procedures for Plasma Collection in Clinical Research

The Early Detection Research Network (EDRN) Standard Operating Procedure (SOP) For Collection of EDTA Plasma.

Reproduced and adapted with permission from [1].

Abstract

The variables surrounding collection and processing of blood specimens affect blood chemistry and the proteome in a way that introduces systematic changes that may be mistakenly attributed to a particular physiopathological condition. A limiting factor of current clinical proteomic studies has been the lack of accepted preanalytical and analytical guidelines. Recent worldwide efforts have been made to standardize blood collection, processing and storage conditions for case and control samples as part of the HUPO Plasma Proteome Project initiative. Given the complexity of the blood proteome, collection and handling present a broad range of specific preanalytical technical challenges from venipuncture to the use of blood derivatives, protease inhibitors, and processing specifications and storage conditions. As the areas of clinical validation of different disease states from blood-derived sources (i.e., disease biomarkers) move toward validation stages, the importance of controlled and standardized protocols are imperative. The establishment of standard operating procedures (SOPs) for the collection and processing of plasma and sera allows for systematic analysis of samples without the potential of bias and variance. This protocol is the SOP for plasma/serum use in clinical proteome research, based on the Early Detection Research Network (EDRN) and Specimen Collection and Handling Committee (SCHC) of the HUPO Plasma Proteome Project [1].

General Requirements

- Gloves must be worn at all times when handling specimens. This includes during the removal of the rubber stopper from the blood tubes, centrifugation, pipetting, disposal of contaminated tubes, and clean up of any spills. Tubes, needles, and pipettes must be properly disposed of in biohazard containers, in accordance with institutional requirements.

Richard J. Simpson and David W. Greening (eds.), *Serum/Plasma Proteomics: Methods and Protocols*, Methods in Molecular Biology, vol. 728, DOI 10.1007/978-1-61779-068-3, © Springer Science+Business Media, LLC 2011

- Universal precautions and Occupational Safety and Health Administration (OSHA) and institutional requirements (http://www.osha.gov/SLTC/biologicalagents/index.html) should be followed, including gloves, eye protection or working in a biosafety cabinet for blood processing.
- All equipment (storage, shipping, and centrifuge) must be labeled as biohazard.
- It is important to take steps to prevent hemolysis in these samples. A vacutainer is recommended. If a needle is used, a 21 gauge needle is recommended.

EDTA Plasma Collection

Supplies

- EDTA Blood Collection Tubes (for example, BD vacutainers catalog #366450)
- Centrifuge with swinging bucket rotor
- 15 mL polypropylene conical tubes (for example, Corning 430052, Fisher catalog #05-538-53D)
- Sterile cryovials with writing surface (for example, Simport T311-2 or Fisher #05-669-57)
- 2, 5 and 10 mL pipettes (for example, Fisher catalog #13-678-11C, 13-678-11D, 13-678-11E)
- Disposable transfer pipettes (for example, Fisher catalog #13-711-20)
- Automatic pipette aid
- Small ice bucket

Plasma Separation Procedure

1. After collection, gently mix the blood by inverting the tube eight to ten times. Store vacutainer tubes upright at 4°C until centrifugation. Blood samples should be centrifuged within 4 h of blood collection.
2. Centrifuge blood samples in a horizontal rotor (swing-out head) for 10–20 min at 1,100–1,300 × *g* at room temperature (20°C).

   Warning: Excessive centrifuge speed (over 2,000 × *g*) may cause tube breakage and exposure to blood and possible injury. If needed, RCF for a centrifuge can be calculated. For an online calculator tool, please refer to: http://www.changbioscience.com/cell/rcf.html.
3. After centrifugation, plasma layer is at the top of the tube. Mononuclear cells and platelets is in a whitish layer, called the "buffy coat," just under the plasma and above the red blood cells (additional processing of these cell fractions is optional).

4. Carefully collect the plasma layer with an appropriate transfer pipette without disturbing the buffy coat layer. If more than one tube is collected, pool the plasma samples from both tubes into a 15 mL conical tube and mix. Pipette the plasma into appropriate sized aliquots in labeled cryovials. Aliquot volume is recommended to be 100 or 250 μL; however, some sites may determine that 1 mL aliquot sizes are needed. Close the caps tightly and place on ice. This process should be completed within 1 h of centrifugation.
5. Check that all aliquot vial caps are secure and that all vials are labeled.
6. Place all aliquots upright in a specimen box or rack in an –80°C or colder freezer. All specimens should remain at –80°C or colder prior to shipping. The samples should not be thawed prior to shipping. (Plasma is shipped on dry ice. Refer to SOP for "Shipping" instructions.)

Data Points

1. Date and time of blood collection
2. Number and volume of aliquots prepared
3. Date and time into –80°C
4. Date and time of shipping
5. Any freeze–thaw that occurs with a sample for any reason
6. Any variations or deviations from the SOP, problems, or issues

Notes

- Sterile, disposable droppers, pipetman, pipette aid, eppendorf repeater are examples of ways to aliquot. Depends on aliquot size and volume and the plasma volume.
- Plasma should not undergo freeze–thaw cycles, so choose aliquot volume carefully.
- Freezers need to have a backup generator or other emergency system options: create emergency management plan, such as moving to a new freezer or adding dry ice in the event of a freezer failure.

References

1. Tuck, M.K., D.W. Chan, D. Chia, et al. (2009), Standard operating procedures for serum and plasma collection: early detection research network consensus statement standard operating procedure integration working group. *J Proteome Res.* **8**, 113–7.

Additional References

Hulmes, J.D., D. Bethea, K. Ho, et al., An investigation of plasma collection, stabilization, and storage procedures for proteomic analysis of clinical samples, in Clinical Proteomics, L.A. Liotta and E. Petricoin, Editors. 2004, Humana Press: Totowa, NJ. Vol. 1, 17–32.

Mischak, H., R. Apweiler, R.E. Banks, et al. (2007), Clinical proteomics: A need to define the field and to begin to set adequate standards. *Proteomics Clin. Appl.* **1**, 148–56.

Rai, A.J., C.A. Gelfand, B.C. Haywood, et al. (2005), HUPO Plasma Proteome Project specimen collection and handling: Towards the standardization of parameters for plasma proteome samples. *Proteomics.* **5**, 3262–77.

Tammen, H. (2008), Specimen collection and handling: standardization of blood sample collection. *Methods Mol Biol.* **428**, 35–42.

Vitzthum, F., G. Siest, D.M. Bunk, et al. (2007), Metrological sharp shooting for plasma proteins and peptides: The need for reference materials for accurate measurements in clinical proteomics and in vitro diagnostics to generate reliable results. *Proteomics Clin. Appl.* **1**, 1016–35.

Appendix B

Standard Operating Procedures for Serum Collection in Clinical Research

The Early Detection Research Network (EDRN) Standard Operating Procedure (SOP) For Collection of Serum.

Reproduced and adapted with permission from [1].

Abstract

The variables surrounding collection and processing of blood specimens affect blood chemistry and the proteome in a way that introduces systematic changes that may be mistakenly attributed to a particular physiopathological condition. A limiting factor of current clinical proteomic studies has been the lack of accepted preanalytical and analytical guidelines. Recent worldwide efforts have been made to standardize blood collection, processing and storage conditions for case and control samples as part of the HUPO Plasma Proteome Project initiative. Given the complexity of the blood proteome, collection and handling present a broad range of specific pre-analytical technical challenges from venipuncture to the use of blood derivatives, protease inhibitors, and processing specifications and storage conditions. As the areas of clinical validation of different disease states from blood-derived sources (i.e., disease biomarkers) move toward validation stages, the importance of controlled and standardized protocols are imperative. The establishment of standard operating procedures (SOPs) for the collection and processing of plasma and sera allows for systematic analysis of samples without the potential of bias and variance. This protocol is the SOP for plasma/serum use in clinical proteome research, based on the Early Detection Research Network (EDRN) and Specimen Collection and Handling Committee (SCHC) of the HUPO Plasma Proteome Project [1].

General Requirements

- Gloves must be worn at all times when handling specimens. This includes during the removal of the rubber stopper from the blood tubes, centrifugation, pipetting, disposal of contaminated tubes, and clean up of any spills. Tubes, needles, and pipettes must be properly disposed of in biohazard containers, in accordance with institutional requirements.
- Universal precautions and Occupational Safety and Health Administration (OSHA) and institutional requirements (http://www.osha.gov/SLTC/biologicalagents/index.html) should

be followed, including gloves, eye protection or working in a biosafety cabinet for blood processing.

- All equipment (storage, shipping, and centrifuge) must be labeled as biohazard.
- It is important to take steps to prevent hemolysis in these samples. A vacutainer is recommended. If a needle is used, a 21 gauge needle is recommended.

Serum Collection

Supplies

- Red Top Vacutainer (NOT SST tubes these contain polymeric gels with several constituents to adjust viscosity, density, and other physical properties) (for example, BD vacutainers catalog #366430)
- Centrifuge with swinging bucket rotor
- 15 mL polypropylene conical tubes (for example, Corning 430052, Fisher catalog #05-538-53D)
- Sterile cryovials with writing surface (for example, Simport T311-2 or Fisher #05-669-57)
- 2, 5 and 10 mL pipettes (for example, Fisher catalog #13-678-11C, 13-678-11D, 13-678-11E)
- Disposable transfer pipettes (for example, Fisher catalog #13-711-20)
- Automatic pipette aid
- Small ice bucket

Serum Separation Procedure

1. Filled red top blood collection tubes ("vacutainers") should sit upright after the blood is drawn at room temperature (20°C) for a minimum of 30 to a maximum of 60 min to allow the clot to form.

   Note: Use red top (serum) tubes (silicon-coated) – no additives and no serum separator tubes (SST). These tubes, without additives, allow the red blood cells to form a clot. The clot also includes white blood cells, platelets etc. After centrifuging, the clot is at the bottom of the tube, and the serum is on top of the clot. The red top tubes do not have to be full to be used.

2. Centrifuge the blood sample at the end of the clotting time (30–60 min) in a horizontal rotor (swing-out bucket) for 20 min at 1,100–1,300 × g at room temperature. If blood is not centrifuged immediately after the clotting time (30–60 min at room temperature), tubes should be refrigerated (4°C) for no longer than 4 h.

Warning: Excessive centrifuge speed (over 2,000 × *g*) may cause tube breakage and exposure to blood and possible injury. If needed, RCF for a centrifuge can be calculated. For an online calculator tool, please refer to: http://www.changbioscience.com/cell/rcf.html.

3. Use pipette to transfer the serum (recommendation: do not pour!). If more than one tube is drawn, pull the serum from both tubes into a 15 mL conical tube and mix. Pipette serum into the labeled cryovials, filling the vials in sequential order. Aliquot volume is recommended to be 100 or 250 μL. Close the caps on the vials tightly. This process should be completed within 1 h of centrifugation.
4. *Note*: Be very careful not to pick up red blood cells when aliquoting. This can be done by keeping the pipette above the red blood cell layer and leaving a small amount of serum in the tube. Check that all aliquot vial caps are secure and that all vials are labeled.
5. Place all aliquots upright in a specimen box or rack in an –80°C or colder freezer. All specimens should remain at –80°C or colder prior to shipping. The samples should not be thawed prior to shipping.

Data Points

1. Is the serum *hemolyzed? If yes, sample cannot be used.*
2. Date and time of blood collection.
3. Number and volume of aliquots prepared.
4. Date and time into –80°C.
5. Date and time of shipping.
6. Any freeze-thaw that occurs with a sample for any reason.
7. Any variations or deviations from the SOP, problems, or issues.

Notes

- Sterile, disposable droppers, pipetman, pipette aid, eppendorf repeater are examples of ways to aliquot. Depends on aliquot size and volume and the sera volume.
- Serum should not undergo freeze–thaw cycles, so choose aliquot volume carefully.
- Freezers need to have a backup generator or other emergency system options: create emergency management plan, such as moving to a new freezer or adding dry ice in the event of a freezer failure.

References

1. Tuck, M.K., D.W. Chan, D. Chia, et al. (2009), Standard operating procedures for serum and plasma collection: early detection research network consensus statement standard operating procedure integration working group. *J Proteome Res.* **8**, 113–7.

Additional References

Hulmes, J.D., D. Bethea, K. Ho, et al., An investigation of plasma collection, stabilization, and storage procedures for proteomic analysis of clinical samples, in Clinical Proteomics, L.A. Liotta and E. Petricoin, Editors. 2004, Humana Press: Totowa, NJ. Vol. **1**, 17–32.

Mischak, H., R. Apweiler, R.E. Banks, et al. (2007), Clinical proteomics: A need to define the field and to begin to set adequate standards. *Proteomics Clin. Appl.* **1**, 148–56.

Rai, A.J., C.A. Gelfand, B.C. Haywood, et al. (2005), HUPO Plasma Proteome Project specimen collection and handling: Towards the standardization of parameters for plasma proteome samples. *Proteomics.* **5**, 3262–77.

Tammen, H. (2008), Specimen collection and handling: standardization of blood sample collection. *Methods Mol Biol.* **428**, 35–42.

Vitzthum, F., G. Siest, D.M. Bunk, et al. (2007), Metrological sharp shooting for plasma proteins and peptides: The need for reference materials for accurate measurements in clinical proteomics and in vitro diagnostics to generate reliable results. Proteomics Clin. Appl. **1**, 1016–1035.

Appendix C

Measured Concentrations of Plasma Proteins from Quantitative Assays

Haab, B.B., B.H. Geierstanger, G. Michailidis, et al. *Immunoassay and antibody microarray analysis of the HUPO Plasma Proteome Project reference specimens: Systematic variation between sample types and calibration of mass spectrometry data*. Proteomics. 2005. **5**, 3278–91.

Information relating to 70 IPI numbers (66 unique analytes) that had a match between the analyte-derived lists and the MS-derived lists is presented.

"Antibody name" = the name that was used in the searches for analyte-associated IPI numbers.

"Name from analyte search" = the name in the IPI database that matched the antibody/analyte name.

"Concentration" = the geometric mean concentration over all specimens as found by immunoassay or antibody microarray.

DB, Dade Behring; **GNF**, Genomics Institute of the Novartis Foundation; **MSI**, Molecular Staging.

Antibody name	Name from analyte search	Concentration, pg/mL	No. of labs	No. of peptides	Laboratory	IPI SET
Albumin	Albumin	4.0E+10	17	201	DB	2
Transferrin	Transferrin	2.3E+09	16	249	DB	2
Apolipoprotein A 1	Apolipoprotein A-1	1.4E+09	17	82	DB	2
α2-macroglobulin	Alpha-2-macroglobulin	1.4E+09	17	211	DB	2
α1-antitrypsin	Serine (or cysteine) proteinase inhibitor, clade A (alpha-1 antiproteinase, antitrypsin), member 1	1.1E+09	15	183	DB	2
C3c	Complement component 3	9.5E+08	5	98	DB	2
Haptoglobin	Haptoglobin	8.8E+08	18	113	DB	2
Hemopexin	Hemopexin	7.5E+08	16	86	DB	2
Apolipoprotein B	Apolipoprotein B (including Ag(x) antigen)	7.2E+08	13	328	DB	2
Fibrinogen	Fibrinogen, gamma polypeptide	6.7E+08	16	66	DB	2
Fibrinogen	Fibrinogen, gamma polypeptide	6.7E+08	12	51	DB	2
α1-acid-glycoprotein	Alpha-1-acid glycoprotein 2 precursor	6.1E+08	14	24	DB	1
α1-acid-glycoprotein	Orosomucoid 1	6.1E+08	16	45	DB	2
Antithrombin III	Serine (or cysteine) proteinase inhibitor, clade C (antithrombin), member 1	3.2E+08	17	70	DB	2
Apolipoprotein A-II	Apolipoprotein A-II	3.0E+08	15	18	DB	2
Prealbumin	Transthyretin (prealbumin, amyloidosis type 1)	2.6E+08	17	27	DB	2
Ceruloplasmin	Ceruloplasmin (ferroxidase)	2.1E+08	15	134	DB	2
C4	Complement C4 precursor (contains: C4A anaphylatoxin)	1.7E+08	17	157	DB	1
Plasminogen	Plasminogen	1.4E+08	12	72	DB	2
Fibronectin	Fibronectin 1	1.1E+08	1	86	DB	2

Apolipoprotein E	Apolipoprotein E	3.4E+07	8	30	DB	2
vWF	Von Willebrand factor	1.3E+06	2	46	GNF	2
β2Microglobulin	Beta 2-microglobulin protein	1.1E+06	1	1	DB	1
β2Microglobulin	Beta-2-microglobulin	1.1E+06	3	1	DB	2
sTfR	Transferrin receptor (p90, CD71)	5.8E+05	1	2	DB	2
VAP-1	Amine oxidase, copper containing 3 (vascular adhesion protein 1)	1.2E+05	2	6	MSI	2
Protein C	Mannose-binding lectin (protein C) 2, soluble (opsonic defect)	9.7E+04	2	7	MSI	2
VCAM-1	Vascular cell adhesion molecule 1	9.4E+04	3	9	MSI/GNF	2
TGFβ1	Transforming growth factor, beta 1 (Camurati-Engelmann disease)	7.5E+04	2	2	GNF	2
IGF-BP3	Insulin-like growth factor binding protein 3	5.9E+04	6	17	MSI/GNF	2
ICAM-1	Intercellular adhesion molecule 1 (CD54), human rhinovirus receptor	4.3E+04	2	4	MSI/GNF	2
MMPg	Matrix metalloproteinase 9 (gelatinase B, 92 kDa gelatinase, 92 kDa type IV collagenase)	4.1E+04	2	5	MSI/GNF	2
VE-cadherin	Cadherin 5, type 2, VE-cadherin (vascular epithelium)	3.0E+04	3	11	MSI	2
M-CSF R	Colony stimulating factor 1 receptor, formerly McDonough feline sarcoma viral (v-fms) oncogene homolog	2.6E+04	3	11	MSI	2
L-Selectin	Selectin L (lymphocyte adhesion molecule 1)	1.7E+04	5	10	MSI	2
ALCAM	Activated leukocyte cell adhesion molecule	1.6E+04	2	5	MSI	2
IGFBP2	Insulin-like growth factor binding protein 2, 36 kDa	1.5E+04	1	3	MSI	2

(continued)

Appendix C (continued)

Antibody name	Name from analyte search	Concentration, pg/mL	No. of labs	No. of peptides	Laboratory	IPI SET
TIMP1	Tissue inhibitor of metalloproteinase 1 (erythroid potentiating activity, collagenase inhibitor)	1.4E+04	1	3	MSI/GNF	2
EGF R1	Epidermal growth factor receptor (erythroblastic leukemia viral (v-erb-b) oncogene homolog, avian)	1.1E+04	3	3	GNF	2
MMP2	Matrix metalloproteinase 2 (gelatinase A, 72 kDa gelatinase, 72 kDa type IV collagenase)	8.8E+03	1	7	MSI/GNF	2
NAP-2	Nucleosome assembly protein 1-like 4	7.5E+03	1	1	MSI	2
LIF Rα	Leukemia inhibitory factor receptor	5.0E+03	2	4	MSI	2
PDGF-Rα	Platelet-derived growth factor receptor, alpha polypeptide	4.6E+03	1	2	MSI	2
MMP1	Matrix metalloproteinase 1 (interstitial collagenase)	2.6E+03	1	1	MSI/GNF	2
FasL	Tumor necrosis factor (ligand) superfamily, member 6	1.5E+03	1	2	MSI/GNF	2
NSE	Enolase 2 (gamma, neuronal)	1.4E+03	1	1	GNF	2
MMP8	Matrix metalloproteinase 8 (neutrophil collagenase)	9.0E+02	1	1	MSI/GNF	2
VEGF-D	C-fos induced growth factor (vascular endothelial growth factor D)	5.0E+02	1	1	MSI/GNF	2
ENA-78	Chemokine (C-X-C motif) ligand 5	3.4E+02	1	1	MSI	2
CD30	Tumor necrosis factor receptor superfamily, member 8	3.3E+02	1	2	MSI/GNF	2
MPIF-1	Chemokine (C-C motif) ligand 23	3.2E+02	1	1	MSI	2

GROb	Chemokine (C-X-C motif) ligand 2	3.0E+02	1	1	MSI	2
BDNF	Brain-derived neurotrophic factor	3.0E+02	1	1	MSI	2
AFP	Alpha-fetoprotein	2.9E+02	2	2	MSI/GNF	2
IGF-IR	Insulin-like growth factor 1 receptor	2.4E+02	1	1	MSI/GNF	2
Calcitonin	Calcitonin/calcitonin-related polypeptide, alpha	1.9E+02	1	1	GNF	2
Calcitonin	Calcitonin gene-related peptide type 1 receptor precursor	1.9E+02	1	1	GNF	1
FGFβ	Fibroblast growth factor-20	1.6E+02	1	1	GNF	1
IL-10Rβ	Interleukin 10 receptor, beta	1.5E+02	1	1	MSI	2
Angp2	Angiopoietin 2	9.7E+01	1	1	GNF	2
MCP-1	Splice isoform A of P15529	7.6E+01	1	1	MSI/GNF	1
SCF	KIT ligand	5.9E+01	1	1	MSI/GNF	2
IFN_γ	Interferon, gamma	5.4E+01	1	1	MSI/GNF	2
OSM	Oncostatin M	4.8E+01	1	1	MSI	2
IL1α	Interleukin 1, alpha	4.5E+01	1	1	MSI/GNF	2
TNFα	Tumor necrosis factor (TNF superfamily, member 2)	3.7E+01	2	1	MSI/GNF	2
AR	Androgen receptor (dihydrotestosterone receptor, testicular feminization; spinal and bulbar muscular atrophy; Kennedy disease)	2.6E+01	1	1	MSI/GNF	2
I-TAC	Chemokine (C-X-C motif) ligand 11	2.3E+01	1	1	MSI	2
CGB	Chorionic gonadotropin, beta polypeptide	1.9E+01	1	1	MSI/GNF	2
IL7	Interleukin 7	7.0E+00	1	1	MSI/GNF	2

Appendix D

Plasma/Serum Proteomics Selected Readings

History

Cohn, Oncley, Strong, Hughes et al. (1944) Chemical, Clinical, and Immunological Studies on the Products of Human Plasma Fractionation. I. the Characterization of the Protein Fractions of Human Plasma. *J Clin Invest*, **23**, 417–32

Cohn. (1950) Demonstration of new processes of blood collection and separation of red blood cells, white blood cells, and platelets; protein glycoprotein, lipoprotein, and other components of plasma. *Science*, **112**, 450–1

Landsteiner. (1961) On agglutination of normal human blood. *Transfusion*, **1**, 5–8

Lewisohn. (1937) Twenty Years' Experience with the Citrate Method of Blood Transfusion. *Ann Surg*, **105**, 602–9

Lewisohn. (1955) Blood transfusion: 50 years ago and today. *Surg Gynecol Obstet*, **101**, 362–8

McCullough. (2005) Transfusion Medicine, 2nd Edition. Churchill Livingstone,

Putnam, F.W., (1975), The Plasma Proteins. 2nd ed. Vol. 1. New York: Academic Press. 57–130.

Plasma/Biofluid Proteomics (Review)

Anderson and Anderson. (2002) The human plasma proteome: history, character, and diagnostic prospects. *Mol Cell Proteomics*, **1**, 845–67

Apweiler, Aslanidis, Deufel, Gerstner et al. (2009) Approaching clinical proteomics: current state and future fields of application in fluid proteomics. *Clin Chem Lab Med*, **47**, 724–44

Liumbruno, D'Alessandro, Grazzini and Zolla. (2010) Blood-related proteomics. *J Proteomics*, **73**, 483–507

Omenn, G.S., D.J. States, M. Adamski, et al. (2005), Overview of the HUPO Plasma Proteome Project: results from the pilot phase with 35 collaborating laboratories and multiple analytical groups, generating a core dataset of 3020 proteins and a publicly-available database. *Proteomics*. **5**, 3226–45.

Thadikkaran, Siegenthaler, Crettaz, Queloz et al. (2005) Recent advances in blood-related proteomics. *Proteomics*, **5**, 3019–34

Preanalytical/Analytical Consideration

Barelli, Crettaz, Thadikkaran, Rubin et al. (2007) Plasma/serum proteomics: pre-analytical issues. *Expert Rev Proteomics*, **4**, 363–70

Di Girolamo, Alessandroni, Somma and Guadagni. (2009) Pre-analytical operating procedures for serum Low Molecular Weight protein profiling. *J Proteomics*, **73**, 667–77

Drake, S.K., R.A. Bowen, A.T. Remaley, et al. (2004), Potential interferences from blood collection tubes in mass spectrometric analyses of serum polypeptides. *Clin Chem*. **50**, 2398–401.

Elliott and Peakman. (2008) The UK Biobank sample handling and storage protocol for the collection, processing and archiving of human blood and urine. *Int J Epidemiol*, **37**, 234–44

Ferguson, Hochstrasser and Banks. (2007) Impact of preanalytical variables on the analysis of biological fluids in proteomic studies. *Proteomics Clin. Appl.*, 739–746

Findeisen, P., D. Sismanidis, M. Riedl, et al. (2005), Preanalytical impact of sample handling on proteome profiling experiments with matrix-assisted laser desorption/ionization time-of-flight mass spectrometry. *Clin Chem.* **51**, 2409–11.

Hortin, G.L. (2006), The MALDI-TOF mass spectrometric view of the plasma proteome and peptidome. *Clin Chem.* **52**, 1223–37.

Hortin, G.L. and D. Sviridov (2010), The dynamic range problem in the analysis of the plasma proteome. *J Proteomics.* **73**, 629–36.

Hulmes, Bethea, Ho, Huang et al. (2004) An investigation of plasma collection, stabilization, and storage procedures for proteomic analysis of clinical samples. *Clinical Proteomics*, Humana Press, **1**, 17–32

Jackson and Banks. (2010) Banking of clinical samples for proteomic biomarker studies: A consideration of logistical issues with a focus on pre-analytical variation. *Proteomics Clin. Appl.*, **4**, 250–270

Rai, Gelfand, Haywood, Warunek et al. (2005) HUPO Plasma Proteome Project specimen collection and handling: Towards the standardization of parameters for plasma proteome samples. *Proteomics*, **5**, 3262–77

Rai and Vitzthum. (2006) Effects of preanalytical variables on peptide and protein measurements in human serum and plasma: implications for clinical proteomics. *Expert Rev Proteomics*, **3**, 409–26

Rodriguez, Tezak, Mesri, Carr et al. (2010) Analytical validation of protein-based multiplex assays: a workshop report by the NCI-FDA interagency oncology task force on molecular diagnostics. *Clin Chem*, **56**, 237–43

Tammen. (2008) Specimen collection and handling: standardization of blood sample collection. *Methods Mol Biol*, **428**, 35–42

Tuck, Chan, Chia, Godwin et al. (2009) Standard operating procedures for serum and plasma collection: early detection research network consensus statement standard operating procedure integration working group. *J Proteome Res*, **8**, 113–7

Vitzthum, Siest, Bunk, Preckel et al. (2007) Metrological sharp shooting for plasma proteins and peptides: The need for reference materials for accurate measurements in clinical proteomics and in vitro diagnostics to generate reliable results. *Proteomics Clin. Appl.*, **1**, 1016–1035

Yi, Kim and Gelfand. (2007) Inhibition of intrinsic proteolytic activities moderates preanalytical variability and instability of human plasma. *J Proteome Res*, **6**, 1768–81

Low-Molecular Weight/Peptidome Analyses

Gatlin, C.L., White, K.Y. Tracy, M.B. et al. Enhancement in MALDI-TOF MS analysis of the low molecular weight human serum proteome. *J Mass Spectrom*, **46**, 85–9.

Greening and Simpson. (2010) A centrifugal ultrafiltration strategy for isolating the low-molecular weight (≤25K) component of human plasma proteome. *J Proteomics* **73**, 637–648

Harper, Workman, Schuetzner, Timperman et al. (2004) Low-molecular-weight human serum proteome using ultrafiltration, isoelectric focusing, and mass spectrometry. *Electrophoresis*, **25**, 1299–1306

Hickman, H.D. and Yewdell J.W. Mining the plasma immunopeptidome for cancer peptides as biomarkers and beyond. *Proc Natl Acad Sci U S A*, **107**, 18747–8.

Richter, Schulz-Knappe, Schrader, Standker et al. (1999) Composition of the peptide fraction in human blood plasma: database of circulating human peptides. *J Chromatogr B Biomed Sci Appl*, **726**, 25–35

Tammen, Schulte, Hess, Menzel et al. (2005) Peptidomic analysis of human blood specimens: comparison between plasma specimens and serum by differential peptide display. *Proteomics*, **5**, 3414–22

Tirumalai, Chan, Prieto, Issaq et al. (2003) Characterization of the low molecular weight human serum proteome. *Mol Cell Proteomics*, **2**, 1096–103

Villanueva, Shaffer, Philip, Chaparro et al. (2006) Differential exoprotease activities confer tumor-specific serum peptidome patterns. *J Clin Invest*, **116**, 271–84

Zheng, Baker and Hancock. (2006) Analysis of the low molecular weight serum peptidome using ultrafiltration and a hybrid ion trap-Fourier transform mass spectrometer. *J Chromatogr A*, **1120**, 173–84

Separation/Enrichment Analyses

Anderson and Anderson. (1977) High resolution two-dimensional electrophoresis of human plasma proteins. *Proc Natl Acad Sci U S A*, **74**, 5421–5

Barnea, Sorkin, Ziv, Beer et al. (2005) Evaluation of prefractionation methods as a preparatory step for multidimensional based chromatography of serum proteins. *Proteomics*, **5**, 3367–75

Berven, Ahmad, Clauser and Carr. Optimizing performance of glycopeptide capture for plasma proteomics. *J Proteome Res*, **9**, 1706–15

Fitzgerald, A. and Walsh B.J. New method for prefractionation of plasma for proteomic analysis. *Electrophoresis*, **31**, 3580–5.

Gundry, Fu, Jelinek, Van Eyk et al. (2007) Investigation of an albumin-enriched fraction of human serum and its albuminome. *Proteomics Clin. Appl.*, **1**, 73–88

Li, Gong, Wang, Wu et al. (2005) Comparison of alternative analytical techniques for the characterisation of the human serum proteome in HUPO Plasma Proteome Project. *Proteomics*, **5**, 3423–41

Mehta, Ross, Lowenthal, Fusaro et al. (2003) Biomarker amplification by serum carrier protein binding. *Dis Markers*, **19**, 1–10

Omenn, States, Adamski, Blackwell et al. (2005) Overview of the HUPO Plasma Proteome Project: results from the pilot phase with 35 collaborating laboratories and multiple analytical groups, generating a core dataset of 3020 proteins and a publicly-available database. *Proteomics*, **5**, 3226–45

Pieper, Gatlin, Makusky, Russo et al. (2003) The human serum proteome: display of nearly 3700 chromatographically separated protein spots on two-dimensional electrophoresis gels and identification of 325 distinct proteins. *Proteomics*, **3**, 1345–64

Whiteaker, Zhang, Eng, Fang et al. (2007) Head-to-head comparison of serum fractionation techniques. *J Proteome Res*, **6**, 828-36

Zhou, Lucas, Chan, Issaq et al. (2004) An investigation into the human serum "interactome". *Electrophoresis*, **25**, 1289–98

Depletion

Bjorhall, Miliotis and Davidsson. (2004) Comparison of different depletion strategies for improved resolution in proteomic analysis of human serum samples. *Proteomics*, **5**, 307–317

Fu, Garnham, Elliott, Bovenkamp et al. (2005) A robust, streamlined, and reproducible method for proteomic analysis of serum by delipidation, albumin and IgG depletion, and two-dimensional gel electrophoresis. *Proteomics*, **5**, 2656–64.

Gong, Li, Yang, Ying et al. (2006) Different immunoaffinity fractionation strategies to characterize the human plasma proteome. *J Proteome Res*, **5**, 1379–87

Pieper, Su, Gatlin, Huang et al. (2003) Multi-component immunoaffinity subtraction chromatography: an innovative step towards a comprehensive survey of the human plasma proteome. *Proteomics*, **3**, 422–32

Quantitative

Addona, Abbatiello, Schilling, Skates et al. (2009) Multi-site assessment of the precision and reproducibility of multiple reaction monitoring-based measurements of proteins in plasma. *Nat Biotechnol*, **27**, 633–41

Campbell, J., Rezai,T., Prakash, A., et al. Evaluation of absolute peptide quantitation strategies using selected reaction monitoring. *Proteomics*, **11**, 1148–52.

Kuzyk, Smith, Yang, Cross et al. (2009) Multiple reaction monitoring-based, multiplexed, absolute quantitation of 45 proteins in human plasma. *Mol Cell Proteomics*, **8**, 1860–77

Database

Deutsch, Eng, Zhang, King et al. (2005) Human Plasma PeptideAtlas. *Proteomics*, **5**, 3497-500

Jeong, Lee, Cho, Lee et al. Data management and functional annotation of the Korean reference plasma proteome. *Proteomics*, **10**, 1250–5

Mathivanan and Pandey. (2008) Human proteinpedia as a resource for clinical proteomics. *Mol Cell Proteomics*, **7**, 2038–47

Muthusamy, Hanumanthu, Suresh, Rekha et al. (2005) Plasma Proteome Database as a resource for proteomics research. *Proteomics*, **5**, 3531–6

Immunoassay/Multiplex Assay

Fu, Schoenhoff, Savage, Zhang et al. (2010) Multiplex assays for biomarker research and clinical application: Translational science coming of age. *Proteomics Clin. Appl.*, **4**, 271–284

Gold, L., Ayers, D., Bertino, J., et al. Aptamer-based multiplexed proteomic technology for biomarker discovery. *PLoS One*. **5**, e15004.

Haab, Geierstanger, Michailidis, Vitzthum et al. (2005) Immunoassay and antibody microarray analysis of the HUPO Plasma Proteome Project reference specimens: Systematic variation between sample types and calibration of mass spectrometry data. *Proteomics*, **5**, 3278–91

Rodriguez, Tezak, Mesri, Carr et al. (2010) Analytical validation of protein-based multiplex assays: a workshop report by the NCI-FDA interagency oncology task force on molecular diagnostics. *Clin Chem*, **56**, 237–43

Search Algorithms

Kapp, Schutz, Connolly, Chakel et al. (2005) An evaluation, comparison, and accurate benchmarking of several publicly available MS/MS search algorithms: sensitivity and specificity analysis. *Proteomics*, **5**, 3475–90

Yu, Taylor, Davis, Bonilla et al. (2010) Maximizing the sensitivity and reliability of peptide identification in large-scale proteomic experiments by harnessing multiple search engines. *Proteomics*, **10**, 1172–89

Plasma Proteome Reference Datasets

Adkins, Varnum, Auberry, Moore et al. (2002) Toward a human blood serum proteome: analysis by multidimensional separation coupled with mass spectrometry. *Mol Cell Proteomics*, **1**, 947–55

Anderson, Polanski, Pieper, Gatlin et al. (2004) The human plasma proteome: a nonredundant list developed by combination of four separate sources. *Mol Cell Proteomics*, **3**, 311–26

Greening, Glenister, Kapp, Moritz et al. (2008) Comparison of human platelet-membrane cytoskeletal proteins with the plasma proteome: Towards understanding the platelet-plasma nexus. *Proteomics Clin. Appl.*, **2**, 63–77

Omenn, States, Adamski, Blackwell et al. (2005) Overview of the HUPO Plasma Proteome Project: results from the pilot phase with 35 collaborating laboratories and multiple analytical groups, generating a core dataset of 3020 proteins and a publicly-available database. *Proteomics*, **5**, 3226–45

States, Omenn, Blackwell, Fermin et al. (2006) Challenges in deriving high-confidence protein identifications from data gathered by a HUPO plasma proteome collaborative study. *Nat Biotechnol*, **24**, 333–8

Schenk, Schoenhals, de Souza and Mann. (2008) A high confidence, manually validated human blood plasma protein reference set. *BMC Med Genomics*, **1**, 41

Other

Anderson. (2010) The clinical plasma proteome: a survey of clinical assays for proteins in plasma and serum. *Clin Chem*, **56**, 177–85

Hortin, G.L. (2005), Can mass spectrometric protein profiling meet desired standards of clinical laboratory practice? *Clin Chem*. **51**, 3–5.

Omenn. (2007) Exploring the Human Plasma Proteome. Wiley-VCH

Omenn, G.S. Data management and data integration in the HUPO plasma proteome project. *Methods Mol Biol*. **696**, 247–57.

Appendix E

Reference Ranges for Blood Tests Sorted by Mass and Molarity

Hormones predominate at the left part of the scale, shown at ng/L or pmol/L, being in very low concentration. There appears to be the greatest cluster of substances at μg/L or nmol/L range, becoming less so toward mg/L or μmol/L. However, there is another cluster containing many metabolic substances like cholesterol and glucose at the limit g/L or mmol/L range.

To translate a substance from the molar to the mass concentration scale above:

- Numerically: *molar concentration* × *molar mass* = *mass concentration*
- Measured directly in distance on the scales:

  Figure link: http://upload.wikimedia.org/wikipedia/commons/c/cb/Blood_values_sorted_by_mass_and_molar_concentration.png

  Author: Mikael Häggström
  Date: Dec 2010
  Website: http://en.wikipedia.org/wiki/Reference_ranges_for_blood_tests

Reference range list from Uppsala University Hospital ("Laborationslista"). Artnr 40284 Sj74a. Issued on April 22, 2008.

Index

A

B

C

D

E

Q

R

S

T

V

X

Z

MIX
Papier aus verantwortungsvollen Quellen
Paper from responsible sources
FSC® C105338

If you have any concerns about our products,
you can contact us on
ProductSafety@springernature.com

In case Publisher is established outside the EU,
the EU authorized representative is:
Springer Nature Customer Service Center GmbH
Europaplatz 3, 69115 Heidelberg, Germany

Printed by Libri Plureos GmbH
in Hamburg, Germany